Universitext

For further volumes:
http://www.springer.com/series/223

Nancy Childress

Class Field Theory

 Springer

Nancy Childress
Department of Mathematics and Statistics
Arizona State University
Tempe, AZ, USA
nc@asu.edu

ISBN: 978-0-387-72489-8   e-ISBN: 978-0-387-72490-4
DOI 10.1007/978-0-387-72490-4

Library of Congress Control Number: 2008935390

Mathematics Subject Classification (2000): 11Rxx, 11Sxx, 11-01

Printed on acid-free paper

springer.com

Preface

In essence, class field theory is the study of the abelian extensions of arbitrary global or local fields. In particular, one is interested in characterizing the abelian extensions of a given field K in terms of the arithmetical data for K. The most basic example of such a characterization is the Kronecker-Weber Theorem, which states that the abelian extensions of the field of rational numbers are subfields of its cyclotomic extensions, so expressible in terms of roots of unity.

Also of interest is to describe how the prime ideals in the ring of integers of a global or local field decompose in its finite abelian extensions. In the case of the quadratic extensions of the field of rational numbers, such a description is obtained through the Law of Quadratic Reciprocity. There are also higher reciprocity laws of course, but all of these are subsumed by what is known as Artin Reciprocity, one of the most powerful results in class field theory.

I have always found class field theory to be a strikingly beautiful topic. As it developed, techniques from many branches of mathematics were adapted (or invented!) for use in class field theory. The interplay between ideas from number theory, algebra and analysis is pervasive in even the earliest work on the subject. And class field theory is still evolving. While it is prerequisite for most any kind of research in algebraic number theory, it also continues to engender active research. It is my hope that this book will serve as a gateway into the subject.

Class field theory has developed through the use of many techniques and points of view. I have endeavored to expose the reader to as many of the different techniques as possible. This means moving between ideal theoretic and idèle theoretic approaches, with L-functions and the Tate cohomology groups thrown in for good measure. I have attempted to include some information about the history of the subject as well. The book progresses from material that is likely more naturally accessible to students, to material that is more challenging.

The global class field theory for number fields is presented in Chapters 2-6, which are intended to be read in sequence. For the most part they are not prerequisite for Chapter 7. (The exceptions to this are in Chapter 6: profinite groups and the theory of infinite Galois extensions in Section 6, and the notion of a ramified prime in an infinite extension from Section 7.) The local material is positioned last primarily because it is somewhat more challenging; for this reason, working through the earlier chapters first may be of benefit.

For students who have completed an introductory course on algebraic number theory, a one-term course on global class field theory might comprise Chapters 2-5 and sections 1–4 of Chapter 6. For more experienced students, some of the material in these chapters may be familiar, e.g., the sections on Dirichlet series and the Theorem on Primes in Arithmetic Progressions. In that case, the remainder of Chapter 6 may be included to produce a course still entirely on global class field theory. For somewhat more sophisticated students, Chapter 7 provides the option of including the local theory.

Facility with abstract algebra and (very) basic topology and complex analysis is assumed. Chapter 1 contains an outline of some of the prerequisite material on number fields and their completions. Nearly all of the results in Chapter 1 appear without proof, but details can be found in Fröhlich and Taylor's *Algebraic Number Theory*, [FT], or (for the global fields) Marcus' *Number Fields*, [Ma].

The level of preparation in abstract algebra that is required increases slightly as one progresses through the book. However, I have included a little background material for certain topics that might not appear in a typical first-year course in abstract algebra. For example there are brief discussions on topological groups, infinite Galois theory, and projective limits. Finite Galois theory is heavily used throughout, and concepts such as modules, exact sequences, the Snake Lemma, etc., play important roles in several places. A small amount of cohomology is introduced, but there is no need for previous experience with cohomology.

The source for the material on Dirichlet characters in Chapter 2 is Washington's *Cyclotomic Fields*, [Wa], while the material on Dirichlet series was adapted primarily from Serre's *A Course in Arithmetic*, [Se1], and the book by Fröhlich and Taylor, [FT]. The section on Dirichlet density is derived mostly from Janusz' *Algebraic Number Fields*, [J], and Lang's *Algebraic Number Theory*, [L1].

I first saw class fields interpreted in terms of Dirichlet density in Sinnott's lectures, [Si], which greatly influenced the organization of the material in Chapters 3 and 4. (This point of view appears also in Marcus' *Number Fields*, [Ma].) Other sources that were particularly valuable in the writing of these two chapters were [J], [L1], and Cassels and Fröhlich's *Algebraic Number Theory*, [CF].

The main source consulted in the preparation of Chapters 5 and 6 is [L1], although [J], [CF], [Si], Neukirch's *Class Field Theory*, [N], and the lecture notes of Artin and Tate, [AT], also were very valuable throughout. For section 7 of Chapter 6, [Wa] is the primary source, and Lang's *Cyclotomic Fields I and II*, [L3], was also consulted.

Other references that proved particularly useful in the preparation of the chapters on global class field theory include Gras' *Class Field Theory*, [G], and Milne's lecture notes, [Mi].

The presentation of local class field theory in Chapter 7 relies mainly on the article by Hazewinkel, [Haz2]. Also very useful were Iwasawa's *Local Class Field Theory*, [I], Neukirch's book, [N], and the seminal article of Lubin and Tate, [LT].

A preliminary version of this book was used by a group of students and faculty at the University of Colorado, Boulder. I am indebted to them for their careful reading of the manuscript, and the many useful comments that resulted. My thanks especially to David Grant, who led the group and kept detailed notes on these comments, and to the members: Suion Ih, Erika Frugoni, Vinod Radhakrishnan, Zachary Strider McGregor-Dorsey and Jonathan Kish.

Several incarnations of the manuscript for this book have been used for courses in class field theory that I have offered periodically. I am grateful to my class field theory students over the past few years, who have participated in these courses using early versions of the manuscript. Among those who have helped in spotting typographical errors and other oddities are Eric Driver, Ahmed Matar, Chase Franks, Rachel Wallington, Michael McCamy and Shawn Elledge. Special thanks also to John Kerl for advice on creating diagrams in LaTeX and to Linda Arneson for her excellent work in typing the first draft of the course outline, which grew into this book.

In completing this book, I am most fortunate to have worked with Mark Spencer, Frank Ganz and David Hartman at Springer, and to have had valuable input from the reviewers. My sincere thanks to them as well.

Tempe, AZ
2007

Contents

1 A Brief Review 1
  1 Number Fields 1
  2 Completions of Number Fields 8
  3 Some General Questions Motivating Class Field Theory 14

2 Dirichlet's Theorem on Primes in Arithmetic Progressions 17
  1 Characters of Finite Abelian Groups 17
  2 Dirichlet Characters 20
  3 Dirichlet Series 30
  4 Dirichlet's Theorem on Primes in Arithmetic Progressions 35
  5 Dirichlet Density 40

3 Ray Class Groups 45
  1 The Approximation Theorem and Infinite Primes 45
  2 Ray Class Groups and the Universal Norm Index Inequality 47
  3 The Main Theorems of Class Field Theory 60

4 The Idèlic Theory 63
  1 Places of a Number Field 64
  2 A Little Topology 66
  3 The Group of Idèles of a Number Field 68
  4 Cohomology of Finite Cyclic Groups and the Herbrand Quotient 75
  5 Cyclic Galois Action on Idèles 83

5 Artin Reciprocity 105
  1 The Conductor of an Abelian Extension of Number Fields and the
     Artin Symbol 105
  2 Artin Reciprocity 111
  3 An Example: Quadratic Reciprocity 128
  4 Some Preliminary Results about the Artin Map on Local Fields 130

6   The Existence Theorem, Consequences and Applications 135
   1   The Ordering Theorem and the Reduction Lemma 136
   2   Kummer n-extensions and the Proof of the Existence Theorem 139
   3   The Artin Map on Local Fields 148
   4   The Hilbert Class Field 153
   5   Arbitrary Finite Extensions of Number Fields 159
   6   Infinite Extensions and an Alternate Proof of the Existence Theorem . . 162
   7   An Example: Cyclotomic Fields 168

7   Local Class Field Theory 181
   1   Some Preliminary Facts About Local Fields 182
   2   A Fundamental Exact Sequence 186
   3   Local Units Modulo Norms 191
   4   One-Dimensional Formal Group Laws 195
   5   The Formal Group Laws of Lubin and Tate 198
   6   Lubin–Tate Extensions 201
   7   The Local Artin Map 210

Bibliography 219

Index 223

Chapter 1
A Brief Review

For the convenience of the reader and to fix notation, in this chapter we recall some basic definitions and theorems for extensions of number fields and their completions. Typically the material discussed in this chapter would be presented in detail in an introductory course on algebraic number theory.

We conclude this chapter with a brief discussion of some questions that arise naturally in the study of algebraic number fields. These questions were important to the development of class field theory. In subsequent chapters, we shall explore some of the mathematics they have inspired. Class field theory provides information on the nature of the abelian extensions of number fields, their ramified primes, their primes that split completely, elements that are norms, etc. We also treat the abelian extensions of local fields in a later chapter, where analogous questions may be asked. The present chapter is intended to be used as a quick reference for the notation, terminology and precursory facts relating to these concepts.

We state nearly all of the results in this chapter without proof. For a more thorough treatment of introductory algebraic number theory, see Fröhlich and Taylor's *Algebraic Number Theory*, [FT], (which includes material on completions), or Marcus' *Number Fields*, [Ma], (which does not). Somewhat more advanced books on the subject include Janusz' *Algebraic Number Fields*, [J], and Lang's *Algebraic Number Theory*, [L1].

1 Number Fields

A *number field* is a finite extension of the field $\mathbb{Q}$ of rational numbers. If F is a number field, denote the ring of algebraic integers in F by $\mathcal{O}_F$. It is well-known that $\mathcal{O}_F$ is a Dedekind domain, so that any ideal of $\mathcal{O}_F$ has a unique factorization into a product of prime ideals. A *fractional ideal* of F is a non-zero finitely generated $\mathcal{O}_F$-submodule of F. The fractional ideals of F form a group $\mathcal{I}_F$ under multiplication; the identity in $\mathcal{I}_F$ is $\mathcal{O}_F$, and for a fractional ideal $\mathfrak{a}$, we have

$$\mathfrak{a}^{-1} = \{x \in F : x\mathfrak{a} \subseteq \mathcal{O}_F\}.$$

N. Childress, *Class Field Theory*, Universitext, DOI 10.1007/978-0-387-72490-4_1,
© Springer Science+Business Media, LLC 2009

The principal fractional ideals of F form a (normal) subgroup of $\mathcal{I}_F$, denoted $\mathcal{P}_F$. The quotient $\mathcal{C}_F = \mathcal{I}_F/\mathcal{P}_F$ is called the *ideal class group* of F. A non-trivial theorem in algebraic number theory says that $\mathcal{C}_F$ is a finite group for any number field F. Its order is the *class number* of F, denoted h_F.

Given a finite extension K/F of algebraic number fields, consider the ideal $\mathfrak{p}\mathcal{O}_K$, where $\mathfrak{p}$ is a non-zero prime ideal of $\mathcal{O}_F$. Using unique factorization of ideals in $\mathcal{O}_K$, we have

$$\mathfrak{p}\mathcal{O}_K = \mathfrak{P}_1^{e_1} \cdots \mathfrak{P}_g^{e_g}$$

where the $\mathfrak{P}_j$ are (distinct) prime ideals of $\mathcal{O}_K$, $g = g(\mathfrak{p})$ is a positive integer and the e_j are positive integers. We call e_j the *ramification index* for $\mathfrak{P}_j/\mathfrak{p}$, denoted $e_j = e(\mathfrak{P}_j/\mathfrak{p})$. If K/F is a Galois extension, then the Galois group permutes the $\mathfrak{P}_j$ transitively, so that $e_1 = \cdots = e_g = e$, say.

Since every non-zero prime ideal is maximal in a Dedekind domain, the quotients $\mathcal{O}_K/\mathfrak{P}_j$ and $\mathcal{O}_F/\mathfrak{p}$ are fields, called *residue fields*. Indeed, they are finite fields of characteristic p, where $\mathfrak{p} \cap \mathbb{Z} = p\mathbb{Z}$. We may view $\mathcal{O}_F/\mathfrak{p}$ as a subfield of $\mathcal{O}_K/\mathfrak{P}_j$. The *residue field degree* is

$$f(\mathfrak{P}_j/\mathfrak{p}) = \left[\mathcal{O}_K/\mathfrak{P}_j : \mathcal{O}_F/\mathfrak{p} \right].$$

If K/F is Galois, then $f(\mathfrak{P}_1/\mathfrak{p}) = \cdots = f(\mathfrak{P}_g/\mathfrak{p}) = f$, say.

In general, we have $\sum_{j=1}^g e(\mathfrak{P}_j/\mathfrak{p})f(\mathfrak{P}_j/\mathfrak{p}) = [K:F]$. When K/F is Galois, this becomes $efg = [K:F]$.

If K/F is an extension of number fields, we say that the prime $\mathfrak{p}$ is *unramified* in K/F if $e(\mathfrak{P}_j/\mathfrak{p}) = 1$ for all j, $\mathfrak{p}$ is *totally ramified* in K/F if there is a unique prime $\mathfrak{P}$ above $\mathfrak{p}$ with $e(\mathfrak{P}/\mathfrak{p}) = [K:F]$, $\mathfrak{p}$ *remains inert* in K/F if $\mathfrak{p}\mathcal{O}_K$ is prime in $\mathcal{O}_K$, and $\mathfrak{p}$ *splits completely* in K/F if $g = [K:F]$.

Given an extension K/F of number fields and a prime ideal $\mathfrak{p}$ of $\mathcal{O}_F$, one approach to finding the factorization of $\mathfrak{p}\mathcal{O}_K$ is the following, sometimes called the *Dedekind-Kummer Theorem*.

Theorem 1.1. *Let K/F be an extension of number fields and suppose $\mathcal{O}_K = \mathcal{O}_F[\alpha]$. Let $f(X) = Irr_F(\alpha, X)$, the irreducible polynomial of α over F, and let $\mathfrak{p}$ be a prime ideal in $\mathcal{O}_F$. Put $\mathbb{F}_\mathfrak{p} = \mathcal{O}_F/\mathfrak{p}$, and denote the image of $f(X)$ in $\mathbb{F}_\mathfrak{p}[X]$ by $\overline{f(X)}$, (reduce the coefficients of f modulo $\mathfrak{p}$). Suppose in $\mathbb{F}_\mathfrak{p}[X]$, the factorization of $\overline{f(X)}$ is given by*

$$\overline{f(X)} = \overline{p_1(X)}^{e_1} \cdots \overline{p_g(X)}^{e_g}$$

where the $\overline{p_j(X)}$ are distinct monic irreducible polynomials in $\mathbb{F}_\mathfrak{p}[X]$. For each j, let $p_j(X)$ be a monic lift of the corresponding $\overline{p_j(X)}$ to $\mathcal{O}_F[X]$, and let $\mathfrak{P}_j$ be the ideal of $\mathcal{O}_K$ generated by $\mathfrak{p}$ and $p_j(\alpha)$. Then

$$\mathfrak{p}\mathcal{O}_K = \mathfrak{P}_1^{e_1} \cdots \mathfrak{P}_g^{e_g}$$

with the $\mathfrak{P}_j$ distinct prime ideals of $\mathcal{O}_K$.   □

The discriminant of an extension of number fields will be of use to us. This can be defined in terms of the discriminants of bases for K as a vector space over F. Recall that if K/F is a finite extension of number fields, with $\{v_1, \ldots, v_n\}$ an F-basis of K, then we define the *discriminant* of this basis to be

$$d(v_1, \ldots, v_n) = \det[\mathrm{Tr}_{K/F}(v_i v_j)] = \det[\sigma_i(v_j)]^2$$

where $\sigma_1, \ldots, \sigma_n : K \hookrightarrow F^{\mathrm{alg}}$ are F-monomorphisms. The relationship between the discriminants of two different bases for K over F can be described in terms of the change of basis matrix between them: If A is an $n \times n$ matrix with $(w_1, \ldots, w_n)^t = A(v_1, \ldots, v_n)^t$, then

$$d(w_1, \ldots, w_n) = (\det A)^2 d(v_1, \ldots, v_n).$$

In the case when $K = F(\alpha)$ where $[K : F] = n$, the matrix $[\sigma_i(\alpha^{j-1})]$ is Vandermonde, so

$$d(1, \ldots, \alpha^{n-1}) = \prod_{1 \leq i < j \leq n} (\sigma_i(\alpha) - \sigma_j(\alpha))^2.$$

Specifically, if $\mathcal{O}_K = \mathcal{O}_F[\alpha]$ and $f(X)$ is the irreducible polynomial of α over F, then $N_{K/F}(f'(\alpha)) = (-1)^{\frac{n(n-1)}{2}} d(1, \alpha, \ldots, \alpha^{n-1})$.

Note that different F-bases for K need not have the same discriminant. Hence the discriminant of the extension K/F must be defined in terms of all the possible bases for K. To do so, we generate a module with all these discriminants.

Suppose M is a non-zero $\mathcal{O}_F$-submodule of K and M contains an F-basis of K. We let $d(M)$ be the $\mathcal{O}_F$-module generated by all $d(v_1, \ldots, v_n)$ where $\{v_1, \ldots, v_n\} \subset M$ varies through the F-bases for K contained in M. Of course if M is a fractional ideal of K then $d(M)$ is a fractional ideal of F. Moreover, if M is a free $\mathcal{O}_F$-module, say $M = \oplus_{i=1}^n \mathcal{O}_F w_i$, then $d(M) = d(w_1, \ldots, w_n)\mathcal{O}_F$.

The *(relative) discriminant* of the extension K/F is $d_{K/F} = d(\mathcal{O}_K)$, where $\mathcal{O}_K$ is considered as a (finitely generated) $\mathcal{O}_F$-module. This makes $d_{K/F}$ an integral ideal of $\mathcal{O}_F$. The *(absolute) discriminant* of K is $d_K = d_{K/\mathbb{Q}}$. Note that $\mathcal{O}_K$ is a free $\mathbb{Z}$-module of rank $n = [K : \mathbb{Q}]$, so $d_K = d_{K/\mathbb{Q}}$ is a (principal) ideal in $\mathbb{Z}$, generated by $d(v_1, \ldots, v_n)$ where $\{v_1, \ldots, v_n\}$ is any integral basis for $\mathcal{O}_K$, (by *integral basis* we mean a $\mathbb{Z}$-basis for $\mathcal{O}_K$).

One of the reasons why discriminants will be useful to us is that they carry information about the primes that ramify in an extension. For a non-zero prime ideal $\mathfrak{p}$ of $\mathcal{O}_F$, we have that $\mathfrak{p}$ is ramified in K/F if and only if $\mathfrak{p} \mid d_{K/F}$.

We shall make heavy use of the notion of the norm of a fractional ideal. We record its definition and a few basic facts here. Let K/F be a finite extension of number fields. Let $\mathfrak{p}$ be a non-zero prime ideal of $\mathcal{O}_F$ and let $\mathfrak{P}$ be a prime of $\mathcal{O}_K$ dividing $\mathfrak{p}\mathcal{O}_K$. Define the *norm* of $\mathfrak{P}$ as $N_{K/F}(\mathfrak{P}) = \mathfrak{p}^{f(\mathfrak{P}/\mathfrak{p})}$. Now extend $N_{K/F}$ to arbitrary fractional ideals of K by multiplicativity, i.e.,

$$N_{K/F}(\mathfrak{P}_1^{a_1} \cdots \mathfrak{P}_t^{a_t}) = N_{K/F}(\mathfrak{P}_1)^{a_1} \cdots N_{K/F}(\mathfrak{P}_t)^{a_t}.$$

Thus the norm of a fractional ideal of K is a fractional ideal of F. Note that if K/F is Galois, then

$$N_{K/F}(\mathfrak{A})\mathcal{O}_K = \prod_{\sigma \in \mathrm{Gal}(K/F)} \sigma(\mathfrak{A}).$$

If $\alpha \in K$, then $N_{K/F}(\alpha\mathcal{O}_K) = N_{K/F}(\alpha)\mathcal{O}_F$, where the norm on the right is the usual element norm. Also, if $F \subseteq E \subseteq K$ are number fields, then

$$N_{K/F} = N_{E/F} \circ N_{K/E}.$$

The specific case when $F = \mathbb{Q}$ gives $N_{K/\mathbb{Q}}(\mathfrak{A}) = a\mathbb{Z}$ for some $a \in \mathbb{Q}$. We shall sometimes write $N\mathfrak{A}$ for $N_{K/\mathbb{Q}}(\mathfrak{A})$, and frequently in our expressions for Dirichlet series we shall also use $N\mathfrak{A}$ to represent the non-negative generator $|a|$ of $a\mathbb{Z}$.

Given a Galois extension of number fields K/F with Galois group G, a non-zero prime ideal $\mathfrak{p}$ of $\mathcal{O}_F$ and a prime ideal $\mathfrak{P}$ of $\mathcal{O}_K$ with $\mathfrak{P}|\mathfrak{p}\mathcal{O}_K$, we define the *decomposition group*

$$Z(\mathfrak{P}/\mathfrak{p}) = \{\sigma \in G : \sigma(\mathfrak{P}) = \mathfrak{P}\}.$$

Note that $Z(\mathfrak{P}/\mathfrak{p})$ acts on the finite field $\mathbb{F}_\mathfrak{P} = {}^{\mathcal{O}_K}/_\mathfrak{P}$, fixing the subfield $\mathbb{F}_\mathfrak{p} = {}^{\mathcal{O}_F}/_\mathfrak{p}$ elementwise, so there is a natural homomorphism of groups

$$Z(\mathfrak{P}/\mathfrak{p}) \to \mathrm{Gal}(\mathbb{F}_\mathfrak{P}/\mathbb{F}_\mathfrak{p}).$$

From algebraic number theory, we have the following.

Theorem 1.2. *Let K/F be a Galois extension of number fields with Galois group G. Let $\mathfrak{p}$ be a non-zero prime ideal of $\mathcal{O}_F$.*

  i. *G acts transitively on the set of prime ideals $\mathfrak{P}$ of $\mathcal{O}_K$ that divide $\mathfrak{p}\mathcal{O}_K$ whence*

$$[G : Z(\mathfrak{P}/\mathfrak{p})] = \#\{primes\ \mathfrak{P}\ of\ \mathcal{O}_K : \mathfrak{P}|\mathfrak{p}\mathcal{O}_K\} = g.$$

   Also, if $\mathfrak{P}$, $\mathfrak{P}'$ are prime ideals of $\mathcal{O}_K$ dividing $\mathfrak{p}\mathcal{O}_K$, then $Z(\mathfrak{P}/\mathfrak{p})$ and $Z(\mathfrak{P}'/\mathfrak{p})$ are G-conjugate.
  ii. *$N\mathfrak{p} = \#\mathbb{F}_\mathfrak{p}$, $N\mathfrak{P} = \#\mathbb{F}_\mathfrak{P}$ and $\mathrm{Gal}(\mathbb{F}_\mathfrak{P}/\mathbb{F}_\mathfrak{p})$ is cyclic, generated by the Frobenius automorphism $\varphi_\mathfrak{p} : x \mapsto x^{N\mathfrak{p}}$.*

iii. *The homomorphism* $Z(\mathfrak{P}/\mathfrak{p}) \to \mathrm{Gal}(\mathbb{F}_{\mathfrak{P}}/\mathbb{F}_{\mathfrak{p}})$ *is surjective; its kernel is called the inertia subgroup, denoted* $T(\mathfrak{P}/\mathfrak{p})$. *Note that* $[Z(\mathfrak{P}/\mathfrak{p}) : T(\mathfrak{P}/\mathfrak{p})] = f$ *and* $T(\mathfrak{P}/\mathfrak{p})$ *has order* e.   □

In the case of a Galois extension K/F of number fields, the decomposition group and inertia subgroup give rise, via the Galois correspondence, to intermediate fields, called the *decomposition field* and *inertia field*, respectively. Let K_Z be the fixed field of $Z(\mathfrak{P}/\mathfrak{p})$ and let K_T be the fixed field of $T(\mathfrak{P}/\mathfrak{p})$. For an abelian extension, the factorization of the ideals generated by $\mathfrak{p}$ in these intermediate fields is given by the following result from algebraic number theory.

Theorem 1.3 (Layer Theorem). *Let* $\mathfrak{p}$ *be a non-zero prime ideal of* $\mathcal{O}_F$, *where* K/F *is an abelian extension of number fields. Then* $\mathfrak{p}$ *splits completely in* K_Z/F. *The primes above* $\mathfrak{p}$ *remain inert in* K_T/K_Z *and ramify totally in* K/K_T.   □

If $e(\mathfrak{P}/\mathfrak{p}) = 1$, then via the natural homomorphism in (*iii*) of Theorem 1.2, we have $Z(\mathfrak{P}/\mathfrak{p}) \cong \mathrm{Gal}(\mathbb{F}_{\mathfrak{P}}/\mathbb{F}_{\mathfrak{p}})$ is cyclic of order f. The Galois group for the residue fields is generated by the Frobenius automorphism $\varphi_{\mathfrak{p}}$, whence there is a unique element $\sigma \in Z(\mathfrak{P}/\mathfrak{p})$ that corresponds to $\varphi_{\mathfrak{p}}$ under the natural isomorphism. We have $Z(\mathfrak{P}/\mathfrak{p}) = \langle\sigma\rangle$. This element σ is called the *Frobenius element at* $\mathfrak{P}$. We denote it $\sigma = \left(\frac{\mathfrak{P}}{K/F}\right) = (\mathfrak{P}, K/F)$.

Proposition 1.4. Let K/F be a Galois extension of number fields, $\mathfrak{p}$ a non-zero prime of $\mathcal{O}_F$ that is unramified in K/F and $\mathfrak{P}$ a prime of $\mathcal{O}_K$ with $\mathfrak{P}|\mathfrak{p}\mathcal{O}_K$. Then the Frobenius element at $\mathfrak{P}$ is the unique element $\sigma \in \mathrm{Gal}(K/F)$ that satisfies $\sigma(\alpha) \equiv \alpha^{N\mathfrak{p}} \pmod{\mathfrak{P}}$ for every $\alpha \in \mathcal{O}_K$.

Proof. Say $\sigma(\alpha) \equiv \alpha^{N\mathfrak{p}} \pmod{\mathfrak{P}}$ for all $\alpha \in \mathcal{O}_K$. From this congruence we see that $\sigma(\mathfrak{P}) \subseteq \mathfrak{P}$, whence $\sigma(\mathfrak{P}) = \mathfrak{P}$, i.e., $\sigma \in Z(\mathfrak{P}/\mathfrak{p})$. Clearly, the isomorphism $Z(\mathfrak{P}/\mathfrak{p}) \cong \mathrm{Gal}(\mathbb{F}_{\mathfrak{P}}/\mathbb{F}_{\mathfrak{p}})$ maps σ to $\varphi_{\mathfrak{p}}$. Thus $\sigma = \left(\frac{\mathfrak{P}}{K/F}\right)$.   □

If we suppose further that G is abelian, then by (*i*) of Theorem 1.2 we know that $Z(\mathfrak{P}/\mathfrak{p})$ depends only on $\mathfrak{p}$ and we may write $Z(\mathfrak{p})$. Also, if $\mathfrak{p}$ is unramified in K/F, then we show in Proposition 1.5 below that the Frobenius element at $\mathfrak{P}$ depends only on $\mathfrak{p}$. In this case, we call it the *Artin automorphism* for $\mathfrak{p}$, denoted $\left(\frac{\mathfrak{p}}{K/F}\right) = (\mathfrak{p}, K/F)$. We may define a map

$$\{\text{primes of } \mathcal{O}_F \text{ that are unramified in } K/F\} \to G$$

given by $\mathfrak{p} \mapsto \sigma_{\mathfrak{p}} = \left(\frac{\mathfrak{p}}{K/F}\right)$.

Proposition 1.5. Let K/F be an abelian extension of number fields, $\mathfrak{p}$ a non-zero prime of $\mathcal{O}_F$ that is unramified in K/F and $\mathfrak{P}$ a prime of $\mathcal{O}_K$ with $\mathfrak{P}|\mathfrak{p}\mathcal{O}_K$. Then $\sigma = \left(\frac{\mathfrak{P}}{K/F}\right)$ does not depend on the choice of the prime $\mathfrak{P}$ above $\mathfrak{p}$.

Proof. To show independence, suppose the primes $\mathfrak{P}$ and $\mathfrak{P}'$ divide $p\mathcal{O}_K$. Write σ, σ' for the Frobenius elements at $\mathfrak{P}$, $\mathfrak{P}'$, respectively. Now by (i) of Theorem 1.2, there exists $\tau \in G$ such that $\tau(\mathfrak{P}) = \mathfrak{P}'$, so

$$\tau\sigma(\alpha) \equiv \tau(\alpha^{N\mathfrak{p}}) \pmod{\mathfrak{P}'}$$
$$\equiv \tau(\alpha)^{N\mathfrak{p}} \pmod{\mathfrak{P}'} \quad \forall \alpha \in \mathcal{O}_K.$$

But $\tau\sigma = \sigma\tau$ since G is abelian, so $\sigma(\tau(\alpha)) \equiv \tau(\alpha)^{N\mathfrak{p}} \pmod{\mathfrak{P}'}$ for all $\alpha \in \mathcal{O}_K$. Since τ is a bijection, we must have $\sigma(\alpha) \equiv \alpha^{N\mathfrak{p}} \pmod{\mathfrak{P}'}$ for all $\alpha \in \mathcal{O}_K$, whence $\sigma = \sigma'$. $\qquad\qquad\qquad\qquad\qquad\qquad\qquad\qquad\qquad\qquad\qquad\qquad\square$

Proposition 1.6. Suppose K/F is an abelian extension with Galois group G and we have a field L with $F \subseteq L \subseteq K$, (so L/F and K/L are also abelian). Let $\mathfrak{p}$ be a prime of $\mathcal{O}_F$ that is unramified in K/F. Then $\left(\frac{\mathfrak{p}}{L/F}\right)$ and $\left(\frac{\mathfrak{p}}{K/F}\right)$ are both defined and

$$\left(\frac{\mathfrak{p}}{L/F}\right) = \left(\frac{\mathfrak{p}}{K/F}\right)\Big|_L .$$

Proof. Let $\sigma = \left(\frac{\mathfrak{p}}{K/F}\right)$, $\sigma' = \left(\frac{\mathfrak{p}}{L/F}\right)$, and let $\mathfrak{P}$ be a prime ideal of $\mathcal{O}_K$ above $\mathfrak{p}$. Then $\sigma(\alpha) \equiv \alpha^{N\mathfrak{p}} \pmod{\mathfrak{P}}$ for all $\alpha \in \mathcal{O}_K$. Let $\mathfrak{P}' = \mathfrak{P} \cap \mathcal{O}_L$. For every $\alpha \in \mathcal{O}_L$ we have $\sigma(\alpha) \equiv \alpha^{N\mathfrak{p}} \pmod{\mathfrak{P}'}$. Thus $\sigma\big|_L = \sigma'$. $\qquad\qquad\qquad\qquad\square$

Exercise 1.1. Suppose K/F is Galois but not necessarily abelian. Let $\mathfrak{P}$ be a prime of $\mathcal{O}_K$ above $\mathfrak{p}$, and suppose $e(\mathfrak{P}/\mathfrak{p}) = 1$.

a. Find and prove a statement similar to Proposition 1.6 for the Frobenius element at $\mathfrak{P}$.

b. Suppose L is an intermediate field in the extension K/F. How are $\left(\frac{\mathfrak{P}}{K/L}\right)$ and $\left(\frac{\mathfrak{P}}{K/F}\right)$ related?

c. Fix the prime ideal $\mathfrak{p}$ of $\mathcal{O}_F$ and let $\mathfrak{P}$ vary through all the prime ideals of $\mathcal{O}_K$ above $\mathfrak{p}$. Show that the set $\left\{\left(\frac{\mathfrak{P}}{K/F}\right) : \mathfrak{P}|p\mathcal{O}_K\right\}$ is a conjugacy class in $G = \mathrm{Gal}\,(K/F)$. $\qquad\qquad\qquad\qquad\qquad\qquad\qquad\qquad\qquad\qquad\diamond$

Example.

1. Let $F = \mathbb{Q}$, $K = \mathbb{Q}(\zeta_m)$; then $G = \mathrm{Gal}\,(K/F) \cong \left(\mathbb{Z}/m\mathbb{Z}\right)^\times$. We may assume that m is either odd or divisible by 4, so that $p|m$ if and only if $p\mathbb{Z}$ ramifies in $K/\mathbb{Q}$. Suppose $p \nmid m$. Let $\sigma = \left(\frac{p\mathbb{Z}}{\mathbb{Q}(\zeta_m)/\mathbb{Q}}\right)$ and suppose $\mathfrak{P}|p\mathbb{Z}$. Then $\sigma(\alpha) \equiv \alpha^p \pmod{\mathfrak{P}}$. In particular, $\sigma(\zeta_m) \equiv \zeta_m^p \pmod{\mathfrak{P}}$. We claim this implies $\sigma(\zeta_m) = \zeta_m^p$. Suppose we have verified this claim. Then $\sigma(\zeta_m) = \zeta_m^p$. Since

$\zeta_m^p = \sigma_p(\zeta_m)$ and ζ_m generates $K/\mathbb{Q}$, it follows that $\sigma = \sigma_p$. The Artin automorphism is the same as the pth power map in this case. It remains to verify the claim. The following proposition resolves this.

Proposition 1.7. If ζ, ζ' are m^{th} roots of unity in K and $\mathfrak{P}|p\mathbb{Z}$ is unramified, with $\zeta \equiv \zeta' \pmod{\mathfrak{P}}$, then $\zeta = \zeta'$.

Proof. Let μ_m denote the set of mth roots of unity and consider the polynomial $X^m - 1 = \prod_{\eta \in \mu_m}(X - \eta)$. Differentiate to obtain:

$$mX^{m-1} = \sum_{\eta \in \mu_m} \prod_{\eta' \neq \eta}(X - \eta').$$

Now evaluate for $X = \zeta$:

$$m\zeta^{m-1} = \sum_{\eta \in \mu_m} \prod_{\eta' \neq \eta}(\zeta - \eta') = \prod_{\eta' \neq \zeta}(\zeta - \eta').$$

Suppose $\zeta \equiv \zeta' \pmod{\mathfrak{P}}$ and $\zeta \neq \zeta'$. Then $\prod_{\eta' \neq \zeta}(\zeta - \eta') \equiv 0 \pmod{\mathfrak{P}}$, which yields $m\zeta^{m-1} \equiv 0 \pmod{\mathfrak{P}}$. It follows that $\mathfrak{P}|m\zeta^{m-1}\mathcal{O}_K$, so $\mathfrak{P}|m\mathcal{O}_K$. We conclude that $p|m$ and thus p is ramified (a contradiction). $\square$

We shall encounter the Artin automorphism again in Chapter V; it plays a central role in our proofs of the main theorems of class field theory. For now, we are content to use it to show the following result on primes that split completely in subextensions of cyclotomic fields. The reader is encouraged to keep this result in mind as we discuss the definition of class field in Chapter 3.

Theorem 1.8. *If $K \subseteq \mathbb{Q}(\zeta_m)$, then identify $\text{Gal}(\mathbb{Q}(\zeta_m)/\mathbb{Q}) \cong \left(\mathbb{Z}/m\mathbb{Z}\right)^\times$ and let $H < \left(\mathbb{Z}/m\mathbb{Z}\right)^\times$ be the subgroup corresponding to $\text{Gal}(\mathbb{Q}(\zeta_m)/K)$. The primes $p \nmid m$ that split completely in $K/\mathbb{Q}$ are those p such that $p \bmod m \in H$.*

Proof. The primes that split completely in $K/\mathbb{Q}$ are precisely the primes with trivial decomposition group, hence precisely the unramified primes with trivial Artin automorphism.

Since $p \nmid m$, we have that p is unramified. Hence p splits completely if and only if its Artin automorphism is trivial: $\left(\frac{p\mathbb{Z}}{K/\mathbb{Q}}\right) = 1$. But $\left(\frac{p\mathbb{Z}}{K/\mathbb{Q}}\right) = \left(\frac{p\mathbb{Z}}{\mathbb{Q}(\zeta_m)/\mathbb{Q}}\right)\Big|_K = \sigma_p\Big|_K$. Thus

$$p \text{ splits completely in } K/\mathbb{Q} \iff \sigma_p|_K = 1$$
$$\iff \sigma_p \in \text{Gal}(\mathbb{Q}(\zeta_m)/K)$$
$$\iff p \bmod m \in H. \qquad \square$$

For example, when $m = 13$, $\mathbb{Q}(\zeta_{13})/\mathbb{Q}$ is of degree 12 and has cyclic Galois group. Let K be the unique subfield of $\mathbb{Q}(\zeta_{13})$ with $[K : \mathbb{Q}] = 3$. In this case, we have $H \cong \mathrm{Gal}\,(\mathbb{Q}(\zeta_{13})/K)$ and H is the unique subgroup of $\left(\mathbb{Z}/13\mathbb{Z}\right)^{\times}$ of order 4. Thus $H = \{1, 5, 12, 8\} = \langle 5 \rangle$. By Theorem 1.8,

$$p \text{ splits completely in } K/\mathbb{Q} \iff p \equiv 1, 5, 8, \text{ or } 12 \pmod{13}.$$

Finally, for a number field F, we shall need to understand the group $\mathcal{O}_F^{\times}$. Let r_1 denote the number of embeddings $F \hookrightarrow \mathbb{R}$ and let r_2 denote the number of conjugate pairs of imaginary embeddings $F \hookrightarrow \mathbb{C}$. (Then $[F : \mathbb{Q}] = r_1 + 2r_2$.) The group $\mathcal{O}_F^{\times}$ is described by the following.

Theorem 1.9 (Dirichlet Unit Theorem). *Let F be a number field and let r_1 and r_2 represent the number of real embeddings and the number of conjugate pairs of imaginary embeddings of F, respectively. There are units $\varepsilon_1, \ldots, \varepsilon_{r_1+r_2-1} \in \mathcal{O}_F^{\times}$ such that*

$$\mathcal{O}_F^{\times} \cong \mathcal{W}_F \times \langle \varepsilon_1 \rangle \times \cdots \times \langle \varepsilon_{r_1+r_2-1} \rangle$$
$$\cong \mathcal{W}_F \times \mathbb{Z}^{r_1+r_2-1},$$

where $\mathcal{W}_F$ is the group of roots of unity in F. The ε_j are called a fundamental system of units for F. □

Among many other things, fundamental units are used to define the *regulator* of F. Let $\{\varepsilon_1, \ldots, \varepsilon_r\}$ be a fundamental system of units for F, where $r = r_1 + r_2 - 1$. Consider the matrix

$$A = [\log |\sigma_j(\varepsilon_i)|_j] \quad \text{for } 1 \le i \le r \text{ and } 1 \le j \le r_1 + r_2$$

where:

$$|x|_j = \begin{cases} |x| & \text{if } 1 \le j \le r_1 \\ |x|^2 & \text{if } r_1 + 1 \le j \le r_1 + r_2. \end{cases}$$

Here $|\cdot|$ is the usual absolute value on $\mathbb{C}$, and $\sigma_1, \ldots, \sigma_{r_1}$ are all the real embeddings of F, while $\sigma_{r_1+1}, \ldots, \sigma_{r_1+r_2}$ are a set of representatives (one from each conjugate pair) of the imaginary embeddings. The regulator R_F is the absolute value of the determinant of any $r \times r$ minor of A. It is independent of the choice of the fundamental system $\{\varepsilon_1, \ldots, \varepsilon_r\}$. The volume of a fundamental parallelopiped in the lattice associated to $\mathcal{O}_F$ may be expressed in terms of the regulator as $\sqrt{r_1 + r_2}\, R_F$.

2 Completions of Number Fields

As our study of class field theory progresses, it will become evident that the various completions of a number field F all play equally important roles. We gather some of the basic facts about completions in this section.

Let F be an algebraic number field. An *absolute value* on the field F is a function $F \to [0, \infty)$, where the image of $x \in F$ is denoted $|x|$, such that

  i. $|x| = 0$ if and only if $x = 0$
 ii. $|xy| = |x||y|$ for all $x, y \in F$
iii. $|x + y| \leq |x| + |y|$ for all $x, y \in F$.

A *non-Archimedean absolute value* is an absolute value that satisfies the stronger condition

iii'. $|x + y| \leq \max\{|x|, |y|\}$ for all $x, y \in F$.

If $| \cdot |$ is an absolute value on F, then it gives rise to a metric on F via $d(x, y) = |x - y|$. We say two absolute values on F are *equivalent* if they give rise to the same metric topology on F. When we discuss places a bit later, we shall relax our definition of absolute value slightly, but it will turn out that this relaxed notion of absolute value does not alter the topological situation: Any such (relaxed) absolute value will induce a topology homeomorphic to the topology induced by an absolute value satisfying (*i*), (*ii*) and (*iii*), above.

To obtain a non-Archimedean absolute value on a number field F, let $\mathfrak{p}$ be a non-zero prime ideal of $\mathcal{O}_F$. For $x \in F^\times$, we may factor the fractional ideal generated by x as a product of prime ideals, paying particular attention to our fixed prime $\mathfrak{p}$; say

$$x\mathcal{O}_F = \mathfrak{p}^\alpha \mathfrak{p}_1^{\alpha_1} \cdots \mathfrak{p}_t^{\alpha_t},$$

where the $\mathfrak{p}_j$ are distinct to $\mathfrak{p}$, and $\alpha, \alpha_j \in \mathbb{Z}$. By unique factorization of fractional ideals in F, we may define

$$\operatorname{ord}_\mathfrak{p}(x) = \alpha.$$

The function $\operatorname{ord}_\mathfrak{p} : F^\times \to \mathbb{Z}$ is what is known as a *discrete valuation* on the field F. (To be a valuation, a function $v : F^\times \to \mathbb{R}$ must satisfy $v(xy) = v(x) + v(y)$ and $v(x + y) \geq \min\{v(x), v(y)\}$. To be discrete, its image $v(F^\times)$ must be a discrete subgroup of $\mathbb{R}$.) In fact, $\operatorname{ord}_\mathfrak{p}$ is *normalized*, meaning that $\operatorname{ord}_\mathfrak{p}(F^\times) = \mathbb{Z}$. Using it, we define the $\mathfrak{p}$-*adic absolute value* on F as follows. Fix a real number c, with $0 < c < 1$ and define

$$|x|_\mathfrak{p} = \begin{cases} 0 & \text{if } x = 0 \\ c^{\operatorname{ord}_\mathfrak{p}(x)} & \text{if } x \neq 0. \end{cases}$$

(Different choices of the real number c yield equivalent absolute values.) It is easy to verify that $| \cdot |_\mathfrak{p}$ is a non-Archimedean absolute value on F. Also, if $\mathfrak{p} \neq \mathfrak{q}$, then $| \cdot |_\mathfrak{p}$ and $| \cdot |_\mathfrak{q}$ are inequivalent absolute values on F. Typical choices for the real number c are $1/p$, where $p\mathbb{Z} = \mathfrak{p} \cap \mathbb{Z}$, or $1/N\mathfrak{p}$, where $N\mathfrak{p}$ is the positive generator of $N_{F/\mathbb{Q}}(\mathfrak{p})$.

One may obtain an Archimedean absolute value on a number field F from any embedding $\sigma : F \hookrightarrow \mathbb{C}$ by setting

$$|x|_\sigma = |\sigma(x)| \qquad \text{where } |\cdot| \text{ is the usual absolute value on } \mathbb{C}.$$

Let $\sigma_1, \ldots, \sigma_{r_1}$ be all the distinct real embeddings of F and let $\sigma_{r_1+1}, \ldots, \sigma_{r_1+r_2}$ be distinct imaginary embeddings of F, chosen so that each conjugate pair of imaginary embeddings is represented exactly once. One may show that the absolute values $|\cdot|_{\sigma_1}, \ldots, |\cdot|_{\sigma_{r_1+r_2}}$ are pairwise inequivalent on F, while the absolute values $|\cdot|_\tau$ and $|\cdot|_{\bar\tau}$ for a conjugate pair of imaginary embeddings of F are equivalent on F.

For example, $F = \mathbb{Q}(\sqrt{3})$ has two inequivalent Archimedean absolute values:

$$|a + b\sqrt{3}|_1 = |a + b\sqrt{3}|_\mathbb{R}$$
$$|a + b\sqrt{3}|_2 = |a - b\sqrt{3}|_\mathbb{R}.$$

By a theorem of Ostrowski, every non-trivial absolute value on a number field F arises (up to equivalence) as a $|\cdot|_\mathfrak{p}$ or a $|\cdot|_\sigma$ for some non-zero prime $\mathfrak{p}$ of $\mathcal{O}_F$ or some embedding $\sigma : F \hookrightarrow \mathbb{C}$.

The *completion* of an algebraic number field F with respect to the absolute value $|\cdot|$ is the field

$$\{\text{Cauchy sequences in } F\} \Big/ \{\text{Null sequences in } F\}.$$

Via constant sequences, we may view F as a subfield of any of its completions. If $|\cdot| = |\cdot|_\mathfrak{p}$ for some non-zero prime ideal $\mathfrak{p}$ of $\mathcal{O}_F$, then we denote the completion of F by $F_\mathfrak{p}$. If $|\cdot| = |\cdot|_\sigma$ for some embedding $\sigma : F \hookrightarrow \mathbb{C}$, then the completion of F is isomorphic either to $\mathbb{R}$ or $\mathbb{C}$, according as $\sigma(F) \subseteq \mathbb{R}$ or not.

Let us examine the $\mathfrak{p}$-adic completion $F_\mathfrak{p}$ in more detail. Let

$$\mathcal{O}_\mathfrak{p} = \{x \in F_\mathfrak{p} : |x|_\mathfrak{p} \leq 1\},$$

the ring of $\mathfrak{p}$-*adic integers*. $\mathcal{O}_\mathfrak{p}$ is a local ring (in fact a discrete valuation ring) with unique maximal ideal

$$\mathcal{P}_\mathfrak{p} = \{x \in F_\mathfrak{p} : |x|_\mathfrak{p} < 1\}.$$

(The units of $\mathcal{O}_\mathfrak{p}$ are precisely those elements having absolute value one.) Note that $\mathcal{O}_F$ is a subring of $\mathcal{O}_\mathfrak{p}$ and is dense in $\mathcal{O}_\mathfrak{p}$. We also have $\mathcal{P}_\mathfrak{p} = \mathfrak{p}\mathcal{O}_\mathfrak{p}$ and

$$\mathcal{O}_\mathfrak{p} \Big/ \mathcal{P}_\mathfrak{p} \cong \mathcal{O}_F \Big/ \mathfrak{p}.$$

Hence $F_\mathfrak{p}$ is a *local field*, i.e., it is complete with respect to a discrete valuation, and it has finite residue field.

Choose $\pi \in \mathfrak{p} - \mathfrak{p}^2$ and view π as an element of $F_\mathfrak{p}$. Observe that $\mathcal{P}_\mathfrak{p} = \langle \pi \rangle$; we say π is a *uniformizer* in $F_\mathfrak{p}$. Every element $x \in F_\mathfrak{p}$ may be written as $x = \varepsilon\pi^t$, where $t \in \mathbb{Z}$ and $\varepsilon \in \mathcal{O}_\mathfrak{p}^\times$. This leads to the $\mathfrak{p}$-adic expansion of x:

$$x = \sum_{j=t}^{\infty} a_j \pi^j$$

where the a_j may be chosen from a (finite) set of distinct representatives for $\mathcal{O}_\mathfrak{p}/\mathcal{P}_\mathfrak{p}$.

In particular, when $F = \mathbb{Q}$, we have $\mathfrak{p} = p\mathbb{Z}$ for some prime number p. Denote the completion by $\mathbb{Q}_p$, and the ring of p-adic integers by $\mathbb{Z}_p$. In this case, we may choose the a_j from the set $\{0, \ldots, p-1\}$.

Again let F be a number field and let $\mathfrak{p}$ be a prime ideal of $\mathcal{O}_F$. The following result allows us to lift a factorization of a polynomial over the residue field $\mathcal{O}_F/\mathfrak{p}$ to a factorization over the completion $F_\mathfrak{p}$. It is one version among many that bear the same title. While similar statements hold in more general settings, this version is all that we require.

Theorem 2.1 (Hensel's Lemma). *Let F be a number field and let $\mathfrak{p}$ be a non-zero prime ideal of $\mathcal{O}_F$. Suppose we are given a monic polynomial $f(X) \in \mathcal{O}_\mathfrak{p}[X]$, say of degree n. Let $\bar{f}(X)$ be the canonical image of $f(X)$ in the ring $\mathcal{O}_\mathfrak{p}/\mathcal{P}_\mathfrak{p}[X]$, (reduce the coefficients modulo $\mathcal{P}_\mathfrak{p}$). If we have a factorization $\bar{f}(X) = \bar{g}(X)\bar{h}(X)$ where $\deg(\bar{g}) = t$ and where $\bar{g}$ and $\bar{h}$ are monic and coprime in $\mathcal{O}_\mathfrak{p}/\mathcal{P}_\mathfrak{p}[X]$, then there is a factorization $f(X) = g(X)h(X)$ in $\mathcal{O}_\mathfrak{p}[X]$, where $\deg g = t$ and g, h are monic polynomials that reduce modulo $\mathcal{P}_\mathfrak{p}$ to $\bar{g}$, $\bar{h}$ respectively.* $\square$

In particular, note that if we apply Hensel's Lemma to the special case where $g(X)$ is linear, we find that a simple zero of $\bar{f}(X)$ lifts to a simple zero of $f(X)$. Thus it is perhaps not surprising that the proof of Hensel's Lemma may be regarded as a generalization of Newton's method for approximating a zero of a polynomial. See Exercise 1.3 for an opportunity to apply Hensel's Lemma.

Observe that $\mathbb{Q}_p$ is not algebraically closed, but unlike the situation for the Archimedean absolute value on $\mathbb{Q}$, when one takes an algebraic closure $\mathbb{Q}_p^{alg}$ for $\mathbb{Q}_p$, it will not be complete! (Finite extensions of $\mathbb{Q}_p$ are complete, but the algebraic closure is an infinite extension.) Luckily, the completion of an algebraic closure of $\mathbb{Q}_p$ is both algebraically closed and complete. Thus we have a field that is analogous to the complex numbers $\mathbb{C}$ in this regard. We denote it $\mathbb{C}_p$. There is a unique absolute value on $\mathbb{C}_p$ that extends the p-adic absolute value of $\mathbb{Q}_p$. We continue to use $|\cdot|_p$ to denote this absolute value on $\mathbb{C}_p$.

At first glance, the Archimedean and non-Archimedean absolute values on a number field seem to arise in very different ways. However, this is something of a misconception. To illustrate, we first introduce the following notational convention. Let $|\cdot|_\infty$ denote the usual Archimedean absolute value on $\mathbb{Q}$. Allowing p to be any fixed prime, or $p = \infty$, observe that if we have an absolute value on a number field F that extends $|\cdot|_p$ from $\mathbb{Q}$, then the completion of F with respect to this absolute value is the compositum $F\mathbb{Q}_p$, (where $\mathbb{Q}_\infty$ is $\mathbb{R}$).

Still taking p to be any fixed prime or $p = \infty$, let F be a number field and suppose $\sigma : F \hookrightarrow \mathbb{Q}_p^{alg}$ is an embedding, (where $\mathbb{Q}_\infty^{alg} = \mathbb{C}$). We may define for $x \in F$,

$$|x|_\sigma = |\sigma(x)|_p.$$

Then $|\cdot|_\sigma$ is readily seen to be an absolute value on F that extends $|\cdot|_p$ on $\mathbb{Q}$. Moreover, every extension of $|\cdot|_p$ to F arises (up to topological equivalence) as some $|\cdot|_\sigma$. When do two embeddings σ_1 and σ_2 of F give rise to equivalent absolute values on F? Precisely when they are conjugate embeddings, i.e., when there is an automorphism φ of $\mathbb{Q}_p^{\mathrm{alg}}$ such that $\varphi \circ \sigma_1 = \sigma_2$. Thus there is a bijective correspondence

{topologically distinct extensions of $|\cdot|_p$ to F}

$$\longleftrightarrow \{\text{conjugacy classes of embeddings } \sigma : F \hookrightarrow \mathbb{Q}_p^{\mathrm{alg}}\}.$$

If we write the number field F as a simple extension of $\mathbb{Q}$, say $F = \mathbb{Q}(\alpha)$, then we may use the monic irreducible polynomial $f(X)$ of α over $\mathbb{Q}$ to learn more about the extensions of $|\cdot|_p$. In $\mathbb{Q}_p[X]$, we factor $f(X)$ into irreducibles, say

$$f(X) = h_1(X) \cdots h_g(X)$$

with the $h_i(X)$ distinct irreducible polynomials over $\mathbb{Q}_p$. Then two embeddings σ_1, σ_2 are conjugate if and only if they send α to roots of the same $h_i(X)$. Thus there are g distinct extensions of $|\cdot|_p$ to F. Moreover, if $\sigma(\alpha)$ is a root of $h_j(X)$ and the completion of F with respect to $|\cdot|_\sigma$ is denoted F_σ, then $\deg h_j = [F_\sigma : \mathbb{Q}_p]$.

Returning exclusively to the non-Archimedean case, suppose F is an algebraic number field and $\mathfrak{p}$ is a non-zero prime ideal of $\mathcal{O}_F$, with $\mathfrak{p} \cap \mathbb{Z} = p\mathbb{Z}$. The restriction of the absolute value $|\cdot|_\mathfrak{p}$ to $\mathbb{Q}$ gives the same topology on $\mathbb{Q}$ as $|\cdot|_p$ does. The completion $F_\mathfrak{p}$ is a finite extension of $\mathbb{Q}_p$ that can be embedded as a subfield of $\mathbb{Q}_p^{\mathrm{alg}}$, hence also of $\mathbb{C}_p$. There is one subtle point. This embedding must preserve the topology, so must be chosen so that elements of $F_\mathfrak{p}$ that are "close" with respect to the $\mathfrak{p}$-adic absolute value are mapped to elements of $\mathbb{C}_p$ that are close with respect to $|\cdot|_p$. By the above, we know that $|\cdot|_\mathfrak{p}$ arises from an embedding $\sigma : F \hookrightarrow \mathbb{Q}_p^{\mathrm{alg}}$. We have $F_\mathfrak{p} = F_\sigma$ and our embedding of $F_\mathfrak{p}$ into $\mathbb{C}_p$ will extend σ.

Suppose K/F is an extension of algebraic number fields and $\mathfrak{p}$ is a non-zero prime ideal of $\mathcal{O}_F$. We may factor

$$\mathfrak{p}\mathcal{O}_K = \mathfrak{P}_1^{e_1} \cdots \mathfrak{P}_g^{e_g}$$

where the $\mathfrak{P}_j$ are distinct prime ideals of $\mathcal{O}_K$ and the e_j are positive integers. The absolute values $|\cdot|_{\mathfrak{P}_1}, \ldots, |\cdot|_{\mathfrak{P}_g}$ are pairwise inequivalent on K, but their restrictions to the subfield F all produce the same topology on F as $|\cdot|_\mathfrak{p}$ does. The absolute values $|\cdot|_{\mathfrak{P}_1}, \ldots, |\cdot|_{\mathfrak{P}_g}$ represent all the topologically inequivalent extensions of $|\cdot|_\mathfrak{p}$ to K. Each completion $K_{\mathfrak{P}_j}$ is a finite extension of $F_\mathfrak{p}$. Conversely, if we begin with any finite extension k of $F_\mathfrak{p}$, there will be some finite extension K of F and some prime ideal $\mathfrak{P}$ of $\mathcal{O}_K$ so that $k = K_\mathfrak{P}$. Thus all the finite extensions of $F_\mathfrak{p}$ arise from prime ideals in finite extensions of F.

Analogous to the number field case, the notion of fractional ideal applies to the completions $F_\mathfrak{p}$. As with number fields, we have unique factorization of fractional ideals ($\mathcal{O}_\mathfrak{p}$ is a p.i.d.). Of course the factorizations are very simple in this case because there is only one prime ideal! For the extension $K_\mathfrak{P}/F_\mathfrak{p}$, we can factor the ideal $\mathcal{P}_\mathfrak{p}\mathcal{O}_\mathfrak{P} = \mathfrak{p}\mathcal{O}_\mathfrak{P}$ as $\mathcal{P}_\mathfrak{P}^e$ for some positive integer $e = e(\mathcal{P}_\mathfrak{P}/\mathcal{P}_\mathfrak{p})$, the *local ramification index*. Putting $f = f(\mathcal{P}_\mathfrak{P}/\mathcal{P}_\mathfrak{p}) = \left[\mathcal{O}_\mathfrak{P}/\mathcal{P}_\mathfrak{P} : \mathcal{O}_\mathfrak{p}/\mathcal{P}_\mathfrak{p} \right]$, (the *local residue field degree*), we have $ef = [K_\mathfrak{P} : F_\mathfrak{p}]$. (Since $\mathcal{O}_\mathfrak{p}$ and $\mathcal{O}_\mathfrak{P}$ are local rings, necessarily we have no splitting in the extension $K_\mathfrak{P}/F_\mathfrak{p}$.) We also use the terminology *totally ramified* ($f = 1$), *ramified* ($e > 1$), and *unramified* ($e = 1$) in the local setting, but we can apply these descriptions to the entire extension, since there is only one prime for each field.

In general, if K/F is an extension of number fields and $\mathfrak{p}$ is a prime ideal of $\mathcal{O}_F$ with $\mathfrak{p}\mathcal{O}_K = \mathfrak{P}_1^{e_1} \cdots \mathfrak{P}_g^{e_g}$, then $e(\mathcal{P}_{\mathfrak{P}_j}/\mathcal{P}_\mathfrak{p}) = e(\mathfrak{P}_j/\mathfrak{p})$, and $f(\mathcal{P}_{\mathfrak{P}_j}/\mathcal{P}_\mathfrak{p}) = f(\mathfrak{P}_j/\mathfrak{p})$, i.e., ramification indices and residue field degrees for the number fields and their completions coincide. Moreover, if K/F is Galois, then so is $K_{\mathfrak{P}_j}/F_\mathfrak{p}$, and

$$\mathrm{Gal}(K_{\mathfrak{P}_j}/F_\mathfrak{p}) \cong \{\sigma \in \mathrm{Gal}(K/F) : |\sigma(x)|_{\mathfrak{P}_j} = |x|_{\mathfrak{P}_j} \ \forall x \in K_{\mathfrak{P}_j}\}$$
$$\cong Z(\mathfrak{P}_j/\mathfrak{p}).$$

Example.

2. Let $K = \mathbb{Q}(i)$, $F = \mathbb{Q}$, $\mathfrak{p} = 5\mathbb{Z}$. We have that $5\mathbb{Z}$ splits completely: $5\mathbb{Z}[i] = \mathfrak{P}_1\mathfrak{P}_2$ where $\mathfrak{P}_1 = \langle 2 - i\rangle$ and $\mathfrak{P}_2 = \langle 2 + i\rangle$. Since we have a prime that splits completely, the ramification index and residue field degree are both trivial, i.e., $e = f = 1$, so the extensions $K_\mathfrak{P}/\mathbb{Q}_5$ are trivial. This means that we have $K_{\mathfrak{P}_1} \cong K_{\mathfrak{P}_2} \cong \mathbb{Q}_5$. However, we must be careful how we distinguish between $K_{\mathfrak{P}_1}$ and $K_{\mathfrak{P}_2}$. On one hand, they are both $\mathbb{Q}_5$, but on the other hand each must contain an isomorphic copy of K that is the image of a *continuous* embedding, and the topology on K is different in each case.

There are two square roots of -1 in $\mathbb{Q}_5$; say α is the one that is congruent to 2 modulo 5. Since $K_{\mathfrak{P}_1} \cong \mathbb{Q}_5$, there is an embedding $\iota_{\mathfrak{P}_1} : K \hookrightarrow \mathbb{Q}_5$. This embedding must preserve absolute values, so that $2 - i \in \mathfrak{P}_1$ is identified with $2 - \alpha \in \mathbb{Q}_5$, while $2 + i$ is identified with a 5-adic unit. On the other hand, the embedding $\iota_{\mathfrak{P}_2} : K \hookrightarrow \mathbb{Q}_5$ maps $2 + i \in \mathfrak{P}_2$ to $2 - \alpha \in \mathbb{Q}_5$ while $2 - i$ is now mapped to a unit. Thus we have found the two distinct embeddings of K into $\mathbb{Q}_5$ (one for each extension of $|\cdot|_5$ to K). As the above illustrates, writing "$2 + i$" for an element of $\mathbb{Q}_5$ is ambiguous unless the embedding of K into $\mathbb{Q}_5$ being used is understood.

Exercise 1.2. Let $F = \mathbb{Q}(\beta)$, where β is a root of $f(X) = X^3 - 3$. Let $p = \infty$. Use $f(X)$ to find the extensions of $|\cdot|_\infty$ to F.   ◊

Exercise 1.3. Let $F = \mathbb{Q}(\beta)$, where β is a root of $f(X) = X^3 - X - 1$. Let $p = 17$. Factor $f(X)$ over $\mathbb{Q}_{17}$ to find the number of extensions of $|\cdot|_{17}$ to F and the degrees of their respective completions over $\mathbb{Q}_{17}$ (Hensel's Lemma may be of use for this).

Find ramification indices and residue field degrees for the primes of $\mathcal{O}_F$ above $17\mathbb{Z}$ and verify that this agrees with the information obtained from the factorization of $f(X)$.   ◊

The following theorem collects some other facts about ramification in extensions of $F_\mathfrak{p}$; these results typically are proved in a first course in algebraic number theory. For example, proofs may be found in the books of Fröhlich and Taylor, [FT], Janusz, [J], and Lang, [L1].

Theorem 2.2. *Let $F_\mathfrak{p}$ denote the completion of a number field F at a non-zero prime ideal $\mathfrak{p}$ of $\mathcal{O}_F$.*

  i. *Every finite totally ramified extension of $F_\mathfrak{p}$ has the form $F_\mathfrak{p}(\pi)$, where π is a zero of some Eisenstein polynomial in $\mathcal{O}_\mathfrak{p}[T]$.*
 ii. *If $\pi \in F_\mathfrak{p}^{alg}$ is a zero of some Eisenstein polynomial in $\mathcal{O}_\mathfrak{p}[T]$, then $F_\mathfrak{p}(\pi)/F_\mathfrak{p}$ is totally ramified.*
iii. *For a given positive integer f, there is a unique unramified extension of $F_\mathfrak{p}$ of degree f, (unique within some fixed algebraic closure). It is Galois with a cyclic Galois group, and is obtained by adjoining to $F_\mathfrak{p}$ a lifting of any primitive element for the degree-f extension of the finite field $\mathcal{O}_\mathfrak{p}/\mathcal{P}_\mathfrak{p}$.*   □

Finally we remark that the norm of an element or of a fractional ideal, the discriminant, etc., may be defined for the extension $K_{\mathfrak{P}_j}/F_\mathfrak{p}$ in exactly the same way as was done for extensions of number fields.

Exercise 1.4. Let K/F be an extension of algebraic number fields, and let $\mathfrak{p}$ be a prime ideal of $\mathcal{O}_F$. For any $\alpha \in K$, show that

$$N_{K/F}(\alpha) = \prod_{\mathfrak{P}|\mathfrak{p}} N_{K_{\mathfrak{P}}/F_\mathfrak{p}}(\iota_{\mathfrak{P}}(\alpha)).$$

(HINT: It may help to study the previous example; if we identify $F_\mathfrak{p}$ and $K_{\mathfrak{P}}$ with extensions of $\mathbb{Q}_p$ (where $p\mathbb{Z} = \mathfrak{p} \cap \mathbb{Z}$), the embeddings $\iota_{\mathfrak{P}}$ of K into its completions must be chosen correctly or the above formula does not hold.)   ◊

3 Some General Questions Motivating Class Field Theory

Let F be a number field, $\mathcal{C}_F$ its ideal class group. (Thus $\mathcal{O}_F$ is a u.f.d. if and only if $\mathcal{C}_F = \{1\}$.) Suppose $\mathcal{C}_F \neq \{1\}$. It would be useful to embed F into a larger number field K, where $\mathcal{O}_K$ is a u.f.d. Can one find a finite extension K/F of class number one, i.e., with $\mathcal{C}_K = \{1\}$? For a long time, the answer to this question was not known. There were known examples of cases where such an extension K exists, but (for good reason) no one had succeeded in proving that K must always exist. As was first shown in the early 1960s by Shafarevich and Golod, there are examples of number fields F having no finite extension K of class number one.

Perhaps we may ask for a bit less. Is there an extension K/F such that any non-principal ideal $\mathfrak{a}$ of $\mathcal{O}_F$ becomes principal in $\mathcal{O}_K$, (i.e., the ideal $\mathfrak{a}\mathcal{O}_K$ is principal in $\mathcal{O}_K$)? Happily, the answer here is yes, by a result called the *Principal Ideal Theorem*, (see Chapter 6). This was conjectured by Hilbert, but not proved until the 1930s by Furtwängler.

The well-known theorem of Kronecker and Weber gives a classificaton of all the finite abelian extensions of $\mathbb{Q}$, proving that any such extension is a subfield of a cyclotomic field. There are also results describing the abelian extensions of imaginary quadratic fields (in terms of special values of elliptic modular functions). It is natural to seek to extend this to the study of all abelian extensions of an arbitrary number field F. This is the heart of class field theory. As we shall discover, the finite abelian extensions K of a number field F correspond to certain subgroups of ideals (or alternately, of idèles), whose factor groups turn out to be isomorphic to the Galois groups $\mathrm{Gal}(K/F)$. (These are the *Existence, Completeness* and *Isomorphy Theorems*.) While these results give a description of the nature of the abelian extensions of an arbitrary number field F, the problem of constructing explicitly the abelian extensions of F remains an open question. Progress in special cases has been made using techniques from computational number theory. For a nice introduction to this facet of the subject, see Cohen's *Advanced Topics in Computational Number Theory*, [Coh].

While we do not treat them in this text, there are results (using algebraic K-theory) that provide information on the abelian extensions of fields of finite transcendence degree over their prime fields. This more general setting has been studied in the work of Bloch, Kato, Saito, et al., beginning in the 1980s (for example, see [Ka], [Bl] and [KS]). Raskind, [R], has gathered much of the material on this topic into an extensive survey. Another area of recent interest (also outside the scope of this text) is non-abelian class field theory, (the Langlands philosophy), much of which is still only conjectured.

Chapter 2
Dirichlet's Theorem on Primes in Arithmetic Progressions

As is well-known, Euclid gave a proof of the existence of infinitely many primes. Centuries later, Euler devised another proof using the Riemann zeta function.

Dirichlet's proof of his theorem on primes in arithmetic progressions is a generalization of Euler's idea, except the zeta function is replaced by the L-function of a Dirichlet character. Dirichlet gave the proof for prime modulus in 1837, finishing the general case in 1840.

Theorem (Dirichlet's Theorem on Primes in Arithmetic Progressions). *If m is a positive integer and a is an integer for which $(a, m) = 1$, then there are infinitely many primes p satisfying $p \equiv a$ (mod m).*

Taking $m = 1$ gives a restatement of Euclid's result, and reduces Dirichlet's proof to the one given by Euler. The theorem on arithmetic progressions was a key ingredient in Legendre's attempted "proof" of Quadratic Reciprocity. Unfortunately, while he relied upon it, Legendre didn't prove it; it remained conjectural until Dirichlet's 1837 paper. The first complete proof of Quadratic Reciprocity (by Gauss, using a different approach) also came later, although it pre-dates Dirichlet's Theorem.

In this chapter, we introduce Dirichlet characters and their L-functions, and give a proof of Dirichlet's Theorem on Primes in Arithmetic Progressions, which motivates our discussion of ray class groups in Chapter 3.

1 Characters of Finite Abelian Groups

We recall some basic facts about characters of finite abelian groups. Given a finite abelian group G, a *character* of G is a multiplicative homomorphism $G \to \mathbb{C}^\times$. Let $\hat{G}$ denote the set of all characters of G.

If $\chi, \psi \in \hat{G}$, then we define their *product* $\chi\psi$ to be the function $\chi\psi : G \to \mathbb{C}^x$ that satisfies $\chi\psi(g) = \chi(g)\psi(g)$. It is straightforward to show that $\chi\psi$ is also a character of G, and that under this multiplication, $\hat{G}$ is a group, called the *character group* of G. For example, the homomorphism $\chi_0 : G \to \mathbb{C}^\times$ given by $\chi_0 : g \mapsto 1$ is called the *trivial character* on G. It serves as the identity in $\hat{G}$. In fact, we have the following.

N. Childress, *Class Field Theory*, Universitext, DOI 10.1007/978-0-387-72490-4_2,
© Springer Science+Business Media, LLC 2009

Proposition 1.1. If G is a finite abelian group, then $\hat{G} \cong G$.

Proof. Since G is abelian, it may be written as a direct sum of cyclic groups of the form $\mathbb{Z}/m\mathbb{Z}$. It follows that $\hat{G}$ is the product of groups of the form $\widehat{\mathbb{Z}/m\mathbb{Z}}$. For any $\chi \in \widehat{\mathbb{Z}/m\mathbb{Z}}$ the complex number $\chi(1)$ determines χ completely, since $\mathbb{Z}/m\mathbb{Z}$ is an additive cyclic group generated by 1. Thus we have an injective homomorphism of groups $\widehat{\mathbb{Z}/m\mathbb{Z}} \to \mathbb{C}^\times$ given by $\chi \mapsto \chi(1)$. Note also that the image of this homomorphism is precisely the set of mth roots of unity in $\mathbb{C}$, which is a cyclic group of order m. Hence $\mathbb{Z}/m\mathbb{Z} \cong \widehat{\mathbb{Z}/m\mathbb{Z}}$, and we conclude $G \cong \hat{G}$. $\qquad\square$

It is clear from the above proposition that $\hat{\hat{G}} \cong G$, but we include additional details because it is possible to give a canonical isomorphism explicitly. Given $g \in G$, let $\tilde{g} : \hat{G} \to \mathbb{C}^\times$ be defined by $\tilde{g} : \chi \mapsto \chi(g)$, (so $\tilde{g} \in \hat{\hat{G}}$).

Exercise 2.1. Show that the map $g \mapsto \tilde{g}$ is a homomorphism $G \to \hat{\hat{G}}$. $\qquad\Diamond$

Proposition 1.2. The map $g \mapsto \tilde{g}$ is an isomorphism $G \to \hat{\hat{G}}$.

Proof. To show injectivity, suppose $g \in G$ satisfies $\chi(g) = 1$ for all $\chi \in \hat{G}$. Let $H = \langle g \rangle < G$. Then the elements of $\hat{G}$ may be viewed as (distinct) characters of G/H. But by Proposition 1.1, there are exactly $\#\left(G/H\right)$ of these. It follows that $\#H = 1$. We have shown that if $\chi(g) = 1$ for all $\chi \in \hat{G}$, then $g = 1$. Hence, we have an injective homomorphism $G \hookrightarrow \hat{\hat{G}}$. Since the orders are equal, it follows that this is an isomorphism $G \xrightarrow{\cong} \hat{\hat{G}}$. $\qquad\square$

Proposition 1.3. Let G be a finite abelian group. For $H < G$, let

$$H^\perp = \{\chi \in \hat{G} : \chi(h) = 1 \text{ for all } h \in H\};$$

i. if $H < G$, then $H^\perp \cong \widehat{G/H}$

ii. if $H < G$, then $\hat{H} \cong \hat{G}/H^\perp$

iii. $(H^\perp)^\perp = H$, (if we identify $\hat{\hat{G}} = G$).

Proof.

(i.) The isomorphism is given by identifying $\chi \in H^\perp$ with $\psi \in \widehat{G/H}$, where $\psi(gH) = \chi(g)$. This mapping is well-defined since χ is trivial on H. It is routine to check that it is an isomorphism of groups.

(ii.) Clearly restriction $\chi \mapsto \chi|_H$ is a homomorphism $\hat{G} \to \hat{H}$ with kernel $H^\perp$. This yields an embedding $\hat{G}/H^\perp \hookrightarrow \hat{H}$. It must also be surjective, since the orders agree:

$$\#H^\perp = \#\left(\widehat{G/H}\right) = \#\left({}^G/_H\right)$$
$$= {}^{\#G}/_{\#H}$$
$$\text{whence } \#\hat{H} = \#H = {}^{\#G}/_{\#H^\perp}$$
$$= {}^{\#\hat{G}}/_{\#H^\perp}.$$

(iii.) By definition, $(H^\perp)^\perp = \{\tilde{g} \in \hat{\hat{G}} : \tilde{g}(\chi) = 1 \text{ for all } \chi \in H^\perp\}$. Considering orders, we get

$$\#(H^\perp)^\perp = \#\widehat{\hat{G}/H^\perp} = {}^{\#\hat{\hat{G}}}/_{H^\perp}$$
$$= {}^{\#\hat{G}}/_{\#H^\perp}$$
$$= {}^{\#G}/_{(\#G/\#H)}$$
$$= \#H.$$

Since the orders agree, it suffices to observe that if $h \in H$, then $\tilde{h} : \chi \mapsto \chi(h)$ satisfies $\tilde{h}(H^\perp) = \{1\}$ from which we deduce that $\hat{\hat{H}} = H \subseteq (H^\perp)^\perp$.   □

Finally, we give a proof of two very useful equations known as the orthogonality relations.

Proposition 1.4 (Orthogonality Relations).

i. Fix a character χ of the finite abelian group G. Then

$$\sum_{g \in G} \chi(g) = \begin{bmatrix} 0 & \text{if } \chi \neq \chi_0 \\ \#G & \text{if } \chi = \chi_0 \end{bmatrix}.$$

ii. Fix an element g of the finite abelian group G. Then

$$\sum_{\chi \in \hat{G}} \chi(g) = \begin{bmatrix} 0 & \text{if } g \neq 1 \\ \#G & \text{if } g = 1 \end{bmatrix}.$$

Proof. For (i), let $h \in G$ and note that $\sum_{g \in G} \chi(g) = \sum_{g \in G} \chi(gh) = \chi(h) \sum_{g \in G} \chi(g)$. This implies that $(1 - \chi(h)) \sum_{g \in G} \chi(g) = 0$, for all $h \in G$. If $\chi \neq \chi_0$, then there is some $h \in G$ for which $\chi(h) \neq 1$. It follows that $\sum_{g \in G} \chi(g) = 0$. Of course, for $\chi = \chi_0$, the expression becomes $\sum_{g \in G} \chi_0(g) = \sum_{g \in G} 1 = \#G$.

For (ii), note that $\displaystyle\sum_{\chi\in\hat{G}}\chi(g) = \sum_{\chi\in\hat{G}}\tilde{g}(\chi)$ and use (i).   □

2 Dirichlet Characters

The notion of Dirichlet character predates considerably the more general ideas discussed in the previous section. However, we shall make use of the more modern terminology to define Dirichlet characters as follows. Let n be a positive integer. A *Dirichlet character modulo n* is a character of the abelian group $\left(\mathbb{Z}/n\mathbb{Z}\right)^{\times}$, i.e., a multiplicative homomorphism

$$\chi : \left(\mathbb{Z}/n\mathbb{Z}\right)^{\times} \to \mathbb{C}^{\times}.$$

We call n the *modulus* of χ.

Examples.

1. Let p be an odd prime, and let $\chi : \left(\mathbb{Z}/p\mathbb{Z}\right)^{\times} \to \mathbb{C}^{\times}$ be the Legendre symbol mod p, i.e., $\chi(a) = \left(\frac{a}{p}\right)$.

2. Let i be the usual complex number, and define $\chi : \left(\mathbb{Z}/5\mathbb{Z}\right)^{\times} \to \mathbb{C}^{\times}$ by $\chi(1) = 1, \chi(2) = i, \chi(3) = -i, \chi(4) = -1$.

   If χ is a Dirichlet character of modulus n and $n|m$, then by using the natural homomorphism $\varphi : \left(\mathbb{Z}/m\mathbb{Z}\right)^{\times} \to \left(\mathbb{Z}/n\mathbb{Z}\right)^{\times}$, we may define $\chi' = \chi \circ \varphi$. Now χ' is also a Dirichlet character, but of modulus m. In this situation, we say that χ' is *induced* by χ.

   Let f_{χ} be the minimal modulus for the Dirichlet character χ, i.e., χ is not induced by any Dirichlet character of modulus smaller than f_{χ}. Call f_{χ} the *conductor* of χ. A Dirichlet character defined modulo its conductor is called *primitive*.

Examples.

3. Let $\chi : \left(\mathbb{Z}/12\mathbb{Z}\right)^{\times} \to \mathbb{C}^{\times}$ be given by $\chi(1) = 1, \chi(5) = -1, \chi(7) = 1, \chi(11) = -1$. Since $\chi(a + 3k) = \chi(a)$ we see that χ is induced by the character $\psi : \left(\mathbb{Z}/3\mathbb{Z}\right)^{\times} \to \mathbb{C}^{\times}$, where $\psi(1) = 1, \psi(2) = -1$. Furthermore ψ is primitive. We conclude that $f_{\chi} = 3$.

4. Let $\chi : \left(\mathbb{Z}/12\mathbb{Z}\right)^{\times} \to \mathbb{C}^{\times}$ be given by $\chi(1) = 1, \chi(5) = -1, \chi(7) = -1, \chi(11) = 1$. It is easy to check that χ is primitive, whence $f_{\chi} = 12$.

   A Dirichlet character χ also may be regarded as a function $\chi : \mathbb{Z} \to \mathbb{C}$ by letting

$$\chi(a) = \begin{cases} \chi(a \bmod f_{\chi}) & \text{if } (a, f_{\chi}) = 1 \\ 0 & \text{if } (a, f_{\chi}) \neq 1. \end{cases}$$

We also refer to this periodic function on $\mathbb{Z}$ as a Dirichlet character, and we do not distinguish notationally between a Dirichlet character as a function on $\left(\mathbb{Z}/f_\chi\mathbb{Z}\right)^\times$ and the periodic function on $\mathbb{Z}$ associated to it.

Let χ_0 denote the trivial character of conductor 1. Let χ, ψ be primitive Dirichlet characters of conductors f_χ, f_ψ, respectively. Let $n = \mathrm{lcm}(f_\chi, f_\psi)$. The function

$$\eta : \left(\mathbb{Z}/n\mathbb{Z}\right)^\times \to \mathbb{C}^\times \text{ defined by}$$

$$\eta : a \mapsto \chi(a)\psi(a)$$

is easily seen to be a (possibly imprimitive) Dirichlet character. Define the product $\chi\psi$ to be the primitive Dirichlet character that induces η. In this way, we are able to define a multiplication on primitive Dirichlet characters that is closed. It is trivial to check that this multiplication is associative and commutative and that χ_0 serves as the identity. Note that the conductor of $\chi\psi$ must be a divisor of the product of the conductors of χ and ψ.

Example.

5. Let

$$\chi : \left(\mathbb{Z}/12\mathbb{Z}\right)^\times \to \mathbb{C}^\times \text{ by } \chi(1) = 1, \ \chi(5) = -1, \ \chi(7) = -1, \ \chi(11) = 1$$

$$\psi : \left(\mathbb{Z}/4\mathbb{Z}\right)^\times \to \mathbb{C}^\times \text{ by } \psi(1) = 1, \ \psi(3) = -1.$$

Then $\eta : \left(\mathbb{Z}/12\mathbb{Z}\right)^\times \to \mathbb{C}^\times$ by $\eta(1) = 1$, $\eta(5) = -1$, $\eta(7) = 1$, $\eta(11) = -1$. The character η is imprimitive, induced by the primitive character $\chi\psi$. We find that $\chi\psi : \left(\mathbb{Z}/3\mathbb{Z}\right)^\times \to \mathbb{C}^\times$ by $\chi\psi(1) = 1$, $\chi\psi(2) = -1$. Note: $\chi(2)\psi(2) = 0 \neq \chi\psi(2)$.

Exercise 2.2. Show: if $(f_\chi, f_\psi) = 1$, then $f_{\chi\psi} = f_\chi f_\psi$.   ◊

Given a Dirichlet character $\chi : \left(\mathbb{Z}/n\mathbb{Z}\right)^\times \to \mathbb{C}^\times$, we can associate to it the map $\bar{\chi} : \left(\mathbb{Z}/n\mathbb{Z}\right)^\times \to \mathbb{C}^\times$ by $\bar{\chi}(a) = (\chi(a))^{-1} = \overline{\chi(a)}$, the complex conjugate. It is straightforward to show that $\bar{\chi}$ is a Dirichlet character, that it has the same conductor as χ, and that $\bar{\chi}\chi = \chi_0$. Thus the (primitive) Dirichlet characters form a group under multiplication. The *order* of a Dirichlet character is its order as an element of this group. Because the image of a Dirichlet character is necessarily a (finite) group of roots of unity in $\mathbb{C}^\times$, its order will always be finite. Indeed, if χ has conductor n, then the order of χ must be a divisor of $\varphi(n)$. A Dirichlet character of order 2 is sometimes called a *quadratic* Dirichlet character.

For any Dirichlet character χ, we must have $\chi(-1) = \pm 1$. If $\chi(-1) = 1$, then χ is called *even*; if $\chi(-1) = -1$, then χ is called *odd*.

Exercise 2.3. Show that the set of all even Dirichlet characters is a subgroup of the group of all Dirichlet characters under multiplication.   ◊

Given a fixed positive integer n, the Dirichlet characters having conductors dividing n form a finite group. In fact, letting ζ_n be a primitive n-th root of unity, we can identify $\mathrm{Gal}(\mathbb{Q}(\zeta_n)/\mathbb{Q}) = G$ with $\left(\mathbb{Z}/n\mathbb{Z}\right)^\times$ whence the group of Dirichlet characters modulo n may be regarded as characters of the Galois group G.

Let χ be a character of $G = \mathrm{Gal}(\mathbb{Q}(\zeta_n)/\mathbb{Q})$. Let K be the fixed field of the kernel of χ. We call K the *field associated to* χ.

Example.

6. Let $\chi : G = \mathrm{Gal}(\mathbb{Q}(\zeta_{12})/\mathbb{Q}) \to \mathbb{C}^\times$ by $\chi(\sigma_1) = 1$, $\chi(\sigma_5) = -1$, $\chi(\sigma_7) = 1$, $\chi(\sigma_{11}) = -1$, where $\sigma_j : \zeta_{12} \mapsto \zeta_{12}^j$. Then $\ker \chi = \{\sigma_1 = \mathrm{id}, \sigma_7\}$.

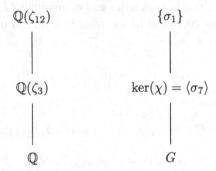

Now $\sigma_7(\zeta_3) = \zeta_3$, so $\ker \chi$ fixes elements of $\mathbb{Q}(\zeta_3)$. Comparing orders and indices, we see that $\mathbb{Q}(\zeta_3)$ is the fixed field of $\ker \chi = \mathrm{Gal}(\mathbb{Q}(\zeta_{12})/\mathbb{Q}(\zeta_3))$. Hence $\mathbb{Q}(\zeta_3)$ is the field associated to χ. It follows that χ is really a character of $G/\ker\chi$, which corresponds to $\mathrm{Gal}(\mathbb{Q}(\zeta_3)/\mathbb{Q}) \cong \left(\mathbb{Z}/3\mathbb{Z}\right)^\times$. Note $f_\chi = 3$.

More generally, let X be a finite group of Dirichlet characters and let n be the least common multiple of all of the conductors of these characters. Then we may view X as a subgroup of $\hat{G}$, where $G = \mathrm{Gal}(\mathbb{Q}(\zeta_n)/\mathbb{Q})$. Let $H = \cap \ker \chi$, where the intersection is over all $\chi \in X$. Then H is a subgroup of G; let K be the fixed field of H. K is called the *field associated to* X.

Note that if $\chi \in X$, then $H < \ker \chi$, so the field associated to χ is a subfield of the field associated to X. Also, if X is cyclic, generated by χ, then the field associated to X is the same as the field associated to χ.

Examples.

7. Let $\chi : \left(\mathbb{Z}/15\mathbb{Z}\right)^\times \to \mathbb{C}^\times$ be given by $\chi(1) = 1$, $\chi(2) = -1$, $\chi(4) = 1$, $\chi(7) = -1$, $\chi(8) = -1$, $\chi(11) = 1$, $\chi(13) = -1$, $\chi(14) = 1$. Consider χ as a character of $\mathrm{Gal}(\mathbb{Q}(\zeta_{15})/\mathbb{Q})$. Now $\ker \chi$ has order 4, so its fixed field K must satisfy $[K : \mathbb{Q}] = 2$. Also K must be real, since its elements must be

fixed by σ_{14}, which is complex conjugation. The quadratic subfields of $\mathbb{Q}(\zeta_{15})$ are diagrammed below.

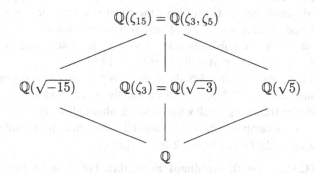

Only one of the quadratic subfields of $\mathbb{Q}(\zeta_{15})/\mathbb{Q}$, is real. Thus $K = \mathbb{Q}(\sqrt{5})$ is the field associated to χ. Note that $d_{K/\mathbb{Q}} = 5$, and χ has conductor 5.

8. Let $\psi : \left(\mathbb{Z}/15\mathbb{Z}\right)^{\times} \to \mathbb{C}^{\times}$ be the Dirichlet character given by $\psi(1) = 1$, $\psi(2) = -1$, $\psi(4) = 1$, $\psi(7) = 1$, $\psi(8) = -1$, $\psi(11) = -1$, $\psi(13) = 1$, $\psi(14) = -1$. Let $\vartheta : \left(\mathbb{Z}/15\mathbb{Z}\right)^{\times} \to \mathbb{C}^{\times}$ be the Dirichlet character given by $\vartheta(1) = 1$, $\vartheta(2) = 1$, $\vartheta(4) = 1$, $\vartheta(7) = -1$, $\vartheta(8) = 1$, $\vartheta(11) = -1$, $\vartheta(13) = -1$, $\vartheta(14) = -1$. Consider ψ and ϑ as characters of $\mathrm{Gal}\,(\mathbb{Q}(\zeta_{15})/\mathbb{Q})$. As in the previous example, each must correspond to a quadratic subextension of $\mathbb{Q}(\zeta_{15})/\mathbb{Q}$. Checking conductors, we see that the conductor of ψ is 3, while the conductor of ϑ is 15. Thus we must have that ψ is a character of $\left(\mathbb{Z}/3\mathbb{Z}\right)^{\times}$, so that the field associated to ψ is $\mathbb{Q}(\zeta_3) = \mathbb{Q}(\sqrt{-3})$. The field associated to ϑ must be the only remaining possibility, i.e., $\mathbb{Q}(\sqrt{-15})$.

9. Let $G = \mathrm{Gal}\,(\mathbb{Q}(\zeta_n)/\mathbb{Q}) \cong \left(\mathbb{Z}/n\mathbb{Z}\right)^{\times}$ and let X be the set of all even characters in $\hat{G}$. Exercise 2.3 tells us that X is a subgroup of $\hat{G}$. Note that X has index 2 in $\hat{G}$ (the product of two odd characters is even, so the only non-trivial coset in $\hat{G}/X$ contains all the odd characters). Since $\chi(-1) = 1$ for all $\chi \in X$, we must have $\sigma_{-1} \in \ker \chi$ for all $\chi \in X$. Now, the automorphism $\sigma_{-1} : \zeta_n \mapsto \zeta_n^{-1}$ is complex conjugation, so the field associated to X must be real. In fact, if χ is any character, then the field associated to χ is real if and only if χ is even. Since X is the largest subgroup of $\hat{G}$ consisting entirely of even characters, one expects that its associated field will be the maximal real subfield of $\mathbb{Q}(\zeta_n)$. Soon we shall have a theoretical result that makes this apparent. For now, however, we simply find $\bigcap_{\chi \in X} \ker \chi$.

We have shown $\bigcap_{\chi \in X} \ker \chi \supseteq \{\sigma_1, \sigma_{-1}\}$. The reverse containment is also true,

for if $\sigma_r \in \bigcap_{\chi \in X} \ker \chi$ is non-trivial, then the canonical isomorphism $G \to \hat{\hat{G}}$

gives that $\tilde{\sigma}_r$ is non-trivial, (see Proposition 1.2). But then there is some $\psi \in \hat{G}$ such that $\psi(\sigma_r) \neq 1$. We cannot have $\psi \in X$, so ψ must be odd. However ψ^2 is even, and hence $\psi^2(\sigma_r) = 1$. It follows that $\psi(\sigma_r) = -1$. Since $[\hat{G} : X] = 2$, any other odd character of G will have the form $\psi \chi$ for some $\chi \in X$, so $\tilde{\sigma}_r$ sends every odd character in $\hat{G}$ to -1 and every even character in $\hat{G}$ to 1. But then $\tilde{\sigma}_r = \tilde{\sigma}_{-1}$. Thus the field associated to X is the fixed field of $\{\sigma_1, \sigma_{-1}\}$, which is $\mathbb{Q}(\zeta_n + \zeta_n^{-1})$, the maximal real subfield of $\mathbb{Q}(\zeta_n)$.

In several of the previous examples, the reader has perhaps noticed a relationship between the conductor of a Dirichlet character and the discriminant of its associated field. We include the following result without proof, although we have seen evidence to support it in our examples, (it may be proved using analytic techniques – see Chapter 7 of Long's *Algebraic Number Theory*, [Lo]).

Theorem 2.1 (Conductor Discriminant Formula). *Let X be a finite group of Dirichlet characters and K its associated (abelian) number field. Then*

$$d_{K/\mathbb{Q}} = (-1)^{r_2} \prod_{\chi \in X} f_\chi$$

where r_2 is the number of pairs of imaginary embeddings of K.   □

Exercise 2.4. Let $p > 2$ be a prime. Use the Conductor Discriminant Formula to find the discriminant of $\mathbb{Q}(\zeta_p + \zeta_p^{-1})$ over $\mathbb{Q}$.   ◇

Exercise 2.5. Let $p > 2$ be a prime and let $n > 0$ be an integer. Use the Conductor Discriminant Formula to find the discriminant of $\mathbb{Q}(\zeta_{p^n})$ over $\mathbb{Q}$.   ◇

We want to describe more precisely the relationship between groups of Dirichlet characters and their associated fields. What results is an enhancement of the Galois correspondence for abelian extensions of $\mathbb{Q}$. Let $\mathbb{Q} \subseteq L \subseteq K$, where $K/\mathbb{Q}$ is finite abelian. Let X_K denote the group of characters of $\mathrm{Gal}(K/\mathbb{Q})$. Then

$$\{\chi \in X_K : \chi(g) = 1 \text{ for all } g \in \mathrm{Gal}(K/L)\} = \mathrm{Gal}(K/L)^\perp$$

$$= \mathrm{Gal}(K/\mathbb{Q})\big/\widehat{\mathrm{Gal}(K/L)}$$

$$= \widehat{\mathrm{Gal}(L/\mathbb{Q})} = X_L$$

where X_L is the group of characters associated to L.

Conversely, beginning with a finite abelian extension $K/\mathbb{Q}$ having group of characters X_K, let Y be any subgroup of X_K and let L be the fixed field of $Y^\perp$. Then

$$Y^\perp = \{g \in \mathrm{Gal}(K/\mathbb{Q}) : \chi(g) = 1 \text{ for all } \chi \in Y\} = \mathrm{Gal}(K/L)$$

and

$$Y = (Y^\perp)^\perp = \mathrm{Gal}(K/L)^\perp = \widehat{\mathrm{Gal}(L/\mathbb{Q})} = X_L.$$

We have shown that there is a bijective correspondence

$$\{\text{subgroups of } X_K\} \longleftrightarrow \{\text{subfields of } K\}$$

given by:

$$Y \longleftrightarrow \text{fixed field of } Y^\perp$$

or by

$$X_L = \text{Gal}(K/L)^\perp \longleftrightarrow L.$$

Note that the above correspondence is order-preserving, i.e., $L_1 \subseteq L_2$ if and only if $X_{L_1} \subseteq X_{L_2}$. The following diagram illustrates this correspondence.

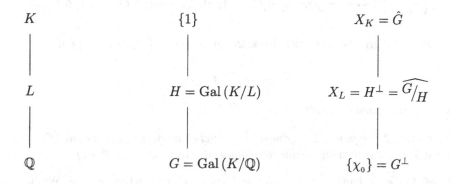

Let us revisit Example 9 momentarily. We consider the group X consisting of all even Dirichlet characters whose conductors divide n. We know that the field associated to X is real. If L is any real subfield of $\mathbb{Q}(\zeta_n)$, then L is fixed by σ_{-1}, so $\sigma_{-1} \in X_L^\perp$, i.e., all the elements of X_L are even. But then $X_L \subseteq X$. It follows that the field associated to X must be the maximal real subfield of $\mathbb{Q}(\zeta_n)$.

Exercise 2.6. Let X_j be the group of Dirichlet characters corresponding to the field L_j $(j = 1, 2)$. Prove or disprove each of the following.

a. The group generated by X_1 and X_2 corresponds to the compositum $L_1 L_2$.
b. The group $X_1 \cap X_2$ corresponds to $L_1 \cap L_2$. $\Diamond$

For abelian extensions of $\mathbb{Q}$, we can compute ramification indices in terms of Dirichlet characters. To describe how, we need to make some preliminary observations. Let n be a positive integer, and suppose $n = \prod_{j=1}^{m} p_j^{a_j}$, where the p_j's are distinct primes and $a_j > 0$. Then

$$\left(\mathbb{Z}/n\mathbb{Z}\right)^\times \cong \prod_{j=1}^{m} \left(\mathbb{Z}/p_j^{a_j}\mathbb{Z}\right)^\times$$

so a Dirichlet character χ defined modulo n may be written as

$$\chi = \prod_{j=1}^{m} \chi_{p_j}$$

where χ_{p_j} is a Dirichlet character modulo $p_j^{a_j}$.

For example, the character $\vartheta : \left(\mathbb{Z}/15\mathbb{Z}\right)^{\times} \to \mathbb{C}^{\times}$ defined by $\vartheta(1) = 1$, $\vartheta(2) = 1$, $\vartheta(4) = 1$, $\vartheta(7) = -1$, $\vartheta(8) = 1$, $\vartheta(11) = -1$, $\vartheta(13) = -1$, $\vartheta(14) = -1$ (from Example 8) may be written as $\vartheta = \vartheta_3 \vartheta_5$ where

$\vartheta_3 : \left(\mathbb{Z}/3\mathbb{Z}\right)^{\times} \to \mathbb{C}$ is given by $\vartheta_3(1) = 1$, $\vartheta_3(2) = -1$

$\vartheta_5 : \left(\mathbb{Z}/5\mathbb{Z}\right)^{\times} \to \mathbb{C}$ is given by $\vartheta_5(1) = 1$, $\vartheta_5(2) = -1$, $\vartheta_5(3) = -1$, $\vartheta_5(4) = 1$.

Let X be a group of Dirichlet characters modulo $n = \prod_{j=1}^{m} p_j^{a_j}$, and put

$$X_{p_j} = \{\chi_{p_j} : \chi \in X\}.$$

We have the following result.

Theorem 2.2. *Suppose X is a group of Dirichlet characters with associated field K. If $p \in \mathbb{Z}$ is prime then the ramification index of p in $K/\mathbb{Q}$ is $e = \#(X_p)$.*

Proof. Let $n = \operatorname{lcm}\{f_\chi : \chi \in X\}$ and say $n = p^a m$ where $p \nmid m$. We have $K \subseteq \mathbb{Q}(\zeta_n)$, and we note that $\mathbb{Q}(\zeta_m) \subseteq L = K(\zeta_m) \subseteq \mathbb{Q}(\zeta_n)$. What is the group of characters with associated field L? Since $\operatorname{Gal}(L/\mathbb{Q})$ is generated by $\operatorname{Gal}(K/\mathbb{Q})$ and $\operatorname{Gal}(\mathbb{Q}(\zeta_m)/\mathbb{Q})$, it is generated by X and the characters modulo m. Since $(p, m) = 1$, $\widehat{\operatorname{Gal}(L/\mathbb{Q})}$ is the direct product of X_p with the characters of $\mathbb{Q}(\zeta_m)$. We have $L = E\mathbb{Q}(\zeta_m)$ where E is the subfield of $\mathbb{Q}(\zeta_{p^a})$ corresponding to X_p.

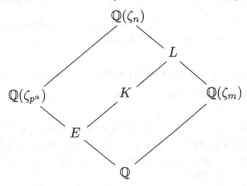

Since p is unramified in $\mathbb{Q}(\zeta_m)/\mathbb{Q}$, the ramification index for p in $L/\mathbb{Q}$ is e. Since p is unramified in L/E, the ramification index for p in $E/\mathbb{Q}$ is e. But p is totally ramified in $\mathbb{Q}(\zeta_{p^a})/\mathbb{Q}$, so also in $E/\mathbb{Q}$. Thus $e = [E : \mathbb{Q}] = \#(X_p)$. $\square$

Corollary 2.3. Let X be a group of Dirichlet characters and let K be its associated field. A prime p is unramified in $K/\mathbb{Q}$ if and only if $\chi(p) \neq 0$ for all $\chi \in X$.

Proof. Suppose p ramifies in $K/\mathbb{Q}$. We have $e = \#X_p \neq 1$, so X_p contains some non-trivial element χ. Thus there is a non-trivial character χ in X with conductor divisible by p, i.e., with $\chi(p) = 0$.

Conversely, if $\chi(p) = 0$ for some element χ of X, then the conductor of χ must be divisible by p. Thus X_p must be non-trivial. But this implies that $e = \#X_p > 1$, whence p is ramified.   □

Example.

10. In $K = \mathbb{Q}(\zeta_{12})$, the subfield $L = \mathbb{Q}(i)$ has associated characters

$$X_L = \{\chi \mod 12 : \chi(\sigma) = 1 \text{ for all } \sigma \in \mathrm{Gal}(K/L)\}.$$

Now $\mathrm{Gal}(K/L) = \{1, \sigma\}$ where σ fixes $i = \zeta_{12}^3$, hence $\sigma : \zeta_{12} \mapsto \zeta_{12}^5$. We find that $\chi(\sigma) = 1$ if and only if $\chi(5) = 1$. Thus $X_L = \{1, \vartheta\}$, where $\vartheta(1) = 1$, $\vartheta(5) = 1$, $\vartheta(7) = -1$, $\vartheta(11) = -1$. Now ϑ has conductor 4, so $\vartheta(p) = 0$ if and only if $p = 2$. This confirms that 2 is the only ramified prime in $\mathbb{Q}(i)/\mathbb{Q}$.

Exercise 2.7. In this exercise we study quadratic Dirichlet characters and their associated fields.

a. Let m be an odd positive integer. How many quadratic Dirichlet characters modulo m are there? How many of them are primitive? (HINT: if p is an odd prime, how many quadratic Dirichlet characters have conductor p? How many have conductor p^2?)

b. What does your answer to part a tell you about the quadratic subfield(s) of $\mathbb{Q}(\zeta_p)$, where p is an odd prime? Does a quadratic subfield always exist? Is it unique? When is it real?

c. Let p be an odd prime. How many quadratic Dirichlet characters modulo $4p$ are there? How many of them are primitive? What does this tell you about the quadratic subfield(s) of $\mathbb{Q}(\zeta_{4p})$, where p is an odd prime?

d. Answer similar questions about the quadratic subfield(s) of $\mathbb{Q}(\zeta_8)$.

e. For any odd prime p, show $\mathbb{Q}(\sqrt{p}) \subseteq \mathbb{Q}(\zeta_m)$ for $m = p$ or $4p$. Use this to show (without Kronecker-Weber) given any integer d, there is some m such that $\mathbb{Q}(\sqrt{d}) \subseteq \mathbb{Q}(\zeta_m)$.   ◊

Theorem 2.4. *Let X be a group of Dirichlet characters with associated field K. Let p be prime, and define the subgroups*

$$Y = \{\chi \in X : \chi(p) \neq 0\}$$
$$Y_1 = \{\chi \in X : \chi(p) = 1\}.$$

Then:

$$X\big/Y \text{ is isomorphic to the inertia subgroup for } p$$

$$X\big/Y_1 \text{ is isomorphic to the decomposition group for } p$$

$$Y\big/Y_1 \text{ is cyclic of order } f.$$

Proof. By Corollary 2.3, the subfield L of K associated to Y must be the largest subfield of K in which p is unramified. Thus $L = K_T$ is the fixed field of the inertia subgroup T_p and $T_p = \mathrm{Gal}\,(K/L)$.

By the bijective correspondence between subgroups of X and subfields of K, we have

$$Y = \mathrm{Gal}\,(K/L)^{\perp}$$

whence

$$X\big/Y = \widehat{\mathrm{Gal}\,(K/\mathbb{Q})}\big/\mathrm{Gal}\,(K/L)^{\perp}$$

$$= \widehat{\mathrm{Gal}\,(K/L)}$$

$$\cong \mathrm{Gal}\,(K/L) = T_p.$$

Thus $e = \#(T_p) = [X : Y]$.

Now consider only the extension $L/\mathbb{Q}$. It is unramified at p and its group of characters is Y. If $n = \mathrm{lcm}\,\{f_\chi : \chi \in Y\}$, then $L \subseteq \mathbb{Q}(\zeta_n)$. Also $p \nmid n$ since p is unramified in $L/\mathbb{Q}$. Recall that the Frobenius automorphism for p in $G = \mathrm{Gal}\,(\mathbb{Q}(\zeta_n)/\mathbb{Q})$ is $\sigma_p : \zeta_n \mapsto \zeta_n^p$ (σ_p is unique since p is unramified). Now $\mathrm{Gal}\,(L/\mathbb{Q}) \cong G\big/\mathrm{Gal}\,(\mathbb{Q}(\zeta_n)/L)$, so the Frobenius automorphism for p in $\mathrm{Gal}\,(L/\mathbb{Q})$ is identified with $\bar\sigma_p$, the coset of σ_p in the quotient. But if $\chi \in Y$, then $\chi(\sigma)$ is trivial on $\mathrm{Gal}\,(\mathbb{Q}(\zeta_n)/L)$ whence $\chi(\bar\sigma_p) = \chi(\sigma_p) = \chi(p)$. Thus $Y_1 = \{\chi \in X : \chi(\bar\sigma_p) = 1\} = \langle\bar\sigma_p\rangle^{\perp}$ and $Y\big/Y_1 \cong \widehat{\langle\bar\sigma_p\rangle} \cong \langle\bar\sigma_p\rangle$, a cyclic group of order f. We have $[Y : Y_1] = f$, $[Y_1 : 1] = g$.

Finally, recall that $\langle\bar\sigma_p\rangle$ is isomorphic to the quotient of the decomposition group by the inertia subgroup. But p is unramified in $L/\mathbb{Q}$, so $\langle\bar\sigma_p\rangle$ is isomorphic to the decomposition group for p in $L/\mathbb{Q}$ (The order of the inertia subgroup is 1). Let F be the fixed field of the decomposition group for p in $L/\mathbb{Q}$. The character group of F is Y_1 (F is the fixed field of $\bar\sigma_p$, so $\chi \in \widehat{\mathrm{Gal}\,(F/\mathbb{Q})}$ if and only if $\chi(\bar\sigma_p) = 1$, so if and only if $\chi \in Y_1$). Hence $Y_1 \cong \mathrm{Gal}\,(F/\mathbb{Q})$.

We return to $K/\mathbb{Q}$. Now p splits completely in $F/\mathbb{Q}$, the primes above p remain inert in L/F, and ramify in K/L.

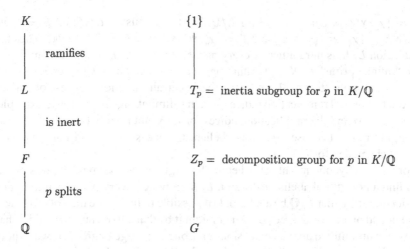

The decomposition group for p in $K/\mathbb{Q}$ is isomorphic to $\mathrm{Gal}\,(K/F)$, since the only splitting of p in $K/\mathbb{Q}$ occurs in $F/\mathbb{Q}$, and p splits completely there. We conclude that $Y_1 = \mathrm{Gal}\,(K/F)^{\perp}$ by the bijective correspondence, and

$$X\big/_{Y_1} = \widehat{\mathrm{Gal}\,(K/\mathbb{Q})}\Big/_{\mathrm{Gal}\,(K/F)^{\perp}} = \widehat{\mathrm{Gal}\,(K/F)}$$

$$\cong \mathrm{Gal}\,(K/F),$$

which is isormorphic to the decomposition group for p in $K/\mathbb{Q}$. $\square$

As mentioned earlier, Dirichlet characters will play an important role in the next section, where we give a proof of Dirichlet's Theorem on Primes in Arithmetic Progressions. But results such as the one above also may be used to study certain finite unramified abelian extensions of a number field K (in the case when K is abelian over $\mathbb{Q}$).

Example.

11. Let χ be a generator of the cyclic group $\widehat{\mathrm{Gal}\,(\mathbb{Q}(\zeta_5)/\mathbb{Q})}$ and let ψ be a generator of $\widehat{\mathrm{Gal}\,(\mathbb{Q}(i)/\mathbb{Q})}$. Let K be the field associated to the character $\chi^2\psi$ and let L be the field associated to the character group $\langle \chi^2, \psi \rangle$. Now $\chi^2\psi$ has conductor 20, so we may view all these characters as elements of $\widehat{\mathrm{Gal}\,(\mathbb{Q}(\zeta_{20})/\mathbb{Q})}$. Writing $\sigma_j : \zeta_{20} \mapsto \zeta_{20}^j$ as usual, we see that $\ker\psi = \langle \sigma_{13} \rangle$, and $\ker\chi^2 = \langle \sigma_9, \sigma_{11} \rangle$, whence L is the fixed field of $\langle \sigma_9 \rangle$. Also, $\ker\chi^2\psi = \langle \sigma_3 \rangle$, so K is the fixed field of $\langle \sigma_3 \rangle$. It is now elementary to verify that $L = \mathbb{Q}(\sqrt{5}, i)$ and $K = \mathbb{Q}(\sqrt{-5})$.

Clearly L/K is unramified at all primes other than 2 and 5. Using the above results on the relationship between Dirichlet characters and ramification, we may find the ramification indices for 2 and 5 in the extensions $L/\mathbb{Q}$ and $K/\mathbb{Q}$ from which the ramification indices in L/K may be deduced. For the extension $K/\mathbb{Q}$ and $p = 2$ or 5, we have $X = \{\chi_0, \chi^2\psi\}$ and $Y = \{\chi_0\}$, whence $e_2 = e_5 = 2$ for $K/\mathbb{Q}$. For the extension $L/\mathbb{Q}$ and $p = 2$, we have $X = \{\chi_0, \chi^2, \psi, \chi^2\psi\}$ and

$Y = \{\chi_0, \chi^2\}$, whence $e_2 = 2$ for $L/\mathbb{Q}$. For the extension $L/\mathbb{Q}$ and $p = 5$, we have $X = \{\chi_0, \chi^2, \psi, \chi^2\psi\}$ and $Y = \{\chi_0, \psi\}$, whence $e_5 = 2$ for $L/\mathbb{Q}$. Thus the extension L/K is unramified at every prime. (In fact, L/K is also unramified at the "infinite prime" of K ... terminology we shall explain in Chapter 3.)

It should be noted that while very effective in certain situations, the use of Dirichlet characters to find ramification indices, etc., has limitations. In the above example, we were able to find the ramification indices for L/K not merely because L/K was abelian, but rather because $L/\mathbb{Q}$ was abelian. Otherwise we could not have used Dirichlet characters for $L/\mathbb{Q}$.

More generally, one might ask whether, for a given number field K, it is possible to find a non-trivial abelian extension L/K in which every prime is unramified (without requiring that $L/\mathbb{Q}$ be abelian). Is it possible to find more than one? Is there a finite bound on the the degree $[L : K]$ or might it be that a fixed number field K has everywhere unramified abelian extensions of arbitrarily large degree? (If we require that $L/\mathbb{Q}$ be abelian, then we leave it as **Exercise 2.8** to show that there is a maximal such L, which has finite degree over K.) We shall return to these kinds of questions (for $L/\mathbb{Q}$ not necessarily abelian) after we have defined the Hilbert class field.

Exercise 2.9. Let K be the field associated to the character $\chi^2\psi$, where χ is a generator of $\mathrm{Gal}\,(\widehat{\mathbb{Q}(\zeta_5)/\mathbb{Q}})$ and ψ is a generator of $\mathrm{Gal}\,(\widehat{\mathbb{Q}(\zeta_3)/\mathbb{Q}})$. Use the above ideas to construct an extension of number fields L/K in which every prime is unramified, and for which $\mathrm{Gal}\,(L/K)$ is cyclic of order 2. (HINT: Let L be the field associated to the character group $\langle \chi^2, \psi \rangle$.)   ◊

Exercise 2.10. Let K be the field associated to the character $\chi^2\psi$, where χ is a generator of $\mathrm{Gal}\,(\widehat{\mathbb{Q}(\zeta_5)/\mathbb{Q}})$ and ψ is a generator of $\mathrm{Gal}\,(\widehat{\mathbb{Q}(\zeta_{13})/\mathbb{Q}})$. Construct an extension of number fields L/K in which every prime is unramified, and for which $\mathrm{Gal}\,(L/K)$ is cyclic of order 2.   ◊

Exercise 2.11. Let K be the field associated to the character $\chi^4\psi$, where χ is a generator of $\mathrm{Gal}\,(\widehat{\mathbb{Q}(\zeta_{13})/\mathbb{Q}})$ and ψ is a generator of $\mathrm{Gal}\,(\widehat{\mathbb{Q}(\zeta_7)/\mathbb{Q}})$. (If you want to express K as a splitting field of some irreducible polynomial over $\mathbb{Q}$, it may be best to employ technology to aid in the computations.) Construct an extension of number fields L/K in which every prime is unramified, and for which $\mathrm{Gal}\,(L/K)$ is cyclic of order 3.   ◊

Exercise 2.12. Let $K = \mathbb{Q}(\sqrt{-21})$. Find an extension L of K so that $L/\mathbb{Q}$ is abelian, $[L : K] = 4$ and every prime is unramified in L/K. (HINT: Use Dirichlet characters modulo 84.)   ◊

3 Dirichlet Series

The Riemann zeta function and the Dirichlet L-functions are examples of Dirichlet series. In this section, we shall discuss a few general properties of Dirichlet series, which we shall need later. This is a very rich subject; much more than what we do here can be said. The reader is encouraged to consult a text on analytic number theory.

A *Dirichlet series* is a series of the form

$$f(s) = \sum_{n=1}^{\infty} \frac{a_n}{n^s}$$

where $a_n \in \mathbb{C}$ for all n, and s is a complex variable. We gather some facts about Dirichlet series in the following discussion. More can be found in Serre's *A Course in Arithmetic*, [Se1].

Exercise 2.13. Prove ABEL'S LEMMA: let (a_n) and (b_n) be sequences, and for $r \geq m$, put $A_{m,r} = \sum_{n=m}^{r} a_n$ and $S_{m,r} = \sum_{n=m}^{r} a_n b_n$. Then

$$S_{m,r} = \sum_{n=m}^{r-1} A_{m,n}(b_n - b_{n+1}) + A_{m,r} b_r.$$

$\Diamond$

Exercise 2.14. Let A be an open subset of $\mathbb{C}$ and let (f_n) be a sequence of holomorphic functions on A that converges uniformly on every compact subset to a function f. Show that f is holomorphic on A and the sequence of derivatives (f_n') converges uniformly on all compact subsets to f'. (HINT: Let D be a closed disk contained in A and let C be its boundary. For s_0 in the interior of D, Cauchy's Formula applies to the $f_n(s_0)$. Let $n \to \infty$ to get a similar formula for $f(s_0)$. For the derivatives, proceed in the same way, using $f'(s_0) = \dfrac{1}{2\pi i} \displaystyle\int_C \frac{f(s)}{(s - s_0)^2}\, ds$.)

$\Diamond$

Lemma 3.1. If $f(s) = \sum_{n=1}^{\infty} \frac{a_n}{n^s}$ converges for $s = s_0$, then it converges uniformly in every domain of the form $\{s : \operatorname{Re}(s - s_0) \geq 0, |\operatorname{Arg}(s - s_0)| \leq \theta\}$ with $\theta < \frac{\pi}{2}$.

Proof. Translating if necessary, we may assume $s_0 = 0$. Then we have that $\sum_{n=1}^{\infty} a_n$ converges and we must show that $f(s)$ converges uniformly in every domain of the form $\{s : \operatorname{Re}(s) \geq 0, |\operatorname{Arg}(s)| \leq \theta\}$ for $\theta < \frac{\pi}{2}$. Equivalently, we must show that $f(s)$ converges uniformly in every domain of the form $\{s : \operatorname{Re}(s) \geq 0, \frac{|s|}{\operatorname{Re}(s)} \leq M\}$.

Let $\varepsilon > 0$ and let $A_{m,r}$ be as in Abel's Lemma (see Exercise 2.13 above). Since $\sum_{n=1}^{\infty} a_n$ converges, there is a sufficiently large number N so that if $r > m \geq N$ then we have $|A_{m,r}| < \varepsilon$. Let $b_n = n^{-s}$ and apply Abel's Lemma to get

$$S_{m,r} = \sum_{n=m}^{r-1} A_{m,n}(n^{-s} - (n+1)^{-s}) + A_{m,r} r^{-s}.$$

Taking absolute values, and noting that

$$|e^{-cs} - e^{-ds}| = |s| \int_c^d e^{-t\operatorname{Re}(s)}\, dt = \frac{|s|}{\operatorname{Re}(s)}(e^{-c\operatorname{Re}(s)} - e^{-d\operatorname{Re}(s)}),$$

we obtain

$$|S_{m,r}| \le \varepsilon \left(1 + \frac{|s|}{\text{Re}(s)} \sum_{n=m}^{r-1} (n^{-\text{Re}(s)} - (n+1)^{-\text{Re}(s)})\right).$$

Hence

$$|S_{m,r}| \le \varepsilon \left(1 + M(m^{-\text{Re}(s)} - r^{-\text{Re}(s)})\right) \le \varepsilon(1 + M).$$

Of course $S_{m,r}$ is just a difference of partial sums of our Dirichlet series, so the uniform convergence of $f(s)$ on the domain $\{s : \text{Re}(s) \ge 0,\ \frac{|s|}{\text{Re}(s)} \le M\}$ follows. $\square$

Theorem 3.2. *If the Dirichlet series $f(s) = \sum_{n=1}^{\infty} \frac{a_n}{n^s}$ converges for $s = s_0$, then it converges (though not necessarily absolutely) for $\text{Re}(s) > \text{Re}(s_0)$ to a function that is holomorphic there.*

Proof. Clear from Lemma 3.1 and Exercise 2.14. $\square$

Corollary 3.3. Let $f(s) = \sum_{n=1}^{\infty} \frac{a_n}{n^s}$ be a Dirichlet series.

i. If the a_n are bounded, then $f(s)$ converges absolutely for $\text{Re}(s) > 1$.
ii. If $A_n = a_1 + \ldots + a_n$ is a bounded sequence, then $f(s)$ converges (though not necessarily absolutely) for $\text{Re}(s) > 0$.
iii. If $f(s) = \sum_{n=1}^{\infty} \frac{a_n}{n^s}$ converges at $s = s_0$, then it converges absolutely for $\text{Re}(s) > \text{Re}(s_0) + 1$.

Proof

i. Suppose the a_n are bounded, say $|a_n| \le M$. Then

$$\left| \sum_{n=1}^{r} \frac{a_n}{n^s} \right| \le M \sum_{n=1}^{r} n^{-\sigma}$$

where $\sigma = \text{Re}(s)$. The result now follows from the convergence of $\sum_{n=1}^{\infty} \frac{1}{n^\sigma}$ for $\sigma > 1$.

ii. For $r > m$ let $A_{m,r} = \sum_{n=m}^{r} a_n$ as in Abel's Lemma. We have that the $A_{m,r}$ are bounded, say $A_{m,r} \le M$. Apply Abel's Lemma with $b_n = n^{-s}$ as in the proof of Lemma 3.1. We get

$$|S_{m,r}| \le M \left(\sum_{n=m}^{r-1} \left| \frac{1}{n^s} - \frac{1}{(n+1)^s} \right| + \left| \frac{1}{r^s} \right| \right).$$

By Theorem 3.2, we may assume that $s = \sigma$ is real. But then we have

$$|S_{m,r}| \le \frac{M}{m^\sigma}$$

from which we see that the partial sums of $f(s)$ comprise a Cauchy sequence when $\text{Re}(s) > 0$.

iii. Let

$$g(s) = f(s + s_0) = \sum (\frac{a_n}{n^{s_0}})(\frac{1}{n^s}).$$

Since $f(s_0)$ converges, $b_n = \frac{a_n}{n^{s_0}} \to 0$ as $n \to \infty$. Hence, $\{b_n\}$ is bounded. By (i), $g(s)$ converges absolutely for $\text{Re}(s) > 1$. Thus $f(s) = g(s - s_0)$ converges absolutely for $\text{Re}(s - s_0) > 1$, i.e., for $\text{Re}(s) > \text{Re}(s_0) + 1$.   □

Let $\chi : (\mathbb{Z}/m\mathbb{Z})^{\times} \to \mathbb{C}^{\times}$, be a Dirichlet character. The *Dirichlet L-function* associated to χ is

$$L(s, \chi) = \sum_{n=1}^{\infty} \frac{\chi(n)}{n^s}.$$

The *Riemann zeta function* is

$$\zeta(s) = \sum_{n=1}^{\infty} \frac{1}{n^s}.$$

Suppose $\chi \neq \chi_0$. Let $A_n = \chi(1) + \ldots + \chi(n)$ and write $n = mk + r$, where $0 \leq r \leq m - 1$. Then

$$\begin{aligned}
A_n &= [\chi(1) + \ldots + \chi(m)] + [\chi(m + 1) + \ldots + \chi(2m)] + \ldots \\
&\quad + [\chi(km + 1) + \ldots + \chi(km + r)] \\
&= \chi(km + 1) + \ldots + \chi(km + r)
\end{aligned}$$

so $|A_n| \leq r < m$.

Now use Corollary 3.3. By (ii), if $\chi \neq \chi_0$, $L(s, \chi)$ is analytic for $\text{Re}(s) > 0$. For any χ (including χ_0), $L(s, \chi)$ converges absolutely for $\text{Re}(s) > 1$ by (iii).

Proposition 3.4. $L(s, \chi)$ has a so-called Euler product:

$$L(s, \chi) = \prod_{p \text{ prime}} (1 - \chi(p)p^{-s})^{-1} \qquad \text{for } \text{Re}(s) > 1.$$

Proof. Fix s, with $\text{Re}(s) > 1$. We want to show:

$$\lim_{N \to \infty} \prod_{p \leq N} \left(1 - \frac{\chi(p)}{p^s}\right)^{-1} = L(s, \chi).$$

Say $p_1, \ldots, p_k$ are all the primes less than N. Then

$$\prod_{i=1}^{k} \left(1 - \frac{\chi(p_i)}{p_i^s}\right)^{-1} = \prod_{i=1}^{k} \left(1 + \frac{\chi(p_i)}{p_i^s} + \ldots + \frac{\chi(p_i^m)}{p_i^{ms}} + \ldots\right)$$

$$= \sum_{m_1,\ldots,m_k \geq 0} \frac{\chi(p_1^{m_1} p_2^{m_2} \cdots p_k^{m_k})}{(p_1^{m_1} p_2^{m_2} \cdots p_k^{m_k})^s}$$

$$= \sum_{n \in \mathcal{J}_N} \frac{\chi(n)}{n^s}$$

where $\mathcal{J}_N = \{n \in \mathbb{Z} : n > 0 \text{ and } n \text{ is not divisible by any prime } p > N\}$.

We have $L(s, \chi) - \prod_{p \leq N} \left(1 - \frac{\chi(p)}{p^s}\right)^{-1} = \sum_{n \in (\mathbb{Z}_+ \setminus \mathcal{J}_N)} \frac{\chi(n)}{n^s}$. Taking absolute values, and applying the triangle inequality, we get

$$\left| \sum_{n \in \mathbb{Z}_+ \setminus \mathcal{J}_N} \frac{\chi(n)}{n^s} \right| \leq \sum_{n \in \mathbb{Z}_+ \setminus \mathcal{J}_N} \frac{1}{n^\sigma} \qquad \text{where } \sigma = \mathrm{Re}\,(s)$$

$$\leq \sum_{n \geq N} \frac{1}{n^\sigma} \to 0 \qquad \text{as } N \to \infty,$$

since $\sum_{n > N} \frac{1}{n^\sigma}$ converges for $\sigma > 1$. The result follows. $\qquad\square$

Note that $L(s, \chi) \neq 0$ for $\mathrm{Re}\,(s) > 1$. We obtain

$$\log L(s, \chi) = -\sum_p \log(1 - \chi(p)p^{-s}),$$

where "log" denotes the branch of the logarithm such that $\log L(s, \chi) \to 0$ as $s \to \infty$.

Using the expansion for $\log(1 + T)$,

$$\log L(s, \chi) = \sum_p -\log(1 - \chi(p)p^{-s})$$

$$= \sum_p \sum_{n \geq 1} \frac{\chi(p)^n p^{-ns}}{n}$$

$$= \sum_p \sum_{n \geq 1} \frac{\chi(p)^n}{np^{ns}}.$$

Now $\left|\frac{\chi(p)^n}{np^{ns}}\right| \leq \left|\frac{1}{p^{ns}}\right| = \frac{1}{p^{n\sigma}}$ where $\sigma = \mathrm{Re}\,(s)$, so

$$\sum_p \sum_{n \geq 1} \left|\frac{\chi(p)^n}{np^{ns}}\right| \leq \sum_p \sum_{n \geq 1} \frac{1}{p^{n\sigma}} \leq \sum_{m \geq 1} \frac{1}{m^\sigma},$$

which converges for $\sigma > 1$. Hence the above expression for $\log L(s, \chi)$ is absolutely convergent for $\mathrm{Re}\,(s) > 1$. This allows us to rearrange the terms to get

$$\log L(s, \chi) = \sum_p \sum_{n \geq 1} \frac{\chi(p)^n}{np^{ns}} = \sum_p \frac{\chi(p)}{p^s} + \sum_p \sum_{n \geq 2} \frac{\chi(p)^n}{np^{ns}}.$$

Let

$$\beta(s, \chi) = \sum_p \sum_{n \geq 2} \frac{\chi(p)^n}{np^{ns}}.$$

Note that $\beta(s, \chi)$ is absolutely convergent for $\mathrm{Re}\,(s) > 1/2$, (so $\beta(1, \chi)$ takes a finite value).

4 Dirichlet's Theorem on Primes in Arithmetic Progressions

In this section, we give a proof of Dirichlet's Theorem on Primes in Arithmetic Progressions. As we mentioned earlier, Dirichlet's proof of this theorem is a generalization of a technique of Euler that used the Riemann zeta function to prove that there are infinitely many primes, (the case $m = 1$). We sketch Euler's proof here.

Recall that the Riemann zeta function is given by

$$\zeta(s) = \sum_{n=1}^{\infty} \frac{1}{n^s} = \prod_{p \text{ prime}} (1 - p^{-s})^{-1}.$$

Suppose there are only finitely many primes $p_1, p_2, \ldots, p_n$ in $\mathbb{Z}$. Then

$$\zeta(s) = \prod_{j=1}^{n} (1 - p_j^{-s})^{-1} = \prod_{j=1}^{n} \left(\frac{1}{1 - \frac{1}{p_j^s}} \right).$$

Taking limits, we have

$$\lim_{s \to 1} \zeta(s) = \prod_{j=1}^{n} \left(\frac{1}{1 - \frac{1}{p_j}} \right),$$

a rational number. But, (from the series for $\zeta(s)$) it is clear that $\lim_{s \to 1} \zeta(s) = \infty$, a contradiction.

In the proof of Dirichlet's Theorem on Primes in Arithmetic Progressions, the zeta function is replaced by Dirichlet L-functions.

Theorem 4.1 (Dirichlet's Theorem on Primes in Arithmetic Progressions). *If m is a positive integer and a is an integer for which $(a, m) = 1$, then there are infinitely many primes p satisfying $p \equiv a \pmod{m}$.*

Proof. Let $(m, a) = 1$ and consider all Dirichlet characters χ that are defined modulo m. Then

$$\sum_{\chi \in (\widehat{\mathbb{Z}/m\mathbb{Z}})^\times} \chi(a)^{-1} \log L(s, \chi) = \sum_\chi \chi(a)^{-1} \left(\sum_p \frac{\chi(p)}{p^s} + \beta(s, \chi) \right)$$

$$= \sum_p \frac{1}{p^s} \sum_\chi \chi(a)^{-1} \chi(p) + \sum_\chi \chi(a)^{-1} \beta(s, \chi)$$

$$= \sum_p \frac{1}{p^s} \sum_\chi \chi(pa^{-1}) + \sum_\chi \chi(a)^{-1} \beta(s, \chi).$$

But, by the orthogonality relations,

$$\sum_\chi \chi(pa^{-1}) = \begin{cases} \varphi(m) & p \equiv a \pmod{m} \\ 0 & \text{otherwise,} \end{cases}$$

so

$$\sum_\chi \chi(a)^{-1} \log L(s, \chi) = \varphi(m) \sum_{p \equiv a \pmod{m}} p^{-s} + \begin{pmatrix} \text{something} \\ \text{abs. conv.} \\ \text{for Re } (s) > \frac{1}{2} \end{pmatrix}. \quad (*)$$

Now let $s \to 1$. For the right side of $(*)$ we get

$$\lim_{s \to 1} \varphi(m) \sum_{p \equiv a \pmod{m}} p^{-s} + \text{(a finite constant)},$$

which would be finite if there were only finitely many primes p with $p \equiv a$ $\pmod{m}$. The proof will be complete if we can show that for the left side of $(*)$ we have $\lim_{s \to 1} \sum_\chi \chi(a)^{-1} \log L(s, \chi) = \infty$.

First, if $\chi = \chi_0$, (with modulus m), then

$$L(s, \chi_0) = \prod_p (1 - \chi_0(p) p^{-s})^{-1} = \zeta(s) \prod_{p \mid m} (1 - p^{-s}) \to \infty \text{ as } s \to 1$$

whence $\log L(s, \chi_0) \to \infty$ as $s \to 1$.

Now, if $\chi \neq \chi_0$, then we have seen by (ii) of Corollary 3.3 that $L(s, \chi)$ is analytic for Re $(s) > 0$. Since $L(1, \chi)$ is defined for $\chi \neq \chi_0$, $\log L(1, \chi)$ will be finite if we can show that $L(1, \chi) \neq 0$ for $\chi \neq \chi_0$. Given this, we'll have $\sum_\chi \chi(a)^{-1} \log L(s, \chi) \to \infty$ as $s \to 1^+$, and the proof will be complete.

Of course, it remains to show that $L(1, \chi) \neq 0$ when $\chi \neq \chi_0$. In 1840, Dirichlet gave an analytic proof of this result. In 1850, Kummer gave an arithmetic proof, which we begin here. We shall need the Dedekind zeta function.

Let K be an algebraic number field and let $\mathfrak{a}$ vary through the nonzero integral ideals of $\mathcal{O}_K$, (so that we may view $N\mathfrak{a}$ as a positive integer). Define the *Dedekind zeta function* of K as

$$\zeta_K(s) = \sum_{\mathfrak{a}} \frac{1}{N\mathfrak{a}^s}.$$

By an argument similar to the one for L-functions, we have

$$\zeta_K(s) = \prod_{\mathfrak{p}} (1 - N\mathfrak{p}^{-s})^{-1},$$

where $\mathfrak{p}$ runs over the prime ideals of $\mathcal{O}_K$. (The proof uses unique factorization of ideals.) It is easy to see that

$$\zeta_K(s) = \sum_{n=1}^{\infty} \frac{\gamma_n}{n^s} \qquad \text{where } \gamma_n = \#\{\mathfrak{a} : N\mathfrak{a} = n\}.$$

Exercise 2.15. Show that $\zeta_K(s)$ is absolutely convergent for $\mathrm{Re}\,(s) > 1$. (Compare this to (i) of Corollary 3.3 in the section on Dirichlet series — are the γ_n bounded?) $\Diamond$

The following theorem comes from the work of Dedekind; we omit the proof.

Theorem 4.2. $\zeta_K(s)$ *can be analytically continued to* $\mathbb{C} - \{1\}$, *with a simple pole at* $s = 1$, *i.e.,*

$$\zeta_K(s) = \frac{\rho(K)}{s-1} + (something\ entire).$$

Moreover

$$\rho(K) = \frac{2^{r_1}(2\pi)^{r_2} h_K R_K}{w_K \sqrt{|d_{K/\mathbb{Q}}|}}$$

where

$$r_1 = \#\ real\ embeddings\ of\ K$$
$$r_2 = \#\ pairs\ of\ imaginary\ embeddings\ of\ K$$
$$h_K = \#\mathcal{C}_K = \ class\ number\ of\ K$$
$$R_K = \ regulator\ of\ K$$
$$w_K = \#\ roots\ of\ unity\ in\ K$$
$$d_{K/\mathbb{Q}} = \ the\ discriminant.$$

$\square$

We shall use the Dedekind zeta function to show that $L(1, \chi) \neq 0$ for $\chi \neq \chi_0$, completing the proof of Dirichlet's Theorem on Primes in Arithmetic Progressions. Take $K = \mathbb{Q}(\zeta_m)$, where m is the modulus in Dirichlet's Theorem, so also the modulus of the characters χ. We have

$$
\begin{aligned}
\zeta_K(s) &= \prod_{\mathfrak{p}}(1 - N\mathfrak{p}^{-s})^{-1} \\
&= \prod_p \prod_{\mathfrak{p}|p}(1 - N\mathfrak{p}^{-s})^{-1} \\
&= \left(\prod_{p|m} \prod_{\mathfrak{p}|p}(1 - N\mathfrak{p}^{-s})^{-1} \right)\left(\prod_{p\nmid m} \prod_{\mathfrak{p}|p}(1 - N\mathfrak{p}^{-s})^{-1} \right).
\end{aligned}
$$

Now $N\mathfrak{p} = p^f$, where f is the residue field degree, and since $K/\mathbb{Q}$ is Galois, we have $efg = \varphi(m)$. If $p \nmid m$, then $e = 1$, $f = \#Z(p) = \mathrm{ord}_m\, p$, $g = \varphi(m)/f$.

Lemma 4.3. If $p \nmid m$, then $(1 - T^f)^{\varphi(m)/f} = \prod_{\chi \bmod m}(1 - \chi(p)T)$.

Proof. Let $G = \mathrm{Gal}(K/\mathbb{Q})$, where $K = \mathbb{Q}(\zeta_m)$ as before. Let $Z = Z(p)$ be the decomposition group for p in $K/\mathbb{Q}$ and define a map $\hat{G} \to \hat{Z}$ by $\chi \mapsto \chi\big|_Z$. Then

$$
\begin{aligned}
\prod_{\chi \in \hat{G}}(1 - \chi(p)T) &= \prod_{\psi \in \hat{Z}} \prod_{\substack{\chi \in \hat{G}, \\ \chi|_Z = \psi}}(1 - \chi(p)T) \\
&= \prod_{\psi \in \hat{Z}}(1 - \psi(p)T)^{\ell(\psi)}
\end{aligned}
$$

where

$$
\begin{aligned}
\ell(\psi) &= \#\{\chi \in \hat{G} : \chi\big|_Z = \psi\} \\
&= \#\ker(\hat{G} \to \hat{Z}) \\
&= {}^{\#\hat{G}}\!\big/_{\#\hat{Z}} \\
&= {}^{\varphi(m)}\!\big/_f = g.
\end{aligned}
$$

Thus

$$
\prod_{\chi \in \hat{G}}(1 - \chi(p)T) = \prod_{\psi \in \hat{Z}}(1 - \psi(p)T)^g.
$$

It remains only to show that $1 - T^f = \prod_{\psi \in \hat{Z}}(1 - \psi(p)T)$. As a subgroup of $\left(\mathbb{Z}/m\mathbb{Z}\right)^\times$, Z is generated by $p \bmod m$. Map $\hat{Z} \to \mu_f = \{f^{\text{th}} \text{ roots of unity}\}$ by

$\psi \mapsto \psi(p)$. Since p is a generator, this is an isomorphism. Thus

$$\prod_{\psi \in \hat{Z}}(1 - \psi(p)T) = \prod_{\eta \in \mu_f}(1 - \eta T)$$

$$= 1 - T^f. \qquad \square$$

Now put $T = p^{-s}$ in Lemma 4.3:

$$(1 - p^{-sf})^{-\varphi(m)/f} = \prod_{\chi}(1 - \chi(p)p^{-s})^{-1}.$$

Take the product over all $p \nmid m$:

$$\prod_{p \nmid m}(1 - p^{-sf})^{-\varphi(m)/f} = \prod_{\chi}\prod_{p \nmid m}(1 - \chi(p)p^{-s})^{-1}.$$

Now, if $p|m$ then $\chi(p) = 0$, so

$$\prod_{p \nmid m}(1 - p^{-sf})^{-\varphi(m)/f} = \prod_{p}(1 - p^{-sf})^{-\varphi(m)/f} = \prod_{\chi} L(s, \chi).$$

On the other hand,

$$\prod_{p \nmid m}(1 - p^{-sf})^{-\varphi(m)/f} = \prod_{p \nmid m}\prod_{\mathfrak{p}|p}(1 - N\mathfrak{p}^{-s})^{-1}$$

$$= \zeta_K(s)\prod_{p|m}\prod_{\mathfrak{p}|p}(1 - N\mathfrak{p}^{-s}).$$

Thus

$$\zeta_K(s)\prod_{p|m}\prod_{\mathfrak{p}|p}(1 - N\mathfrak{p}^{-s}) = \prod_{\chi} L(s, \chi)$$

$$= L(s, \chi_0) \prod_{\chi \neq \chi_0} L(s, \chi)$$

$$= \zeta(s) \prod_{p|m}(1 - p^s) \prod_{\chi \neq \chi_0} L(s, \chi).$$

We get

$$\frac{\left(\prod_{p|m}\prod_{\mathfrak{p}|p}(1 - N\mathfrak{p}^{-s})\right)\zeta_K(s)}{\left(\prod_{p|m}(1 - p^{-s})\right)\zeta(s)} = \prod_{\chi \neq \chi_0} L(s, \chi).$$

Now $\prod_{p|m} \prod_{\mathfrak{p}|p} (1 - N\mathfrak{p}^{-s})$ and $\prod_{p|m} (1 - p^{-s})$ are non-zero constants, while each of $\zeta_K(s)$ and $\zeta(s)$ has a simple pole at $s = 1$. Letting $s \to 1$, the expression on the left side approaches a constant, hence $\prod_{\chi \neq \chi_0} L(s, \chi)$ does too. This shows that $L(1, \chi) \neq 0$ for all $\chi \neq \chi_0$ and our proof of Dirichlet's Theorem is complete. $\square$

5 Dirichlet Density

Let $f(s)$, $g(s)$ be defined for $s \in \mathbb{R}$, $s > 1$. Write $f(s) \sim g(s)$ to signify that $f(s) - g(s)$ is bounded as $s \to 1^+$.

We may reformulate our proof of Dirichlet's Theorem using this notation. Recall that for any χ,

$$\log L(s, \chi) = \sum_p \frac{\chi(p)}{p^s} + \{\text{Dirichlet series converging for Re}\,(s) > 1/2\},$$

so

$$\log L(s, \chi) \sim \sum_p \frac{\chi(p)}{p^s}.$$

Thus (assuming $L(1, \chi) \neq 0$ if $\chi \neq \chi_0$)

$$\sum_\chi \chi(a)^{-1} \log L(s, \chi) \sim \sum_{p \equiv a \,(\text{mod } m)} \frac{\varphi(m)}{p^s}$$

$$\sim \log L(s, \chi_0).$$

Now

$$L(s, \chi_0) = \prod_p \left(1 - \frac{\chi_0(p)}{p^s}\right)^{-1} = \left(\prod_{p|m}(1 - p^{-s})\right)\zeta(s),$$

so $\log L(s, \chi_0) \sim \log \zeta(s)$ and we get

$$\sum_{p \equiv a \,(\text{mod } m)} \frac{1}{p^s} \sim \frac{1}{\varphi(m)} \log \zeta(s).$$

Letting $s \to 1^+$, we find that $\sum_{p \equiv a \,(\text{mod } m)} \frac{1}{p^s}$ diverges, which gives Dirichlet's Theorem.

The above reformulation leads naturally to the notion of Dirichlet density. Note that $\lim_{s \to 1^+} (s-1)\zeta(s) = 1$, and

$$\frac{1}{\varphi(m)} \log \zeta(s) = \frac{1}{\varphi(m)} \left(\log(s-1)\zeta(s) + \log\left(\frac{1}{s-1}\right) \right).$$

Hence

$$\sum_{p \equiv a \,(\mathrm{mod}\ m)} p^{-s} \sim \frac{1}{\varphi(m)} \log\left(\frac{1}{s-1}\right).$$

Indeed

$$\lim_{s \to 1^+} \frac{\sum_{p \equiv a \,(\mathrm{mod}\ m)} p^{-s}}{\log\left(\frac{1}{s-1}\right)} = \frac{1}{\varphi(m)}.$$

This motivates the following definition.

Let S be any set of primes. If

$$\lim_{s \to 1^+} \frac{\sum_{p \in S} p^{-s}}{\log\left(\frac{1}{s-1}\right)} = \delta \quad \text{exists},$$

then we say that S has *Dirichlet density* $\delta = \delta(S)$.

Examples.

12. If $S = \{\text{primes } p : p \equiv a \ (\mathrm{mod}\ m)\}$, then $\delta(S) = \frac{1}{\varphi(m)}$.
13. If S is a finite set of primes, then $\delta(S) = 0$.
14. If $S = \{\text{primes } p : \text{ the first digit of } p \text{ is "1"}\}$, (e.g., 11, 17, 103, etc.), then $\delta(S) = \log_{10} 2$. (This example is due to Bombieri; it illustrates the distinction between Dirichlet density and natural density. See 6.4.5 of Serre's *A Course in Arithmetic*, [Se1].)

Exercise 2.16. Is the converse to Example 13 also true?   ◇

Exercise 2.17. Show that if $S = \{\text{all primes of } \mathbb{Z}\}$, then $\delta(S) = 1$.   ◇

Exercise 2.18. Let H be a subgroup of $\mathbb{Z}/n\mathbb{Z}$ and let

$$S = \{\text{primes } p \text{ of } \mathbb{Z} : p + n\mathbb{Z} \in H\}.$$

Find $\delta(S)$ in terms of $\#H$ and prove that your answer is correct.   ◇

Exercise 2.19. Suppose S and T are sets of primes with $S \cap T = \emptyset$. Show that if any two of $\delta(S)$, $\delta(T)$, $\delta(S \cup T)$ are finite then so is the third, and that in this case $\delta(S \cup T) = \delta(S) + \delta(T)$.   ◇

Theorem 5.1. *Let $K/\mathbb{Q}$ be Galois, and let*

$$\mathcal{S}_K = \{p \in \mathbb{Z} : p \text{ splits completely in } K/\mathbb{Q}\}.$$

Then $\delta(\mathcal{S}_K) = {}^1/{[K:\mathbb{Q}]}$.

Proof. Let $\zeta_K(s) = \prod_{\mathfrak{p}}(1 - N\mathfrak{p}^{-s})^{-1}$ for Re $(s) > 1$, be the Dedekind zeta function for K. Consider $s \in \mathbb{R}, s > 1$. We have

$$\log \zeta_K(s) = -\sum_{\mathfrak{p}} \log(1 - N\mathfrak{p}^{-s})$$

$$= \sum_{\mathfrak{p}} \sum_{n=1}^{\infty} \frac{1}{n} N\mathfrak{p}^{-ns}.$$

Now $\log \zeta_K(s) = \log((s-1)\zeta_K(s)) + \log(1/(s-1))$, so

$$\log \zeta_K(s) \sim \log(1/(s-1))$$

$$\sim \sum_{\mathfrak{p}} \sum_{n=1}^{\infty} \frac{1}{n} N\mathfrak{p}^{-ns}$$

$$\sim \sum_{\mathfrak{p}} N\mathfrak{p}^{-s} + \sum_{\mathfrak{p}} \sum_{n=2}^{\infty} \frac{1}{n} N\mathfrak{p}^{-ns}$$

$$\sim \sum_{\mathfrak{p}} N\mathfrak{p}^{-s}$$

since $\sum_{\mathfrak{p}} \sum_{n=2}^{\infty} \frac{1}{n} N\mathfrak{p}^{-ns}$ is bounded as $s \to 1^+$. Hence

$$\log \zeta_K(s) \sim \log\left(\frac{1}{s-1}\right) \sim \sum_{\mathfrak{p}} N\mathfrak{p}^{-s}$$

$$\sim \sum_{\substack{\mathfrak{p} \\ f(\mathfrak{p}/p)=1=e(\mathfrak{p}/p)}} p^{-s} + \sum_{\substack{\mathfrak{p} \\ f(\mathfrak{p}/p)>1}} p^{-f(\mathfrak{p}/p)s} + \sum_{\substack{\mathfrak{p} \\ f(\mathfrak{p}/p)=1 \\ e(\mathfrak{p}/p)>1}} p^{-s}.$$

Note that the second series is bounded as $s \to 1^+$. The third is also, since the number of ramified primes is finite. Hence,

$$\log \zeta_K(s) \sim \log\left(\frac{1}{s-1}\right) \sim \sum_{p \in \mathcal{S}_K} g(p)p^{-s}.$$

If $p \in \mathcal{S}_K$, then $g(p) = [K:\mathbb{Q}]$, so

$$\log \zeta_K(s) \sim \log\left(\frac{1}{s-1}\right) \sim \sum_{p \in \mathcal{S}_K} [K:\mathbb{Q}]\, p^{-s}$$

whence

$$\log\left(\frac{1}{s-1}\right) = [K : \mathbb{Q}] \sum_{p \in S_K} p^{-s} + b(s)$$

where $b(s)$ is bounded as $s \to 1^+$. We may now compute $\delta(S_K)$.

$$\begin{aligned}
\delta(S_K) &= \lim_{s \to 1^+} \frac{\sum_{p \in S_K} p^{-s}}{\log\left(\frac{1}{s-1}\right)} \\
&= \lim_{s \to 1^+} \left(\frac{[K : \mathbb{Q}]\sum_{p \in S_K} p^{-s} + b(s)}{\sum_{p \in S_K} p^{-s}}\right)^{-1} \\
&= [K : \mathbb{Q}]^{-1}. \qquad \qquad \square
\end{aligned}$$

More generally, we may also define Dirichlet density on sets of prime ideals in a number field F. If S is a set of prime ideals of $\mathcal{O}_F$, and

$$\lim_{s \to 1^+} \frac{\sum_{\mathfrak{p} \in S} N\mathfrak{p}^{-s}}{\log(\frac{1}{s-1})} = \delta \quad \text{exists,}$$

then we say that S has *Dirichlet density* $\delta = \delta_F(S)$.

The previous theorem holds in this more general setting:

Corollary 5.2. Let K/F be Galois, and let

$$S_{K/F} = \{\mathfrak{p} \in \mathcal{O}_F : \mathfrak{p} \text{ splits completely in } K/F\}.$$

Then $\delta_F(S_{K/F}) = {}^1/_{[K \,:\, F]}.$

Proof. **Exercise 2.20.**

$\square$

Let $S, \mathcal{T}$ be sets of primes in $\mathcal{O}_F$, where F is a number field. We define the following notation.

Write $S \prec \mathcal{T}$ to mean $\delta_F(S \setminus \mathcal{T}) = 0$.
Write $S \approx \mathcal{T}$ if $S \prec \mathcal{T} \prec S$.

Exercise 2.21. Let F be a number field.

a. Compute $\delta_F(S)$, where $S = \{\text{primes } \mathfrak{p} \text{ of } \mathcal{O}_F : f(\mathfrak{p}/\mathfrak{p} \cap \mathbb{Z}) = 1\}$.

b. Let S and $\mathcal{T}$ be arbitrary sets of primes in $\mathcal{O}_F$. Prove or disprove: $S \approx \mathcal{T}$ if and only if S and $\mathcal{T}$ differ by finitely many elements. $\diamond$

Theorem 5.3. *Let E and K be number fields, each of which is Galois over $\mathbb{Q}$. Then $S_K \prec S_E$ if and only if $E \subseteq K$.*

Proof. "$\Longleftarrow$" is clear. For "$\Longrightarrow$" suppose $\mathcal{S}_K \prec \mathcal{S}_E$. Note that $\mathcal{S}_{KE} = \mathcal{S}_E \cap \mathcal{S}_K$. We have $\delta(\mathcal{S}_K \setminus \mathcal{S}_E) = 0$, so (using Exercise 2.19),

$$\delta(\mathcal{S}_{KE}) = \delta(\mathcal{S}_E \cap \mathcal{S}_K) = \delta(\mathcal{S}_K \setminus \mathcal{S}_E) + \delta(\mathcal{S}_E \cap \mathcal{S}_K) = \delta(\mathcal{S}_K).$$

Theorem 5.1 gives $[KE : \mathbb{Q}] = [K : \mathbb{Q}]$, whence $KE = K$ and $E \subseteq K$.   □

Exercise 2.22. Can the hypothesis that $E/\mathbb{Q}$ is Galois be omitted in Theorem 5.3?$\Diamond$

Exercise 2.23. Generalize Theorem 5.3 to two extensions E and K of an arbitrary number field F using δ_F and $\mathcal{S}_{K/F}$, $\mathcal{S}_{E/F}$. $\Diamond$

Chapter 3
Ray Class Groups

As we have seen in the previous chapter, there are infinitely many primes of $\mathbb{Z}$ in an arithmetic progression $\{a + jm : j \in \mathbb{N}\}$ whenever $(a, m) = 1$. This is a theorem about primes of $\mathbb{Z}$, but one may hope to generalize it to prime ideals of $\mathcal{O}_F$ where F is an algebraic number field.

Are there infinitely many primes $\mathfrak{p}$ of $\mathcal{O}_F$ in an "arithmetic progression"? What might one mean by an "arithmetic progression"? Perhaps we could interpret it as a question about ideal classes: Given an ideal class $\mathfrak{c} \in \mathcal{C}_F$, are there infinitely many primes of $\mathcal{O}_F$ in $\mathfrak{c}$? (Recall $\mathcal{C}_F = {}^{\mathcal{I}_F}\!/_{\mathcal{P}_F}$ where $\mathcal{I}_F = \{$nonzero fractional ideals of $F\}$, $\mathcal{P}_F = \{$principal fractional ideals in $\mathcal{I}_F\}$.) But where does the modulus m enter into this?

If we hope to follow Dirichlet, we must replace m by an ideal $\mathfrak{m}$ of $\mathcal{O}_F$, and we must consider "congruences" modulo $\mathfrak{m}$. This leads naturally to the idea of generalized ideal class groups, defined for each such $\mathfrak{m}$, called *ray class groups*. We shall also need to expand our notions of Dirichlet character and L-function if an analogue of Dirichlet's argument is to apply in this more general setting.

In this chapter, we pursue these ideas, following the framework of Dirichlet's argument, as Weber did. This will lead us to the notion of class field, and the proof of the Universal Norm Index Inequality. First, however, we shall need a brief discussion of absolute values and the Approximation Theorem.

1 The Approximation Theorem and Infinite Primes

Theorem 1.1 (Approximation Theorem). *Let $|\cdot|_1, \cdots |\cdot|_n$ be non-trivial pairwise inequivalent absolute values on a number field F, and let $\beta_1, \ldots, \beta_n$ be non-zero elements of F. For any $\varepsilon > 0$, there is an element $\alpha \in F$ such that $|\alpha - \beta_j|_j < \varepsilon$, for each $j = 1, \ldots, n$.*

Proof. First, we show that there are elements $x_1, \ldots, x_n \in F$ such that for every $j = 1, \ldots, n$

$$|x_j|_j > 1, \quad \text{and } |x_i|_j < 1 \text{ for any } i \neq j$$

N. Childress, *Class Field Theory*, Universitext, DOI 10.1007/978-0-387-72490-4_3,
© Springer Science+Business Media, LLC 2009

by induction on n.

Let $j = 1$, (other values of j are handled similarly). For $n = 2$, since $|\cdot|_1$ and $|\cdot|_2$ are inequivalent, we must have that there are elements $y, z \in F$ such that $|y|_1 > 1, |y|_2 \leq 1$, while $|z|_1 \leq 1, |z|_2 > 1$. Take $x_1 = \frac{y}{z}$.

Now suppose there is some $x \in F$ that satisfies $|x|_1 > 1$ and $|x|_j < 1$ for all $j = 2, \ldots, n - 1$. By the $n = 2$ case, there is some $v \in F$ with $|v|_1 > 1$ and $|v|_n < 1$. Choose x_1 so that

$$
x_1 = \begin{cases} x & \text{if } |x|_n < 1 \\ x^r v & \text{if } |x|_n = 1 \\ \frac{x^r v}{1 + x^r} & \text{if } |x|_n > 1, \end{cases}
$$

where $r \in \mathbb{Z}$ will be determined as follows. In the case where $|x|_n = 1$, note that while $|v|_j$ may be larger than 1, we still have $|x^r v|_j < 1$ for all $j = 2, \ldots, n$, when r is sufficiently large. In the case where $|x|_n > 1$, we have

$$
|x_1|_j = \frac{|x|_j^r |v|_j}{|1 + x^r|_j} = \frac{|v|_j}{|x^{-r} + 1|_j}.
$$

When $2 \leq j \leq n - 1$, this yields $|x_1|_j \leq \frac{|v|_j}{||x|_j^{-r} - 1|} \to 0$ as $r \to \infty$. Also, $|x_1|_1 \leq \frac{|v|_1}{||x|_1^{-r} - 1|} \to |v|_1$ as $r \to \infty$, while $|x_1|_n \leq \frac{|v|_n}{||x|_n^{-r} - 1|} \to |v|_n < 1$ as $r \to \infty$. Thus, again, if r is sufficiently large, then we have $|x_1|_j < 1$ for all $j = 2, \ldots, n$, while $|x_1|_1 > 1$.

Our induction argument shows that we have $x_1 \in F$ that is large at $|\cdot|_1$ and small at all other $|\cdot|_j$. By symmetry, we may find $x_2, \ldots, x_n$ similarly.

Now let $\alpha = \sum_j \frac{x_j^\ell}{1 + x_j^\ell} \beta_j$, where ℓ will be determined below. We have (by the triangle inequality)

$$
|\alpha - \beta_j|_j \leq \left| \frac{\beta_j}{1 + x_j^\ell} \right|_j + \sum_{i \neq j} \left| \frac{x_i^\ell}{1 + x_i^\ell} \beta_i \right|_j.
$$

Choose ℓ to be a sufficiently large integer so that the above expression is smaller than ε for every j. $\square$

Note that when $\mathfrak{p}$ is a prime ideal of $\mathcal{O}_F$ and $c = |\pi|_\mathfrak{p}$ for $\pi \in \mathfrak{p} - \mathfrak{p}^2$, the statement $\alpha\beta \neq 0$ and $|\alpha - \beta|_\mathfrak{p} < \varepsilon$ gives $\text{ord}_\mathfrak{p}(\frac{\alpha}{\beta} - 1) > n$, where n is given by $\frac{\varepsilon}{|\beta|_\mathfrak{p}} < c^n$. If α and β are $\mathfrak{p}$-adic units, then this just means $\alpha \equiv \beta \pmod{\mathfrak{p}^n}$. In particular, if each of the absolute values in the Approximation Theorem is $\mathfrak{p}$-adic for some $\mathfrak{p}$, we get the Chinese Remainder Theorem.

Note also that when $|\cdot|_j = |\cdot|_\sigma$, where $\sigma : F \hookrightarrow \mathbb{R}$, the statement $\alpha\beta \neq 0$ and $|\alpha - \beta|_\sigma < \varepsilon$ for small ε means that $\sigma(\alpha/\beta) > 0$.

The Approximation Theorem is yet one more result that suggests that it would be advantageous to have some kind of unifying notation that would allow us to treat simultaneously the $\mathfrak{p}$-adic absolute values on a number field F and the absolute values arising from embeddings of F into $\mathbb{C}$.

In the $\mathfrak{p}$-adic situation above, we were able to write $\alpha \equiv \beta$ (mod $\mathfrak{p}^n$). We can write something similar in the case of the real embeddings σ of F if we make the following convention. When $\sigma : F \hookrightarrow \mathbb{R}$, we associate to σ a formal object that we call an *infinite real prime*, which we denote by $\mathfrak{p}_\sigma$. We may then define

$$\alpha \equiv \beta \quad (\text{mod } \mathfrak{p}_\sigma) \text{ if and only if } \sigma(\alpha/\beta) > 0.$$

(We may also define *infinite imaginary primes*: We associate an object $\mathfrak{p}_\sigma$ to each conjugate pair $\sigma, \bar{\sigma} : F \hookrightarrow \mathbb{C}$. We don't use the congruence notation with infinite imaginary primes however.)

Other language used for prime ideals can be adapted to infinite primes as well. In particular, if K/F is an extension of number fields, we say that an infinite prime $\mathfrak{p}_\sigma$ *ramifies* in K/F if and only if $\sigma(F) \subseteq \mathbb{R}$, but for some extension of σ to K we have $\sigma(K) \not\subseteq \mathbb{R}$.

Using the infinite real primes (and the usual primes), we may also define what is known as a *divisor*, (or *modulus*) for F as a formal product $\prod_{\mathfrak{p}} \mathfrak{p}^{t(\mathfrak{p})}$, where $t(\mathfrak{p}) \in \mathbb{N}$ is non-zero for only finitely many $\mathfrak{p}$, and can only take a value of 0 or 1 when $\mathfrak{p}$ is an infinite real prime. (We may consider the notion of an infinite imaginary prime, but if we do, we must take $t(\mathfrak{p}) = 0$ for all infinite imaginary primes $\mathfrak{p}$.) Specifically, we shall denote the product of all the infinite real primes by $\mathfrak{m}_\infty = \prod_{\sigma \text{ real}} \mathfrak{p}_\sigma$.

In the next section, we avoid the use of these infinite primes at first, but at the end we discuss how one may rewrite what we have done in terms of them. It is recommended that the reader consider this question while progressing through the section.

In the next chapter, we shall present the notion of *places*, which is a somewhat different way to treat these ideas, and which will be the language we use in our discussion of idèles.

2 Ray Class Groups and the Universal Norm Index Inequality

If an element $\alpha \in F$ satisfies $\sigma(\alpha) > 0$ for every real embedding σ of F, we say that α is *totally positive*, and write $\alpha \gg 0$. Let $\mathfrak{m}$ be a non-zero integral ideal of $\mathcal{O}_F$. Define $\mathcal{P}_{F,\mathfrak{m}}^+$ to be the subgroup of $\mathcal{P}_F$ generated by

$$\{\langle \alpha \rangle : \alpha \in \mathcal{O}_F, \alpha \equiv 1 \quad (\text{mod } \mathfrak{m}), \text{ and } \alpha \gg 0\}.$$

How do we characterize a fractional ideal in $\mathcal{P}_{F,\mathfrak{m}}^+$? The following exercise provides an answer. Write $\alpha \overset{\times}{\equiv} 1$ (mod $\mathfrak{m}$) when $\alpha \equiv 1$ (mod $\mathfrak{p}^{\text{ord}_\mathfrak{p}(\mathfrak{m})}$) in the completion $F_\mathfrak{p}$ for every $\mathfrak{p} | \mathfrak{m}$. (We are abusing notation slightly here; when writing congruences modulo powers of $\mathfrak{p}$ in the completion, we really mean congruences modulo powers of the unique maximal ideal in the ring of integers of $F_\mathfrak{p}$.)

Exercise 3.1. Show that

$$\mathcal{P}_{F,\mathfrak{m}}^{+} = \{\langle \alpha \rangle : \alpha \in F, \alpha \gg 0, \alpha \overset{\times}{\equiv} 1 \pmod{\mathfrak{m}}\}$$
$$= \{\langle \frac{\alpha}{\beta} \rangle : \frac{\alpha}{\beta} \gg 0; \alpha, \beta \in \mathcal{O}_F \text{ prime to } \mathfrak{m}; \alpha \equiv \beta \pmod{\mathfrak{m}}\}.$$

$\Diamond$

Let $\mathcal{I}_F(\mathfrak{m})$ be the group of fractional ideals of F whose factorizations do not contain a non-trivial power of any prime ideal dividing $\mathfrak{m}$:

$$\mathcal{I}_F(\mathfrak{m}) = \{\mathfrak{a} \in \mathcal{I}_F : \operatorname{ord}_\mathfrak{p} \mathfrak{a} = 0 \text{ for all } \mathfrak{p} | \mathfrak{m}\}.$$

The *strict (narrow) ray class group* (or *generalized ideal class group*) of F for $\mathfrak{m}$, is

$$\mathcal{R}_{F,\mathfrak{m}}^{+} = \mathcal{I}_F(\mathfrak{m}) \big/ \mathcal{P}_{F,\mathfrak{m}}^{+}.$$

Example.

1. Let $F = \mathbb{Q}$, $\mathfrak{m} = m\mathbb{Z}$, where $m \geq 1$. If $\langle r \rangle \in \mathcal{I}(\mathfrak{m})$, then we may suppose $r > 0$ and $r = a/b$, where $(a, m) = (b, m) = 1$. The map

$$\mathcal{I}_\mathbb{Q}(\mathfrak{m}) \longrightarrow \left(\mathbb{Z} \big/ m\mathbb{Z}\right)^{\times}$$

given by $\langle r \rangle \mapsto ab^{-1} \pmod{m}$ is then well-defined. It is clearly surjective and its kernel is $\{\langle r \rangle : r > 0, r = a/b, (a, m) = (b, m) = 1, a \equiv b \pmod{m}\} = \mathcal{P}_{\mathbb{Q},\mathfrak{m}}^{+}$. Hence for $F = \mathbb{Q}$, $\mathfrak{m} = m\mathbb{Z}$, we have

$$\mathcal{I}_\mathbb{Q}(\mathfrak{m}) \big/ \mathcal{P}_{\mathbb{Q},\mathfrak{m}}^{+} \cong \left(\mathbb{Z} \big/ m\mathbb{Z}\right)^{\times}.$$

For a non-zero integral ideal $\mathfrak{m}$ of $\mathcal{O}_F$ we define the *ray modulo* $\mathfrak{m}$ as

$$\mathcal{P}_{F,\mathfrak{m}} = \{\langle \alpha \rangle : \alpha \overset{\times}{\equiv} 1 \pmod{\mathfrak{m}}\}.$$

The *ray class group* of F for $\mathfrak{m}$ is

$$\mathcal{R}_{F,\mathfrak{m}} = \mathcal{I}_F(\mathfrak{m}) \big/ \mathcal{P}_{F,\mathfrak{m}}.$$

(The strict ray class group $\mathcal{R}_{F,\mathfrak{m}}^{+}$ may also be viewed as a ray class group in the above sense if one views $\mathcal{P}_{F,\mathfrak{m}}^{+}$ as a ray modulo the divisor $\mathfrak{m}\mathfrak{m}_\infty$. We discuss this briefly a bit later.)

Returning to Example 1, for $F = \mathbb{Q}$, $\mathfrak{m} = m\mathbb{Z}$, we have $\mathcal{R}^+_{\mathbb{Q},\mathfrak{m}} \cong \left(\mathbb{Z}/m\mathbb{Z}\right)^\times$ and $\mathcal{R}_{\mathbb{Q},\mathfrak{m}} \cong \left(\mathbb{Z}/m\mathbb{Z}\right)^\times \Big/ \{\pm 1\}$.

These ideas are consistent with the original notion of an ideal class group of F, for when $\mathfrak{m} = \mathcal{O}_F$, we have

$$\mathcal{R}_{F,\mathfrak{m}} = {}^{\mathcal{I}_F}\big/\mathcal{P}_F = \mathcal{C}_F, \text{ the ordinary ideal class group}$$

$$\mathcal{R}^+_{F,\mathfrak{m}} = {}^{\mathcal{I}_F}\big/\mathcal{P}^+_F, \text{ the } \textit{strict (narrow) ideal class group.}$$

Exercise 3.2. Let $F = \mathbb{Q}(\sqrt{m})$ where $m > 1$ is a square-free integer. Let $\mathfrak{m} = \mathcal{O}_F$ and let ε be a fundamental unit in $\mathcal{O}_F$.

a. Suppose $N_{F/\mathbb{Q}}(\varepsilon) = -1$. Show that the ideal class group of F and the strict ideal class group of F are isomorphic.

b. Suppose $N_{F/\mathbb{Q}}(\varepsilon) = 1$. Show that the strict ideal class group of F is twice as large as the ideal class group of F. $\Diamond$

Our question about a possible generalization of Dirichlet's Theorem on Primes in Arithmetic Progressions may be formulated as follows.

Fix a non-zero integral ideal $\mathfrak{m}$ of $\mathcal{O}_F$, where F is an algebraic number field. Are there infinitely many prime ideals $\mathfrak{p}$ of $\mathcal{O}_F$ in each class of $\mathcal{R}^+_{F,\mathfrak{m}}$?

In the hope that the answer is yes, we shall attempt (as Weber did) to follow the framework set by Dirichlet. Define a *generalized Dirichlet character* (or *Weber character*) of modulus $\mathfrak{m}$ as a homomorphism of groups $\chi : \mathcal{R}^+_{F,\mathfrak{m}} \to \mathbb{C}^\times$. Recall from our example in the case $F = \mathbb{Q}$, $\mathfrak{m} = m\mathbb{Z}$, we have $\mathcal{R}^+_{\mathbb{Q},\mathfrak{m}} \cong \left(\mathbb{Z}/m\mathbb{Z}\right)^\times$ so that this truly is a consistent generalization of the classical Dirichlet characters. We may also define an L-function for generalized Dirichlet characters: let

$$L_\mathfrak{m}(s, \chi) = \sum_{\substack{\text{integral ideals} \\ \mathfrak{a} \text{ of } \mathcal{O}_F, \\ (\mathfrak{a}, \mathfrak{m})=1}} \chi(\mathfrak{a})N\mathfrak{a}^{-s},$$

where by $\chi(\mathfrak{a})$, we really mean χ of the image of $\mathfrak{a}$ in $\mathcal{R}^+_{F,\mathfrak{m}}$. These L-functions are sometimes called *Weber L-functions*. (In the case $F = \mathbb{Q}$, we recover the Dirichlet L-functions.)

As with Dirichlet L-functions, we have an Euler product for $L_\mathfrak{m}(s, \chi)$:

$$L_\mathfrak{m}(s, \chi) = \prod_{\mathfrak{p} \nmid \mathfrak{m}}(1 - \chi(\mathfrak{p})N\mathfrak{p}^{-s})^{-1}.$$

Taking logs and proceeding as before, we find that there are infinitely many primes $\mathfrak{p}$ in each class, provided that $L_{\mathfrak{m}}(1, \chi)$ is defined, and $L_{\mathfrak{m}}(1, \chi) \neq 0$ for every $\chi \neq \chi_0$.

If we wish to continue to follow Dirichlet's argument, we must find an extension K/F such that $L_{\mathfrak{m}}(s, \chi)$ occurs as a factor of $\zeta_K(s)$. Class field theory will establish this field K.

First we want to study the strict ray class groups $\mathcal{R}_{F,\mathfrak{m}}^+$. There is a well-known result that the ordinary class group is finite; its order has been the subject of much study. We may ask whether the strict ray class groups are also finite groups. The answer is provided by the following proposition.

Proposition 2.1. $\mathcal{R}_{F,\mathfrak{m}}^+$ is a finite group, with

$$\#\mathcal{R}_{F,\mathfrak{m}}^+ = \frac{h_F \, 2^{r_1} \varphi(\mathfrak{m})}{[\mathcal{U}_F : \mathcal{U}_{F,\mathfrak{m}}^+]}$$

where

$$h_F = \#\mathcal{C}_F$$
$$r_1 = \# \text{ of real embeddings of } F$$
$$\varphi(\mathfrak{m}) = \# \left(\mathcal{O}_F / \mathfrak{m} \right)^{\times} = \prod_{\mathfrak{p} | \mathfrak{m}} N\mathfrak{p}^{e_\mathfrak{p}-1}(N\mathfrak{p} - 1), \text{ where } \mathfrak{m} = \prod_{\mathfrak{p}|\mathfrak{m}} \mathfrak{p}^{e_\mathfrak{p}}$$
$$\mathcal{U}_F = \mathcal{O}_F^{\times}, \text{ the units of } \mathcal{O}_F$$
$$\mathcal{U}_{F,\mathfrak{m}}^+ = \{\varepsilon \in \mathcal{U}_F : \varepsilon \gg 0, \, \varepsilon \equiv 1 \pmod{\mathfrak{m}}\}.$$

$\#\mathcal{R}_{F,\mathfrak{m}}^+$ is called the *strict ray class number modulo* $\mathfrak{m}$ or the *ray class number modulo* $\mathfrak{m}\mathfrak{m}_\infty$.

Proof. Let $\mathcal{P}_F(\mathfrak{m}) = \{\text{principal fractional ideals in } \mathcal{I}_F(\mathfrak{m})\} = \mathcal{I}_F(\mathfrak{m}) \cap \mathcal{P}_F$. We divide the proof into four steps.

STEP 1. $\mathcal{I}_F(\mathfrak{m}) \big/ \mathcal{P}_F(\mathfrak{m}) \cong \mathcal{I}_F \big/ \mathcal{P}_F = \mathcal{C}_F$.
Proof of step 1. We must show $\mathcal{I}_F = \mathcal{I}_F(\mathfrak{m})\mathcal{P}_F$ (i.e., for each $\mathfrak{a} \in \mathcal{I}_F$ there exists $\alpha \in F$ such that $\mathfrak{a} = \langle\alpha\rangle\mathfrak{b}$, for some $\mathfrak{b} \in \mathcal{I}_F(\mathfrak{m})$).

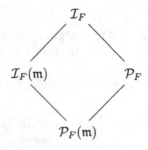

Let $\mathfrak{p}_1, \ldots, \mathfrak{p}_r$ be the primes dividing $\mathfrak{m}$. For each j, let $\pi_j \in \mathfrak{p}_j - \mathfrak{p}_j^2$. Let $\mathfrak{a} \in \mathcal{I}_F$, $e_j = \mathrm{ord}_{\mathfrak{p}_j}(\mathfrak{a}) \in \mathbb{Z}$. Choose $\alpha \in F$ satisfying:

$$|\alpha - \pi_j^{e_j}|_{\mathfrak{p}_j} < |\pi_j^{e_j}|_{\mathfrak{p}_j}$$

for all $i = 1, \ldots, r$ (possible by the Approximation Theorem).

It follows that $|\alpha|_{\mathfrak{p}_j} = |\pi_j^{e_j}|_{\mathfrak{p}_j}$, whence $\mathrm{ord}_{\mathfrak{p}_j}(\alpha) = e_j = \mathrm{ord}_{\mathfrak{p}_j}(\mathfrak{a})$. Thus $\mathrm{ord}_{\mathfrak{p}_j}(\alpha^{-1}\mathfrak{a}) = 0$ for all $j = 1, \ldots, r$.

Let $\mathfrak{b} = \alpha^{-1}\mathfrak{a}$; by the above, it will be in $\mathcal{I}_F(\mathfrak{m})$, and clearly we have $\mathfrak{a} = \langle \alpha \rangle \mathfrak{b}$.

STEP 2. $\mathcal{P}_F(\mathfrak{m}) / \mathcal{P}_{F,\mathfrak{m}}^+ \cong F(\mathfrak{m}) / \mathcal{U}_F F_{\mathfrak{m}}^+$ where

$$F(\mathfrak{m}) = \{\alpha \in F^\times : \langle \alpha \rangle \in \mathcal{I}_F(\mathfrak{m})\}$$

$$F_{\mathfrak{m}}^+ = \{\alpha \in F^\times : \alpha \gg 0,\ \alpha \overset{\times}{\equiv} 1 \pmod{\mathfrak{m}}\}.$$

Proof of step 2. Consider the epimorphism $F(\mathfrak{m}) \to \mathcal{P}_F(\mathfrak{m}) / \mathcal{P}_{F,\mathfrak{m}}^+$ that is given by $\alpha \mapsto \langle \alpha \rangle \mathcal{P}_{F,\mathfrak{m}}^+$. Its kernel is

$$\{\alpha \in F(\mathfrak{m}) : \langle \alpha \rangle \in \mathcal{P}_{F,\mathfrak{m}}^+\}$$
$$= \{\alpha \in F(\mathfrak{m}) : \exists \beta \in F_{\mathfrak{m}}^+ \text{ such that } \langle \alpha \rangle = \langle \beta \rangle\}$$
$$= \{\alpha \in F(\mathfrak{m}) : \exists \beta \in F_{\mathfrak{m}}^+ \text{ such that } \beta = \alpha\varepsilon \text{ for some } \varepsilon \in \mathcal{U}_F\}$$
$$= \mathcal{U}_F F_{\mathfrak{m}}^+.$$

Thus

$$F(\mathfrak{m}) / \mathcal{U}_F F_{\mathfrak{m}}^+ \cong \mathcal{P}_F(\mathfrak{m}) / \mathcal{P}_{F,\mathfrak{m}}^+.$$

STEP 3. $F(\mathfrak{m}) / F_{\mathfrak{m}}^+ \cong (\pm 1)^{r_1} \left(\mathcal{O}_F / \mathfrak{m}\right)^\times$.

Proof of step 3. Map $F(\mathfrak{m}) \to \{\pm 1\}^{r_1} \left(\mathcal{O}_F / \mathfrak{m}\right)^\times$ by

$$\alpha \mapsto (\mathrm{sign}\, \sigma_1(\alpha), \ldots, \mathrm{sign}\, \sigma_{r_1}(\alpha))(\alpha + \mathfrak{m})$$

where $\sigma_1, \ldots, \sigma_{r_1}$ are the real embeddings of F.

We leave it as **Exercise 3.3** to show this map is an epimorphism. Its kernel is

$$\{\alpha \in F(\mathfrak{m}) : \alpha \gg 0,\ \alpha \equiv 1 \pmod{\mathfrak{m}}\} = F_{\mathfrak{m}}^+.$$

STEP 4. $\mathcal{U}_F F_{\mathfrak{m}}^+ / F_{\mathfrak{m}}^+ \cong \mathcal{U}_F / \mathcal{U}_{F,\mathfrak{m}}^+$.

Proof of step 4.

$$\mathcal{U}_F F_{\mathfrak{m}}^+ \Big/ F_{\mathfrak{m}}^+ \cong \mathcal{U}_F \Big/ \mathcal{U}_F \cap F_{\mathfrak{m}}^+ = \mathcal{U}_F \Big/ \mathcal{U}_{F,\mathfrak{m}}^+.$$

Now we combine the information from all four steps. Consider the following diagram.

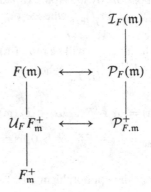

We have

$$
\begin{aligned}
\#R_{F,\mathfrak{m}}^+ &= [\mathcal{I}_F(\mathfrak{m}) : \mathcal{P}_{F,\mathfrak{m}}^+] = [\mathcal{I}_F(\mathfrak{m}) : \mathcal{P}_F(\mathfrak{m})][\mathcal{P}_F(\mathfrak{m}) : \mathcal{P}_{F,\mathfrak{m}}^+] \\
&= [\mathcal{I}_F(\mathfrak{m}) : \mathcal{P}_F(\mathfrak{m})][F(\mathfrak{m}) : F_{\mathfrak{m}}^+] \Big/ [\mathcal{U}_F F_{\mathfrak{m}}^+ : F_{\mathfrak{m}}^+] \\
&= h_F 2^{r_1} \varphi(\mathfrak{m}) \Big/ [\mathcal{U}_F : \mathcal{U}_{F,\mathfrak{m}}^+].
\end{aligned}
$$

$\square$

Example.

2. Let $F = \mathbb{Q}(\sqrt{3})$, $\mathfrak{m} = \mathcal{O}_F$. As we have seen previously, $R_{F,\mathfrak{m}}^+ = \mathcal{I}_F \big/ \mathcal{P}_F^+$. With Proposition 2.1, we obtain $\#R_{F,\mathfrak{m}}^+ = \frac{h_F 2^{r_1} \varphi(\mathfrak{m})}{[\mathcal{U}_F : \mathcal{U}_{F,\mathfrak{m}}^+]}$. For this particular field F, the quantities in the numerator of this formula are well-known: $r_1 = 2$, $h_F = 1$, and $\varphi(\mathfrak{m}) = \varphi(\mathcal{O}_F) = \#\left(\mathcal{O}_F \big/ \mathcal{O}_F\right)^\times = 1$. To find the denominator, note that by Dirichlet's Unit Theorem,

$$\mathcal{U}_F \cong \{\pm 1\} \times \mathbb{Z} \cong \{\pm 1\} \times \langle \varepsilon \rangle$$

where ε is a fundamental unit. The fundamental units for $\mathbb{Q}(\sqrt{3})$ are $2 \pm \sqrt{3}$, both of which are positive. Take $\varepsilon = 2 + \sqrt{3}$; then we have $\varepsilon \gg 0$, and

$$\mathcal{U}_{F,\mathfrak{m}}^+ = \{u \in \mathcal{U}_F : u \gg 0, u \equiv 1 \pmod{\mathcal{O}_F}\} = \langle \varepsilon \rangle.$$

Hence $[\mathcal{U}_F : \mathcal{U}_{F,\mathfrak{m}}^+] = 2$, and we conclude

$$\#\mathcal{R}_{F,\mathfrak{m}}^+ = \frac{1 \cdot 2^2 \cdot 1}{2} = 2.$$

Finally, note that in this case $\mathcal{R}_{F,\mathfrak{m}} = \mathcal{I}_F/\mathcal{P}_F$, so we obtain $\#\mathcal{R}_{F,\mathfrak{m}} = \#\mathcal{C}_F = h_F = 1$.

Exercise 3.4. Let $F = \mathbb{Q}(\sqrt{5})$, $\mathfrak{m} = \mathcal{O}_F$. Find the fundamental units for F, and determine $\mathcal{R}_{F,\mathfrak{m}}^+$ (up to isomorphism of groups). $\qquad\qquad\qquad \Diamond$

The definition of class field is due to Weber ([We2], 1897-1898). Earlier, Kronecker (e.g., [K1], 1853) had observed that every abelian extension of $\mathbb{Q}$ is cyclotomic. (In 1886-1887 Weber gave the first complete proof, [We1].) Kronecker had also observed ([K3], 1883-1890) that the transformation and division equations of modular and elliptic functions generated abelian extensions of imaginary quadratic fields. (He had hoped to prove that every abelian extension of an imaginary quadratic field can be obtained thus. Weber, [We3], partially succeeded in doing this in 1908, but the first complete proof did not come until Takagi, [T], in 1920. See also Hilbert's twelfth problem, [Hi3].) In these examples, Weber observed that the primes that split completely in these abelian extensions seemed to be related to the Galois groups. (Compare Theorem 1.8 in Chapter 1.) These ideas led Weber to his definition of class field.

Let K/F be Galois, and let $\mathfrak{m}$ be an integral ideal of $\mathcal{O}_F$. We say that K is the *class field* over F of $\mathcal{P}_{F,\mathfrak{m}}^+$ if

$$\mathcal{S}_{K/F} = \{\text{primes } \mathfrak{p} \text{ of } \mathcal{O}_F : \mathfrak{p} \text{ splits completely in } K/F\}$$
$$\approx \{\text{primes } \mathfrak{p} \text{ of } \mathcal{O}_F : \mathfrak{p} \in \mathcal{P}_{F,\mathfrak{m}}^+\}.$$

(Recall that $\mathcal{S} \approx \mathcal{T}$ if and only if they differ by a set with Dirichlet density zero.) Example.

3. For $F = \mathbb{Q}$ and $\mathfrak{m} = m\mathbb{Z}$, we have

$$\{p\mathbb{Z} : p\mathbb{Z} \in \mathcal{P}_{\mathbb{Q},\mathfrak{m}}^+\} = \{p\mathbb{Z} : p \equiv 1 \pmod{m}, \ p > 0\}$$
$$= \{p\mathbb{Z} : p\mathbb{Z} \text{ splits completely in } \mathbb{Q}(\zeta_m)/\mathbb{Q}\}$$
$$= \mathcal{S}_{\mathbb{Q}(\zeta_m)/\mathbb{Q}}.$$

Thus $K = \mathbb{Q}(\zeta_m)$ is the class field over $\mathbb{Q}$ of $\mathcal{P}_{\mathbb{Q},\mathfrak{m}}^+$.

More generally, we may define the notion of class field for subgroups of $\mathcal{I}_F(\mathfrak{m})$ that contain $\mathcal{P}_{F,\mathfrak{m}}^+$. If $\mathfrak{m}$ is a non-zero integral ideal of $\mathcal{O}_F$, and $\mathcal{H}$ satisfies

$$\mathcal{P}_{F,\mathfrak{m}}^+ < \mathcal{H} < \mathcal{I}_F(\mathfrak{m}),$$

then we say K is the *class field* over F of $\mathcal{H}$ if K/F is Galois and

$$\mathcal{S}_{K/F} \approx \{\text{primes } \mathfrak{p} \text{ of } \mathcal{O}_F : \mathfrak{p} \in \mathcal{H}\}.$$

What Weber had observed in the cases $F = \mathbb{Q}$ and F equal to an imaginary quadratic extension of $\mathbb{Q}$, was that the Galois groups for K/F were isomorphic to the associated factor groups $\mathcal{I}_F(\mathfrak{m})\big/\mathcal{H}$ (so clearly abelian). The isomorphism with the Galois group in these examples is an illustration of the *Isomorphy Theorem*, which we state shortly (its proof will be a consequence of Artin Reciprocity, see Chapter 5).

It will be some time before we can address the question of the existence of a class field for any such $\mathcal{H}$, but the issue of uniqueness can be settled easily.

Theorem 2.2 (Weber). *If the class field K of $\mathcal{H}$ exists, then it is unique.*

Proof. Recall the Dirichlet density of a set $\mathcal{S}$ of primes of F is

$$\delta_F(\mathcal{S}) = \lim_{s \to 1^+} \frac{\sum_{\mathfrak{p} \in \mathcal{S}} \frac{1}{N\mathfrak{p}^s}}{\log(\frac{1}{s-1})}.$$

We have shown that $\delta_F(\mathcal{S}_{K/F}) = \frac{1}{[K:F]}$. If K_1, K_2 are two class fields for $\mathcal{H}$, then let $K = K_1 K_2$. We find

$$\mathcal{S}_{K/F} = \mathcal{S}_{K_1/F} \cap \mathcal{S}_{K_2/F}$$
$$\approx \{\mathfrak{p} \text{ of } F : \mathfrak{p} \in \mathcal{H}\},$$

i.e.,

$$\mathcal{S}_{K/F} \approx \mathcal{S}_{K_1/F} \approx \mathcal{S}_{K_2/F}.$$

Thus

$$\frac{1}{[K : F]} = \frac{1}{[K_1 : F]} = \frac{1}{[K_2 : F]}$$

and we must have $K = K_1 = K_2$. $\square$

Exercise 3.5. Let F be a number field and let $\mathfrak{n}$, $\mathfrak{m}$ be (not necessarily distinct) ideals of $\mathcal{O}_F$. Let $\mathcal{P}_{F,\mathfrak{n}}^+ < \mathcal{H}_1 < \mathcal{I}_F(\mathfrak{n})$, and $\mathcal{P}_{F,\mathfrak{m}}^+ < \mathcal{H}_2 < \mathcal{I}_F(\mathfrak{m})$. If $\mathcal{H}_1 \neq \mathcal{H}_2$, is it possible for them to have the same class field over F? $\Diamond$

Let us return to our efforts to generalize the techniques of Dirichlet in his proof of the Theorem on Primes in Arithmetic Progressions. Let F, $\mathfrak{m}$ be as before. Recall that a group homomorphism $\chi : \mathcal{I}_F(\mathfrak{m})\big/\mathcal{P}_{F,\mathfrak{m}}^+ \longrightarrow \mathbb{C}^\times$ is called a generalized Dirichlet character, and we have defined the Weber L-function $L_{\mathfrak{m}}(s, \chi) = \sum_{\mathfrak{a} \in \mathcal{I}(\mathfrak{m})} \frac{\chi(\mathfrak{a})}{N\mathfrak{a}^s}$, which has Euler product $L_{\mathfrak{m}}(s, \chi) = \prod_{\mathfrak{p} \nmid \mathfrak{m}}(1 - \chi(\mathfrak{p})N\mathfrak{p}^{-s})^{-1}$ for $\mathrm{Re}(s) > 1$. We collect a few facts about Weber L-functions below, but omit the proofs, which are generalizations of the arguments used to prove the analogous facts about Dirichlet L-functions.

Some Facts About $L_{\mathrm{m}}(s, \chi)$.

1. For $\chi \neq \chi_0$, $L_{\mathrm{m}}(s, \chi)$ can be analytically continued to the entire complex plane.
2. $L_{\mathrm{m}}(s, \chi_0)$ can be analytically continued to the entire complex plane except for a simple pole at $s = 1$.
3. For χ_0, we have:

$$L_{\mathrm{m}}(s, \chi_0) = \prod_{\mathfrak{p} \nmid \mathfrak{m}} (1 - N\mathfrak{p}^{-s})^{-1}$$

$$= \left(\prod_{\mathfrak{p} \mid \mathfrak{m}} (1 - N\mathfrak{p}^{-s}) \right) \zeta_F(s).$$

Recall that if $\mathcal{S}$ is finite, then it has Dirichlet density $\delta_F(\mathcal{S}) = 0$; if $\mathcal{S}$ consists of *all* the primes of F, then $\delta_F(\mathcal{S}) = 1$.

For $\mathfrak{a} \in \mathcal{I}_F(\mathfrak{m})$ we may define

$$\mathcal{S}_{\mathfrak{a},\mathfrak{m}} = \{ \text{primes } \mathfrak{p} \text{ of } \mathcal{O}_F : \mathfrak{p} \equiv \mathfrak{a} \text{ in } \mathcal{R}_{F,\mathfrak{m}}^+ \} = \{ \text{primes } \mathfrak{p} \in \mathfrak{a}\,\mathcal{P}_{F,\mathfrak{m}}^+ \}.$$

We claim that if for every character $\chi \neq \chi_0$ of $\mathcal{R}_{F,\mathfrak{m}}^+$ we have $L_{\mathrm{m}}(1, \chi) \neq 0$, then

$$\delta_F(\mathcal{S}_{\mathfrak{a},\mathfrak{m}}) = \frac{1}{\#\mathcal{R}_{F,\mathfrak{m}}^+}.$$

Of course, this says $\mathcal{S}_{\mathfrak{a},\mathfrak{m}}$ contains infinitely many prime ideals. Thus the generalization of the Theorem on Primes in Arithmetic Progressions will follow once we have shown that $L_{\mathrm{m}}(1, \chi) \neq 0$ when $\chi \neq \chi_0$. This turns out to be related to Weber's notion of class field, as we shall soon discover.

First we need to prove the claim about $\delta_F(\mathcal{S}_{\mathfrak{a},\mathfrak{m}})$. More generally, we shall prove the following.

Proposition 2.3. Let $\mathfrak{a} \in \mathcal{I}_F(\mathfrak{m})$. Suppose $\mathcal{P}_{F,\mathfrak{m}}^+ < \mathcal{H} < \mathcal{I}_F(\mathfrak{m})$. If for all characters $\chi \neq \chi_0$ of $\mathcal{I}_F(\mathfrak{m})$ that are trivial on $\mathcal{H}$, we have $L_{\mathrm{m}}(1, \chi) \neq 0$, then

$$\delta_F(\{ \text{primes } \mathfrak{p} \text{ of } \mathcal{O}_F : \mathfrak{p} \in \mathfrak{a}\mathcal{H} \}) = \frac{1}{[\mathcal{I}_F(\mathfrak{m}) : \mathcal{H}]}.$$

Proof.

$$L_{\mathrm{m}}(s, \chi) = \prod_{\mathfrak{p} \nmid \mathfrak{m}} (1 - \chi(\mathfrak{p}) N\mathfrak{p}^{-s})^{-1}$$

$$\log L_{\mathrm{m}}(s, \chi) = -\sum_{\mathfrak{p} \nmid \mathfrak{m}} \log(1 - \chi(\mathfrak{p}) N\mathfrak{p}^{-s})$$

$$= \sum_{\mathfrak{p} \nmid \mathfrak{m}} \sum_{n=1}^{\infty} \frac{\chi(\mathfrak{p})^n N\mathfrak{p}^{-ns}}{n}$$

$$= \sum_{\mathfrak{p} \nmid \mathfrak{m}} \chi(\mathfrak{p}) N\mathfrak{p}^{-s} + \sum_{\mathfrak{p} \nmid \mathfrak{m}} \sum_{n=2}^{\infty} \frac{\chi(\mathfrak{p})^n N\mathfrak{p}^{-ns}}{n}$$

$$\sim \sum_{\mathfrak{p} \nmid \mathfrak{m}} \chi(\mathfrak{p}) N\mathfrak{p}^{-s}.$$

Now if χ is a character of $\mathcal{I}_F(\mathfrak{m})$ that is trivial on $\mathcal{H}$, then χ may be viewed as a character of $\mathcal{I}_F(\mathfrak{m})/\mathcal{H}$. For a fixed prime $\mathfrak{p}$ of $\mathcal{O}_F$, we have

$$\sum_{\chi} \chi(\mathfrak{a})^{-1} \chi(\mathfrak{p}) = \begin{cases} 0 & \text{if } \mathfrak{p} \notin \mathfrak{a}\mathcal{H} \\ [\mathcal{I}(\mathfrak{m}) : \mathcal{H}] & \text{if } \mathfrak{p} \in \mathfrak{a}\mathcal{H} \end{cases}$$

where the sum is over $\chi \in \widehat{\mathcal{I}_F(\mathfrak{m})}/\mathcal{H}$. Taking $\beta_\chi(s) = \sum_{\mathfrak{p} \nmid \mathfrak{m}} \sum_{n=2}^{\infty} \frac{\chi(\mathfrak{p})^n N\mathfrak{p}^{-ns}}{n}$, we get

$$\sum_{\chi} \chi(\mathfrak{a})^{-1} \log L_\mathfrak{m}(s, \chi) = \sum_{\chi} \chi(\mathfrak{a})^{-1} \left[\sum_{\mathfrak{p} \nmid \mathfrak{m}} \chi(\mathfrak{p}) N\mathfrak{p}^{-s} + \beta_\chi(s) \right]$$

$$\sum_{\chi} \chi(\mathfrak{a})^{-1} [\log L_\mathfrak{m}(s, \chi) - \beta_\chi(s)] = \sum_{\mathfrak{p} \in \mathfrak{a}\mathcal{H}} [\mathcal{I}_F(\mathfrak{m}) : \mathcal{H}] N\mathfrak{p}^{-s}. \tag{*}$$

Let $\mathcal{S} = \{\text{primes } \mathfrak{p} \text{ of } \mathcal{O}_F : \mathfrak{p} \in \mathfrak{a}\mathcal{H}\}$. We must show that

$$\lim_{s \to 1^+} \frac{\sum_{\mathfrak{p} \in \mathcal{S}} N\mathfrak{p}^{-s}}{\log(\frac{1}{s-1})}$$

converges to the desired limit. From $(*)$,

$$\sum_{\mathfrak{p} \in \mathcal{S}} [\mathcal{I}_F(\mathfrak{m}) : \mathcal{H}] N\mathfrak{p}^{-s} = \sum_{\chi \neq \chi_0} \chi(\mathfrak{a})^{-1} (\log L_\mathfrak{m}(s, \chi) - \beta_\chi(s))$$

$$+ \log((s-1) L_\mathfrak{m}(s, \chi_0)) - \log(s-1) - \beta_{\chi_0}(s)$$

Letting $h = [\mathcal{I}_F(\mathfrak{m}) : \mathcal{H}]$, this becomes

$$\sum_{\mathfrak{p} \in \mathcal{S}} N\mathfrak{p}^{-s} = \frac{1}{h} \sum_{\chi \neq \chi_0} \chi(\mathfrak{a})^{-1} [\log L_\mathfrak{m}(s, \chi) - \beta_\chi(s)]$$

$$+ \frac{1}{h} \log[(s-1) L_\mathfrak{m}(s, \chi_0)] + \frac{1}{h} \log\left(\frac{1}{s-1}\right) - \frac{1}{h} \beta_{\chi_0}(s).$$

Rearranging, we get

$$\sum_{\mathfrak{p} \in \mathcal{S}} N\mathfrak{p}^{-s} - \frac{1}{h} \log\left(\frac{1}{s-1}\right) = \frac{1}{h} \sum_{\chi \neq \chi_0} \chi(\mathfrak{a})^{-1}[\log L_{\mathfrak{m}}(s, \chi) - \beta_\chi(s)]$$

$$+ \frac{1}{h} \log[(s-1)L_{\mathfrak{m}}(s, \chi_0)] - \frac{1}{h}\beta_{\chi_0}(s).$$

Note that the right side of the above equation is bounded as $s \rightarrow 1^+$ (since $L_{\mathfrak{m}}(1, \chi) \neq 0$ for all $\chi \neq \chi_0$, and $L_{\mathfrak{m}}(s, \chi_0)$ has a simple pole at $s = 1$). Therefore

$$\frac{\sum_{\mathfrak{p} \in \mathcal{S}} N\mathfrak{p}^{-s}}{\log(\frac{1}{s-1})} - \frac{1}{h} \longrightarrow 0 \quad \text{as} \quad s \longrightarrow 1^+.$$

But then

$$\delta_F(\mathcal{S}) = \frac{1}{h} = \frac{1}{[\mathcal{I}_F(\mathfrak{m}) : \mathcal{H}]}.$$

$\square$

Using Artin Reciprocity, it is possible to prove other results similar in nature to Proposition 2.3. See the homework problems in Chapter 5 for some examples.

We can now begin to address the remaining ingredient in our attempt to generalize the Theorem on Primes in Arithmetic Progressions. Recall that we need to show $L_{\mathfrak{m}}(1, \chi) \neq 0$ whenever $\chi \neq \chi_0$ is trivial on $\mathcal{H}$. The following theorem nearly accomplishes this.

Theorem 2.4. *Suppose K/F is Galois, and $\mathcal{P}_{F,\mathfrak{m}}^+ < \mathcal{H} < \mathcal{I}_F(\mathfrak{m})$. Suppose there is some set of primes $\mathcal{T} \subseteq \mathcal{H}$ with $\mathcal{S}_{K/F} \approx \mathcal{T}$. Then*

$$[\mathcal{I}_F(\mathfrak{m}) : \mathcal{H}] \leq [K : F],$$

and $L_{\mathfrak{m}}(1, \chi) \neq 0$ whenever $\chi \neq \chi_0$ and χ is trivial on $\mathcal{H}$.

Proof. Let $m(\chi) = \text{ord}_{s=1}(L_{\mathfrak{m}}(s, \chi))$. For $\chi \neq \chi_0$, we know $m(\chi) \geq 0$, while $m(\chi_0) = -1$. Since χ is trivial on $\mathcal{H}$, we may view χ as a character of $\mathcal{I}_F(\mathfrak{m})/\mathcal{H}$.

There is some constant a such that

$$\prod_{\chi \in \widehat{\mathcal{I}(\mathfrak{m})/\mathcal{H}}} L_{\mathfrak{m}}(s, \chi) = a(s-1)^{\sum_\chi m(\chi)} + \cdots$$

Taking logs, we get

$$\sum_\chi \log L_{\mathfrak{m}}(s, \chi) \sim \left(\sum_\chi m(\chi)\right) \log(s-1)$$

$$= -\left(\sum_\chi m(\chi)\right) \log\left(\frac{1}{s-1}\right).$$

Now

$$\log L_{\mathfrak{m}}(s, \chi) = \sum_{\mathfrak{p} \nmid \mathfrak{m}} \sum_{n=1}^{\infty} \frac{\chi(\mathfrak{p})^n}{n N \mathfrak{p}^{ns}}$$

$$\sim \sum_{\mathfrak{p} \nmid \mathfrak{m}} \chi(\mathfrak{p}) N \mathfrak{p}^{-s}$$

as before, so

$$\sum_{\chi} \log L_{\mathfrak{m}}(s, \chi) \sim \sum_{\mathfrak{p} \in \mathcal{H}} [\mathcal{I}_F(\mathfrak{m}) : \mathcal{H}] N \mathfrak{p}^{-s}$$

as before. Hence,

$$\sum_{\mathfrak{p} \in \mathcal{H}} [\mathcal{I}_F(\mathfrak{m}) : \mathcal{H}] N \mathfrak{p}^{-s} \sim -\left(\sum_{\chi} m(\chi) \right) \log \left(\frac{1}{s-1} \right).$$

But

$$\sum_{\mathfrak{p} \in \mathcal{H}} [\mathcal{I}_F(\mathfrak{m}) : \mathcal{H}] N \mathfrak{p}^{-s} = [\mathcal{I}_F(\mathfrak{m}) : \mathcal{H}] \left(\sum_{\mathfrak{p} \in \mathcal{J}} N \mathfrak{p}^{-s} + \sum_{\mathfrak{p} \in \mathcal{H} \setminus \mathcal{J}} N \mathfrak{p}^{-s} \right).$$

Dividing by $\log(\frac{1}{s-1})$, and letting $s \to 1^+$, we get

$$-\left(\sum_{\chi} m(\chi) \right) = \lim_{s \to 1^+} \frac{[\mathcal{I}_F(\mathfrak{m}) : \mathcal{H}] \sum_{\mathfrak{p} \in \mathcal{S}_{K/F}} N \mathfrak{p}^{-s}}{\log(\frac{1}{s-1})}$$

$$+ \lim_{s \to 1^+} \frac{[\mathcal{I}_F(\mathfrak{m}) : \mathcal{H}] \sum_{\mathfrak{p} \in \mathcal{H} \setminus \mathcal{S}_{K/F}} N \mathfrak{p}^{-s}}{\log(\frac{1}{s-1})}$$

$$= [\mathcal{I}_F(\mathfrak{m}) : \mathcal{H}] \delta_F(\mathcal{S}_{K/F}) + [\mathcal{I}_F(\mathfrak{m}) : \mathcal{H}] \text{ (a finite nonnegative constant)}$$

$$\geq [\mathcal{I}_F(\mathfrak{m}) : \mathcal{H}] \Big/ [K : F].$$

(Note that the first term on the right above converges since $\delta_F(\mathcal{S}_{K/F})$ exists, whence the second term on the right side must also converge because the left side is finite.) Recalling that $m(\chi) \geq 0$ for all $\chi \neq \chi_0$ and that $m(\chi_0) = -1$, we have

$$1 \geq 1 - \sum_{\chi \neq \chi_0} m(\chi) \geq [\mathcal{I}_F(\mathfrak{m}) : \mathcal{H}] \Big/ [K : F] > 0$$

whence

$$0 \geq - \sum_{\chi \neq \chi_0} m(\chi) > -1.$$

But this forces $m(\chi) = 0$ for all $\chi \neq \chi_0$. Hence $-\left(\sum_{\chi} m(\chi) \right) = 1$ and

$$1 \geq [\mathcal{I}_F(\mathfrak{m}) : \mathcal{H}] \Big/ [K : F], \quad \text{i.e.,}$$

$$[K : F] \geq [\mathcal{I}_F(\mathfrak{m}) : \mathcal{H}].$$

Finally, whenever $\chi \neq \chi_0$, the fact that $m(\chi) = 0$ gives that $L_\mathfrak{m}(s, \chi)$ has a nonzero constant term when expanded in powers of $s - 1$. But this implies that $L_\mathfrak{m}(1, \chi) \neq 0$ for all $\chi \neq \chi_0$. □

Corollary 2.5. If K/F is Galois and K is the class field for $\mathcal{H}$ where $\mathcal{P}_{F,\mathfrak{m}}^+ < \mathcal{H} < \mathcal{I}_F(\mathfrak{m})$, then

$$[\mathcal{I}_F(\mathfrak{m}) : \mathcal{H}] = [K : F].$$

Proof. Say K is the class field for $\mathcal{H}$, i.e., {primes $\mathfrak{p}$ of $\mathcal{O}_F : \mathfrak{p} \in \mathcal{H}$} $\approx \mathcal{S}_{K/F}$. Then

$$\delta_F(\{\text{primes } \mathfrak{p} \in \mathcal{H}\} \setminus \mathcal{S}_{K/F}) = \lim_{s \to 1^+} \frac{\sum\limits_{\mathfrak{p} \in \mathcal{H} \setminus \mathcal{S}_{K/F}} N\mathfrak{p}^{-s}}{\log(\frac{1}{s-1})} = 0,$$

so proceeding as in the proof of Theorem 2.4, we get

$$1 = - \sum_{\chi} m(\chi) = [\mathcal{I}_F(\mathfrak{m}) : \mathcal{H}] \Big/ [K : F].$$

□

By Theorem 2.4, we shall have concluded the proof of the generalization of the Theorem on Primes in Arithmetic Progressions as soon as we verify that there is always a class field for $\mathcal{H} = \mathcal{P}_{F,\mathfrak{m}}^+$. This issue will be settled in Chapter 6. With what we have done thus far, we may readily obtain the following result first proved by Weber ([We2], 1897-1898).

Theorem 2.6 (Universal Norm Index Inequality). *(Historically, this was called the First Fundamental Inequality of Class Field Theory. Later it was called the Second Fundamental Inequality.) Let K/F be a Galois extension of number fields and let $\mathcal{H} = \mathcal{P}_{F,\mathfrak{m}}^+ \mathcal{N}_{K/F}(\mathfrak{m})$ where*

$$\mathcal{N}_{K/F}(\mathfrak{m}) = \{\mathfrak{a} \in \mathcal{I}_F(\mathfrak{m}) : \mathfrak{a} = N_{K/F}(\mathfrak{A}) \text{ for some } \mathfrak{A} \text{ in } \mathcal{I}_K\}.$$

(Note that the factorization of the fractional ideal $\mathfrak{A}$ of K cannot contain a non-trivial power of any prime ideal that divides $\mathfrak{m}\mathcal{O}_K$, i.e., $\mathfrak{A} \in \mathcal{I}_K(\mathfrak{m}\mathcal{O}_K)$.) Then

$$[\mathcal{I}_F(\mathfrak{m}) : \mathcal{H}] \leq [K : F].$$

Proof. If $\mathfrak{p} \in \mathcal{S}_{K/F}$ and $\mathfrak{P}|\mathfrak{p}$, where $\mathfrak{P}$ is a prime of $\mathcal{O}_K$, then $N_{K/F}\mathfrak{P} = \mathfrak{p}$ (since $\mathfrak{p} \in \mathcal{S}_{K/F}$, it splits completely in K/F giving $f(\mathfrak{P}/\mathfrak{p}) = 1$). Thus $\mathcal{S}_{K/F} \approx \mathcal{T}$ for some $\mathcal{T} \subseteq \mathcal{H}$ and we may apply the previous theorem to get $[\mathcal{I}_F(\mathfrak{m}) : \mathcal{H}] \leq [K : F]$. $\qquad\qquad\square$

It is natural to ask whether one may better specify the relationship between the two indices $[\mathcal{I}_F(\mathfrak{m}) : \mathcal{H}]$ and $[K : F]$. The "Global Cyclic Norm Index Inequality" (historically, the Second Fundamental Inequality of Class Field Theory, later the First Fundamental Inequality) says that if K/F is cyclic and $\mathfrak{m}$ is divisible by a sufficiently high power of each $\mathfrak{p}$ that ramifies in K/F, then

$$[\mathcal{I}_F(\mathfrak{m}) : \mathcal{P}_{F,\mathfrak{m}}^{+}\mathcal{N}_{K/F}(\mathfrak{m})] \geq [K : F].$$

It then follows that these indices are equal in the cyclic case. The proof of the Global Cyclic Norm Index Inequality uses techniques that are entirely different to those used in this chapter, and must be delayed for now (see Chapter 4). We shall be able to say more about the non-cyclic abelian case when we study Artin Reciprocity in Chapter 5.

As was mentioned earlier, it is possible to rephrase what we have done in terms of divisors. For a divisor $\mathfrak{m} = \prod_{\mathfrak{p}} \mathfrak{p}^{a(\mathfrak{p})}$ of F, we shall write $\mathfrak{m}_0 = \prod_{\text{finite } \mathfrak{p}} \mathfrak{p}^{a(\mathfrak{p})}$ and $\mathfrak{m}_{\text{re}} = \prod_{\text{real } \mathfrak{p}} \mathfrak{p}^{a(\mathfrak{p})}$. Of course, if $\mathfrak{p}$ is real, then $a(\mathfrak{p})$ is either 0 or 1, and in general, we have $a(\mathfrak{p}) = 0$ for all but finitely many $\mathfrak{p}$.

Given a divisor $\mathfrak{m}$ of F, we write $\alpha \equiv 1 \pmod{\mathfrak{m}}$ to denote that $\alpha \overset{\times}{\equiv} 1 \pmod{\mathfrak{m}_0}$ (i.e., $\text{ord}_{\mathfrak{p}}(\alpha - 1) \geq \text{ord}_{\mathfrak{p}}(\mathfrak{m}_0)$ for all $\mathfrak{p}$ dividing $\mathfrak{m}_0$), and that $\sigma(\alpha) > 0$ whenever σ is a real embedding with $\mathfrak{p}_{\sigma}$ dividing $\mathfrak{m}_{\text{re}}$.

Remembering that $\mathfrak{m}$ is a divisor of F (and not necessarily an ideal), we let $\mathcal{P}_{F,\mathfrak{m}}$ denote the set of principal fractional ideals of F that have a generator α with $\alpha \equiv 1 \pmod{\mathfrak{m}}$. ($\mathcal{P}_{F,\mathfrak{m}}$ is sometimes called the *ray modulo the divisor* $\mathfrak{m}$.) Also, set $\mathcal{I}_F(\mathfrak{m}) = \mathcal{I}_F(\mathfrak{m}_0)$. We call $\mathcal{R}_{F,\mathfrak{m}} = {\mathcal{I}_F(\mathfrak{m})}\big/{\mathcal{P}_{F,\mathfrak{m}}}$ the *ray class group modulo the divisor* $\mathfrak{m}$.

Comparing this with what we did before, note that for an *ideal* $\mathfrak{m}_0$, $\mathcal{P}_{F,\mathfrak{m}_0}^{+} = \mathcal{P}_{F,\mathfrak{m}_0\mathfrak{m}_{\infty}}$, where $\mathfrak{m}_{\infty} = \prod_{\text{real } \mathfrak{p}} \mathfrak{p}$. Similarly, we have $\mathcal{R}_{F,\mathfrak{m}_0}^{+} = \mathcal{R}_{F,\mathfrak{m}_0\mathfrak{m}_{\infty}}$. This notation is consistent, for in the case of $\mathcal{P}_{F,\mathfrak{m}_0}$ and $\mathcal{R}_{F,\mathfrak{m}_0}$, the ideal $\mathfrak{m}_0$ may also be viewed as a divisor with no infinite factors.

3 The Main Theorems of Class Field Theory

In this chapter, we have seen from a historical perspective the origins of some of the central ideas of Class Field Theory. Many of these ideas first surfaced in the work of Kronecker, Weber and Hilbert. However, their proofs often remained elusive until the 20th century; the majority of these proofs were first given in generality by Takagi, ([T], 1920).

In subsequent chapters, we shall discuss the theorems in detail, introducing some of the techniques that contributed to the discovery (or the reformulation) of their proofs. For now, we simply give an outline of the results themselves. The first two results give us a bijective correspondence between the finite abelian extensions of a number field F and the groups $\mathcal{H}$ that satisfy $\mathcal{P}^+_{F,\mathrm{m}} < \mathcal{H} < \mathcal{I}_F(\mathrm{m})$ for some m. The third result tells us that the Galois group of such an extension is related to the group $\mathcal{H}$. Moreover, this relationship can be described in terms of a canonical map (the Artin map).

Existence Theorem. For any $\mathcal{H}$, with $\mathcal{P}^+_{F,\mathrm{m}} < \mathcal{H} < \mathcal{I}_F(\mathrm{m})$, there is a class field K/F associated to $\mathcal{H}$.

In fact, the theorem holds if we replace $\mathcal{P}^+_{F,\mathrm{m}}$ above with $\mathcal{P}_{F,\mathrm{m}}$ where m is a divisor. In particular, if we take $\mathcal{H} = \mathcal{P}^+_{F,\mathrm{m}}$, then the Existence Theorem implies that $L_{\mathrm{m}}(1, \chi) \neq 0$ for all χ except $\chi = \chi_0$, so provides the missing step to complete the proof of the generalization of Dirichlet's Theorem on Primes in Arithmetic Progressions.

Completeness Theorem. For any abelian extension K/F, there is some m and some $\mathcal{H}$ with $\mathcal{P}^+_{F,\mathrm{m}} < \mathcal{H} < \mathcal{I}_F(\mathrm{m})$ such that K is the class field over F of $\mathcal{H}$.

Isomorphy Theorem. When $\mathcal{P}^+_{F,\mathrm{m}} < \mathcal{H} < \mathcal{I}_F(\mathrm{m})$, and K is the class field over F of $\mathcal{H}$, we have

$$\mathrm{Gal}\,(K/F) \cong {\mathcal{I}_F(\mathrm{m})}\big/{\mathcal{H}}$$

with the isomorphism being induced by the Artin map.

In particular, if $\mathrm{m} = \mathcal{O}_F$, and $\mathcal{H} = \mathcal{P}_{F,\mathrm{m}}$, then we get a class field K, for $\mathcal{H}$, called the *Hilbert class field*. By the Isomorphy Theorem, $\mathrm{Gal}\,(K/F) \cong {\mathcal{I}_F}\big/{\mathcal{P}_F} = \mathcal{C}_F$. We shall also see that K is the maximal unramified abelian extension of F (every prime is unramified including the infinite ones).

Chapter 4
The Idèlic Theory

Idèles were introduced by Chevalley ([Ch2], 1940); the modern definition is due to Weil. They were used by Chevalley as an alternative to the approach of Takagi using ray class groups and L-functions. For example, with idèles, he was able to give a proof of the Universal Norm Index Inequality that did not rely on L-functions ([Ch2], 1940), and he was able to consider infinite Galois extensions, as we discuss in Chapter 6. Chevalley's idèle class groups will play a role similar to that played by the ray class groups of Chapter 3. One of our tasks in the present chapter will be to describe precisely the relationship between idèle class groups and ray class groups.

In order to define idèles, we shall need places. In about 1900, Hensel introduced the p-adic numbers for p a prime in $\mathbb{Z}$. This can be generalized as in Chapter I to $\mathfrak{p}$-adic numbers, where $\mathfrak{p}$ is a prime ideal of $\mathcal{O}_F$ for some number field F. To define idèles, we consider simultaneously all of the Archimedean absolute values on F, together with the $\mathfrak{p}$-adic absolute values on F for *all* the primes $\mathfrak{p}$ of $\mathcal{O}_F$.

Each place is a collection of topologically equivalent absolute values, treated as a single entity. The idèles are then defined in terms of all of the places of the number field F. Because the idèles of F carry global information about F in terms of local information at each of its places, they are a successful implementation of the *local global principle*, which is a recurring theme in algebraic number theory.

In the first section, we use a definition of absolute value that is slightly less restrictive than the definition we gave in Chapter 1. This allows us to normalize our absolute values in a particular way, so that certain formulas will hold once we begin working with idèles in the third section. Fortunately, it does not affect things topologically.

In the fourth section, we pause to study a small amount of cohomology, so that we may make use of the Herbrand quotient when we return to idèles in Section 5. The cohomological approach to class field theory was developed in the 1950s, (e.g., Artin's notion of "class formations" as discussed in [AT]; see also [HN] and [Tat]). These ideas give rise to alternate proofs of many of the results for number fields included in this text, and also can be used to treat local class field theory. They are important to the proofs of other related results, (e.g., one can use higher cohomology groups to get information on the norms in non-cyclic abelian extensions, as we mention at the end of Section 5).

N. Childress, *Class Field Theory*, Universitext, DOI 10.1007/978-0-387-72490-4_4,

1 Places of a Number Field

Let F be an algebraic number field. An *absolute value* on F is a mapping

$$\| \cdot \| : F \to [0, \infty)$$

that satisfies $\|0\| = 0$, whose restriction to $F^\times$ is a homomorphism of multiplicative groups $F^\times \to \mathbb{R}_+^\times$, and that satisfies

$$\|1 + x\| \leq c \quad \text{whenever } \|x\| \leq 1$$

(for some suitable constant c).

Such an absolute value induces a (metric) topology on F via fundamental systems of neighborhoods of the form:

$$\{x \in F : \|x - a\| < \varepsilon\}, \ \varepsilon > 0.$$

Note that we must have $c \geq 1$.

Exercise 4.1. For any such absolute value $\| \cdot \|$, show that there exists a positive real number λ such that the absolute value $\| \cdot \|^\lambda$ satisfies the triangle inequality (i.e., is an absolute value in the stricter sense of Chapter 1).   ◊

We say that two absolute values are *equivalent* if they induce the same topology. Exercise 4.1 gives that any absolute value $\| \cdot \|$ is topologically equivalent to an absolute value that satisfies the triangle inequality.

A *place* of F is an equivalence class of non-trivial absolute values on F. Denote the set of places of F by V_F. By a theorem of Ostrowski, each of the places of F falls into one of the following three categories.

1. Places that contain one of the $\mathfrak{p}$-adic absolute values given by $\|\alpha\|_\mathfrak{p} = N\mathfrak{p}^{-\mathrm{ord}_\mathfrak{p}(\alpha)}$, for a non-zero prime ideal $\mathfrak{p}$ of $\mathcal{O}_F$. These are the *finite* (or *non-Archimedean*, or *discrete*) places of F.
2. Places that contain one of the absolute values $\|\alpha\|_\sigma = |\sigma(\alpha)|_\mathbb{R}$, for some real embedding $\sigma : F \hookrightarrow \mathbb{R}$ of F. These are the *infinite real* (or *real Archimedean*) places of F.
3. Places that contain one of the absolute values $\|\alpha\|_\sigma = |\sigma(\alpha)|_\mathbb{C}^2$, for some $\sigma : F \hookrightarrow \mathbb{C}$, an imaginary embedding of F. These are the *infinite imaginary* (or *imaginary Archimedean*) places of F.

Note that two distinct non-zero prime ideals of $\mathcal{O}_F$ cannot produce absolute values that are equivalent, so there is a distinct finite place for each non-zero prime ideal of $\mathcal{O}_F$. Similarly, distinct real embeddings produce inequivalent absolute values so are associated to distinct places of F. In the case of imaginary embeddings, we have that each place contains the two (equivalent) absolute values corresponding to a conjugate pair of embeddings. On the other hand, if two imaginary embeddings of

F are not conjugate, then they give rise to inequivalent absolute values. Thus there is a single place for each conjugate pair of imaginary embeddings of F.

For a number field F, there are a finite number of infinite places. Also, given $x \in F^\times$, there can be only finitely many prime ideals $\mathfrak{p}$ of $\mathcal{O}_F$ for which $\|x\|_\mathfrak{p} \neq 1$, (namely those $\mathfrak{p}$ that appear in the factorization of the fractional ideal $x\mathcal{O}_F$).

For a non-zero prime ideal $\mathfrak{p}$ of $\mathcal{O}_F$, we let $v_\mathfrak{p}$ denote the place containing $\|\cdot\|_\mathfrak{p}$. For an embedding $\sigma : F \hookrightarrow \mathbb{C}$, we let v_σ denote the place containing $\|\cdot\|_\sigma$.

Conversely, for a finite place $v \in V_F$, we let $\mathfrak{p}_v$ denote the associated prime ideal of $\mathcal{O}_F$. To simplify notation, we write ord_v instead of $\mathrm{ord}_{\mathfrak{p}_v}$.

For a place $v \in V_F$, we let $\|\cdot\|_v$ denote the specific absolute value described in (1), (2) or (3) above (and not merely an arbitrary absolute value from the place v). These particular absolute values satisfy the *product formula*: For any $x \in F^\times$

$$\prod_{v \in V_F} \|x\|_v = 1.$$

It is important to note that if v is an infinite imaginary place, the absolute value $\|\cdot\|_v$ does *not* satisfy the triangle inequality. However, by Exercise 4.1 there is some λ such that $\|\cdot\|_v^\lambda$ does. Of course, we know well that $\lambda = 1/2$.

Examples.

1. Let $F = \mathbb{Q}(\sqrt{3})$ and $K = \mathbb{Q}(\zeta_{12}) = \mathbb{Q}(i, \sqrt{3})$. Now F has two embeddings (both real):

$$\sigma_1 : a + b\sqrt{3} \mapsto a + b\sqrt{3} \text{ and } \sigma_2 : a + b\sqrt{3} \mapsto a - b\sqrt{3}.$$

Each of these may be extended to K in two ways:

$$\sigma_{1,1} : \sqrt{3} \mapsto \sqrt{3}, i \mapsto i \text{ and } \sigma_{1,2} : \sqrt{3} \mapsto \sqrt{3}, i \mapsto -i \qquad \text{extend } \sigma_1,$$
$$\sigma_{2,1} : \sqrt{3} \mapsto -\sqrt{3}, i \mapsto i \text{ and } \sigma_{2,2} : \sqrt{3} \mapsto -\sqrt{3}, i \mapsto -i \text{ extend } \sigma_2.$$

The real places of F are v_{σ_1} and v_{σ_2}, (F has no imaginary places). There is one place of K above v_{σ_1}, namely the imaginary place $v_{\sigma_{1,1}} = v_{\sigma_{1,2}}$, and one place of K above v_{σ_2}, namely the imaginary place $v_{\sigma_{2,1}} = v_{\sigma_{2,2}}$.

2. Let $F = \mathbb{Q}(\sqrt{3})$ and $K = \mathbb{Q}(\sqrt[8]{3})$. The places of K are:

$$v_{\sigma_{1,1}}, \ v_{\sigma_{1,2}} \text{ and } v_{\sigma_{1,3}} = v_{\bar{\sigma}_{1,3}} \text{ above } v_{\sigma_1}, \text{ and}$$
$$v_{\sigma_{2,1}} = v_{\bar{\sigma}_{2,1}} \text{ and } v_{\sigma_{2,2}} = v_{\bar{\sigma}_{2,2}} \text{ above } v_{\sigma_2}$$

where

$$\sigma_{1,1} : \sqrt[8]{3} \mapsto \sqrt[8]{3},$$

$$\sigma_{1,2} : \sqrt[8]{3} \mapsto -\sqrt[8]{3},$$

$$\sigma_{1,3} : \sqrt[8]{3} \mapsto i\sqrt[8]{3},$$

$$\sigma_{2,1} : \sqrt[8]{3} \mapsto \xi\sqrt[8]{3},$$

$$\sigma_{2,2} : \sqrt[8]{3} \mapsto \xi^3\sqrt[8]{3},$$

for ξ a primitive 8th root of unity.

3. Let $F = \mathbb{Q}(\zeta_{12})$. Note that 3 ramifies in $\mathbb{Q}(\zeta_3)/\mathbb{Q}$ with ramification index 2, while 3 is inert in $\mathbb{Q}(i)/\mathbb{Q}$. Thus we have $e = f = 2$ for the prime 3 in $F/\mathbb{Q}$, $\mathfrak{p} = \langle\sqrt{-3}\rangle$ is prime in $\mathcal{O}_F$, and $3\mathcal{O}_F = \mathfrak{p}^2$. We may compute $\|3\|_{\mathfrak{p}} = N\mathfrak{p}^{-2} = 9^{-2}$. Meanwhile, as we saw in Example 1, F has two imaginary places v_1 and v_2, where v_1 contains the absolute values arising from the identity and complex conjugation, while v_2 contains the absolute values arising from the maps $\zeta_{12} \mapsto \zeta_{12}^5$ and $\zeta_{12} \mapsto \zeta_{12}^7$. We have $\|3\|_{v_j} = |3|^2 = 9$, for $j = 1, 2$. Consequently, we compute $\prod_{v \in V_F} \|3\|_v = \|3\|_{\mathfrak{p}} \|3\|_{v_1} \|3\|_{v_2} = 1$ as the product rule predicts.

Exercise 4.2. Let $F = \mathbb{Q}(\sqrt{2})$ and $K = \mathbb{Q}(\sqrt[4]{2}, \sqrt{3})$. Find all of the infinite places of F and K, grouping the places of K according to which place of F they extend. ◇

Exercise 4.3. Let $F = \mathbb{Q}(i)$ and $x = 2 - i$. Compute $\|x\|_v$ for all of the places $v \in V_F$ and verify that the product formula holds for x. ◇

Exercise 4.4. Let K/F be a Galois extension of number fields, and let v be an infinite real place of F. Show that the places of K above v are either all real, or all imaginary. ◇

For a number field F and a place $v \in V_F$, we may complete F with respect to (any of the absolute values in) v. Denote the completion by F_v. Note that if v is a finite place, then $F_v = F_{\mathfrak{p}}$ for some non-zero prime ideal $\mathfrak{p}$ of $\mathcal{O}_F$. If v is an infinite real place, then $F_v \cong \mathbb{R}$, while if v is an infinite imaginary place, then $F_v \cong \mathbb{C}$.

We may embed F into F_v for each place v; write $\iota_v : F \hookrightarrow F_v$ for the embedding. Note that ι_v is continuous if F is given the topology from $\|\cdot\|_v$ and that $\iota_v(F)$ is dense in F_v. If $\alpha \in F$ is non-zero, then $\iota_v(\alpha)$ is a non-zero element of F_v for every v. Furthermore, $\iota_v(\alpha)$ is a unit of F_v for all but finitely many of the places v. (For an infinite place v, we understand the "units" of F_v to be the group $F_v^\times$; for a finite place v, we understand the "units" of F_v to be the group $\mathcal{U}_v = \mathcal{O}_v^\times$, the elements having absolute value one.) These embeddings ι_v will be important to our discussion of idèles.

2 A Little Topology

A *topological group* is a group G that is also a topological space, for which multiplication $\mu : G \times G \to G$, given by $\mu : (a, b) \mapsto ab$, and inversion $\rho : G \to G$, given by $\rho : a \mapsto a^{-1}$ are continuous.

Proposition 2.1. Let G be a topological group and fix $g \in G$. If $f_g : G \to G$ is given by $f_g(x) = gx$ (left multiplication by g), then f_g is a homeomorphism.

Proof. First note that for any subset A of G, we have

$$f_g^{-1}(A) = \{g^{-1}a : a \in A\} = f_{g^{-1}}(A)$$

so that it suffices to show that $f_g^{-1}(A)$ is open in G for any open subset A of G and for any $g \in G$. Let $\eta_g : \{g\} \times G \to G \times G$ be inclusion. Then $\mu \circ \eta_g : \{g\} \times G \to G$ is continuous. But $(\mu \circ \eta_g)^{-1}(A) = \{g\} \times f_g^{-1}(A)$. Hence $f_g^{-1}(A)$ is open in G. $\square$

Similarly right multiplication by g is a homeomorphism, so that a topological group G is necessarily a homogeneous space (i.e., for any two elements of G there is a homeomorphism from G onto G that carries one element to the other).

If G is a topological group and H is a subgroup of G, then we may give the set of left cosets of H in G the quotient topology. Since we may also define left multiplication by g on the cosets, it is easy to see that the set of left cosets G/H is a homogeneous space. If H is normal in G, then G/H is also a topological group.

Exercise 4.5. Note that the additive group of integers $\mathbb{Z}$ is a normal subgroup of the additive group of real numbers $\mathbb{R}$. If $\mathbb{R}$ is given its usual topology, then it is a topological group. Describe the topological group $\mathbb{R}/\mathbb{Z}$. $\Diamond$

Exercise 4.6. Show that if H is a subgroup of the topological group G, then the closure of H is also a subgroup of G. $\Diamond$

Exercise 4.7. Let G be a topological group with identity 1, and suppose there is some compact neighborhood A of 1 in G. Show that G is locally compact. $\Diamond$

Exercise 4.8. Let H be a subgroup of the topological group G.

a. Show that if H is open in G, then H is also closed.

b. Show that if H is closed with finite index in G, then H is also open.

c. Show that if G is compact and H is open in G, then $[G : H]$ is finite. $\Diamond$

Exercise 4.9. Let G be a topological group with identity 1. Prove or disprove and salvage:

a. If H is a subgroup of G that contains an open neighborhood of 1, then H is open in G.

b. If A is an open neighborhood of 1 in G, then A is a subgroup of G. $\Diamond$

Exercise 4.10. For each of the following, determine whether it is a topological group. If so, is it compact, locally compact or neither? Is it connected? totally disconnected?

a. The p-adic integers $\mathbb{Z}_p$ under addition, with its usual p-adic metric topology.

b. The p-adic units $\mathbb{Z}_p^\times$ under multiplication, with the usual p-adic metric topology.

c. The additive group $\mathbb{Q}_p$, with its usual p-adic metric topology.

d. The complex numbers $\mathbb{C}$ under addition, with its usual metric topology.

e. The torus ${}^{\mathbb{C}}\!/\!_{\mathcal{L}}$ under addition, with the quotient topology, where $\mathcal{L}$ is a lattice in $\mathbb{C}$. $\diamond$

3 The Group of Idèles of a Number Field

An *idèle* of a number field F is an "infinite vector" $\mathbf{a} = (\ldots, a_v, \ldots)_{v \in V_F}$ where each a_v is an element of its corresponding $F_v^\times$, and where $a_v \in \mathcal{U}_v$ for all but finitely many v.

The idèles of F form a multiplicative group, denoted $J_F = \prod_v F_v^\times$, (the symbol $\prod$ denotes a so-called "restricted topological product"— see below).

We let $\mathcal{E}_F = \prod_{v \in V_F} \mathcal{U}_v$, (clearly a subgroup of J_F). We may give $\mathcal{E}_F$ the product topology, where each $\mathcal{U}_v$ has its metric topology.

We want to put a topology on J_F that will make it a locally compact topological group. The challenge is to make the operations in J_F continuous and to have the subspace topology on $\mathcal{E}_F$ agree with the product topology. To do so, we require $\mathbf{a}\mathcal{E}_F$ to be an open subset of J_F for every $\mathbf{a} \in J_F$, and also require that the map $\mathcal{E}_F \mapsto \mathbf{a}\mathcal{E}_F$ (multiplication by $\mathbf{a}$) be a homeomorphism for every $\mathbf{a} \in J_F$. These requirements already are sufficient to determine the topology on J_F. Specifically, J_F must be the restricted topological product of the $F_v^\times$ with respect to the $\mathcal{U}_v$. We define this next. Then Exercise 4.11 shows that as a subspace of J_F, $\mathcal{E}_F$ will have the product topology as desired. Note however that the topology on J_F will *not* be the product topology.

In general, restricted topological product is defined as follows. If $\{B_i : i \in \mathfrak{I}\}$ is a family of topological spaces, and if for each i (or for all but finitely many of the i), we are given an open subset $A_i \subseteq B_i$, we may form

$$B = \{\{x_i\}_{i \in \mathfrak{I}} : x_i \in B_i \text{ for all } i; \ x_i \in A_i \text{ for all but finitely many } i\}.$$

Give B a topology by taking

$$\{\prod C_i : C_i \text{ is an open subset of } B_i \text{ for all } i; \ C_i = A_i \text{ for all but finitely many } i\}$$

as a basis of open sets. The space B with this topology is called the *restricted topological product* of the B_i with respect to the A_i. Note that for any finite subset $\mathfrak{I}$ of $\mathfrak{I}$, this makes $\prod_{i \in \mathfrak{I}} B_i \prod_{i \notin \mathfrak{I}} A_i$ an open set in B.

Exercise 4.11. Show that for J_F, a basis of open sets may be given by

$$\{\mathbf{a}A : \mathbf{a} \in J_F, \text{ and } A \text{ is an open subset of } \mathcal{E}_F\}.$$

Hence the subspace topology on $\mathcal{E}_F$ is the product topology, as desired.   ◊

Exercise 4.12. Show that J_F with this topology is a topological group.   ◊

Note that for a finite place v, $\mathcal{U}_v$ is compact, so

$$\mathcal{E}_F = \prod_{v \text{ infinite}} \mathcal{U}_v \prod_{v \text{ finite}} \mathcal{U}_v \cong (\mathbb{R}^\times)^{r_1} \times (\mathbb{C}^\times)^{r_2} \times \{\text{a compact set}\}.$$

It is easy to show that in the restricted topological product, if the B_i are locally compact and the A_i are compact, then B is locally compact. We give a proof for the special case $B = J_F$ next.

Proposition 3.1. J_F is a locally compact topological group.

Proof. By Exercise 4.12, J_F is a topological group. Thus, it suffices to find a compact neighborhood of 1. For v infinite, let

$$A_v = \{x \in F_v^\times : \|x - 1\|_v \leq 1/2\}$$

(a compact neighborhood of 1 in $F_v^\times$). Now let $A = \prod_{v \text{ infinite}} A_v \prod_{v \text{ finite}} \mathcal{U}_v$. Clearly A is a compact neighborhood of 1 in $\mathcal{E}_F$, as $\mathcal{E}_F$ has the product topology. Thus A is a compact neighborhood of 1 in J_F as well.   □

Exercise 4.13. What happens if instead of J_F we consider $\prod_v F_v^\times$ with the product topology?   ◊

Exercise 4.14. For an idèle $\mathbf{a} = (\ldots, a_v, \ldots) \in J_F$, we define the *content* of $\mathbf{a}$ to be content$(\mathbf{a}) = \prod_{v \in V_F} \|a_v\|_v$. Show that the map $\mathbf{a} \mapsto$ content$(\mathbf{a})$ is a continuous homomorphism $J_F \to \mathbb{R}_+^\times$.   ◊

Proposition 3.2. The quotient group $J_F \big/ \mathcal{E}_F$ is isomorphic to $\mathcal{I}_F$, the group of fractional ideals of F.

Proof. Define a map $\eta : J_F \to \mathcal{I}_F$ by

$$\eta : \mathbf{a} = (\ldots, a_v, \ldots) \mapsto \langle \mathbf{a} \rangle = \prod_{v \text{ finite}} \mathfrak{p}_v^{\text{ord}_v(a_v)}.$$

Note that the product on the right is actually a finite product, since $\text{ord}_v(a_v) = 0$ for all but finitely many v. Now η is clearly a surjective homomorphism of groups, and

$$\ker \eta = \{(\ldots, a_v, \ldots) : \text{ord}_v(a_v) = 0 \text{ for all finite } v\} = \mathcal{E}_F.$$ □

We may view $\alpha \in F^\times$ as an idèle $(\ldots, \iota_v(\alpha), \ldots)$, where $\iota_v : F \hookrightarrow F_v$ is an embedding of F into its completion at v. This gives an embedding, called the *diagonal embedding,*

$$\iota : F^\times \hookrightarrow J_F, \quad \text{where } \iota(\alpha) = (\ldots, \iota_v(\alpha), \ldots).$$

Usually it will do no harm to identify α and $\iota(\alpha)$, and we shall often write $F^\times$ when we really mean $\iota(F^\times)$. If we do, we find that $\eta(\alpha) = \prod_{v \text{ finite}} \mathfrak{p}_v^{\mathrm{ord}_v(\alpha)} = \alpha \mathcal{O}_F$ and $\eta(F^\times) = \mathcal{P}_F$. This observation gives us:

Proposition 3.3. For a number field F, we have

$$J_F \Big/ {}_{F^\times \mathcal{E}_F} \cong \mathcal{C}_F = {}^{\mathcal{I}_F} \Big/ \mathcal{P}_F.$$

$\square$

The embeddings $\iota_v : F \hookrightarrow F_v$ mentioned above are important. A given place v of F will lie above either the infinite real place ∞ of $\mathbb{Q}$ or above a finite place of $\mathbb{Q}$ corresponding to a prime p of $\mathbb{Z}$, (abusing the language slightly, we say "above p"). Above the place ∞, we choose the ι_v from the set of embeddings $F \hookrightarrow F_v \subseteq \mathbb{C}$, so that each infinite place of F is represented exactly once. Analogously, for the finite places v above p, we want to choose the ι_v from the set of embeddings $F \hookrightarrow F_v \subseteq \mathbb{C}_p$ so that each place of F above p is represented exactly once.

Let $| \cdot |$ denote the usual absolute value on $\mathbb{C}$. For the infinite places and their embeddings, we have $\|x\|_v = |\iota_v(x)|^d$ (where $d = 1$ if v is real and $d = 2$ if v is imaginary, i.e., $d = [F_v : \mathbb{R}]$). The same occurs for the finite places: for a finite place $v = v_{\mathfrak{p}}$, where $\mathfrak{p}$ lies above the prime p of $\mathbb{Z}$, the embedding $\iota_v : F \hookrightarrow F_v \subseteq \mathbb{C}_p$ satisfies $\|x\|_v = |\iota_v(x)|_p^d$, where $d = [F_v : \mathbb{Q}_p]$ and $| \cdot |_p$ is the p-adic absolute value on $\mathbb{C}_p$, normalized so that $|p|_p = \frac{1}{p}$.

Example.

4. If $F = \mathbb{Q}(i, \sqrt{3}) = \mathbb{Q}(\zeta)$, where ζ is a primitive 12th root of unity, we find that there are two places of F above $p = 5$. Say they are associated to the distinct prime ideals $\mathfrak{p}_5$ and $\mathfrak{p}_5'$ of $\mathcal{O}_F$. Let θ_5 be one of the two solutions to $X^2 - 3 = 0$ in $\mathbb{C}_5$. Note that $[\mathbb{Q}_5(\theta_5) : \mathbb{Q}_5] = 2$ and (identifying the completions of F with extensions of $\mathbb{Q}_p$), $F_{\mathfrak{p}_5} = \mathbb{Q}_5(\theta_5) = F_{\mathfrak{p}_5'}$. There are two solutions to $X^2 + 1 = 0$ in $\mathbb{Q}_5$; call them κ_5 and $\tilde{\kappa}_5$. Note that $\kappa_5 \tilde{\kappa}_5 = 1$, and without loss of generality we may choose $\kappa_5 \equiv 2 \pmod 5$, (so $\tilde{\kappa}_5 \equiv 3 \pmod 5$). If we look at $i = \zeta^3 \in F$, we find that $i - 2$ is in one of $\mathfrak{p}_5$, $\mathfrak{p}_5'$, while $i - 3$ is in the other. Suppose $i - 2 \in \mathfrak{p}_5$. Now $|\kappa_5 - 2|_5 = |\tilde{\kappa}_5 - 3|_5 < 1$, so the embeddings $\iota_{\mathfrak{p}_5}, \iota_{\mathfrak{p}_5'} : F \hookrightarrow \mathbb{Q}_5(\theta_5)$ must satisfy $\iota_{\mathfrak{p}_5}(i) = \kappa_5$, $\iota_{\mathfrak{p}_5'}(i) = \tilde{\kappa}_5$. We may then take $\iota_{\mathfrak{p}_5}(\sqrt{3}) = \theta_5 = \iota_{\mathfrak{p}_5'}(\sqrt{3})$ to complete the definitions of $\iota_{\mathfrak{p}_5}$ and $\iota_{\mathfrak{p}_5'}$.

Let us look also at the case $p = 3$. There is a single prime ideal $\mathfrak{p}_3$ of F above the prime $p = 3$, and we have $[F_{\mathfrak{p}_3} : \mathbb{Q}_3] = 4$. In this case, we have only one place above our prime, and we may choose θ_3 to be either of the solutions to $X^2 - 3 = 0$ and κ_3 to be either of the solutions to $X^2 + 1 = 0$ in $\mathbb{C}_3$. We have $F_{\mathfrak{p}_3} = \mathbb{Q}_3(\theta_3, \kappa_3)$ and we let $\iota_{\mathfrak{p}_3} : F \hookrightarrow \mathbb{Q}_3(\theta_3, \kappa_3)$ be given by $\iota_{\mathfrak{p}_3}(i) = \kappa_3$ and $\iota_{\mathfrak{p}_3}(\sqrt{3}) = \theta_3$.

Finally, let us look at the infinite places of F. There are two imaginary places (one above each real place of $\mathbb{Q}(\sqrt{3})$); say v_1 is above the real place of $\mathbb{Q}(\sqrt{3})$ that corresponds to the embedding that sends $\sqrt{3} \mapsto \sqrt{3}$, and v_2 is above the real place of $\mathbb{Q}(\sqrt{3})$ that corresponds to the embedding that sends $\sqrt{3} \mapsto -\sqrt{3}$. We have $F_{v_1} = F_{v_2} = \mathbb{C}$ and we let $\iota_{v_1}, \iota_{v_2} : F \hookrightarrow \mathbb{C}$ be given by $\iota_{v_1}(\sqrt{3}) = \sqrt{3}$, $\iota_{v_2}(\sqrt{3}) = -\sqrt{3}$, and $\iota_{v_1}(i) = i = \iota_{v_2}(i)$.
Say $\alpha = \sqrt{-3} \in F$. The idèle in J_F associated to α is

$$\iota(\alpha) = (\underbrace{\sqrt{-3}, -\sqrt{-3}}_{\text{above } \infty}, \ldots, \underbrace{\theta_3 \kappa_3}_{\text{above } 3}, \underbrace{\theta_5 \kappa_5, \theta_5 \tilde{\kappa}_5}_{\text{above } 5}, \ldots).$$

Exercise 4.15. Show that $F^\times$, (viewed as a subset of J_F, so identified with $\iota(F^\times)$), is a discrete subgroup of J_F. It is called the subgroup of *principal idèles* and $C_F = J_F \big/ F^\times$ is called the *group of idèle classes*. $\diamond$

Exercise 4.16. The following steps may be used to give a proof of the finiteness of the ideal class group $\mathcal{C}_F$.

a. Denote the kernel of the content map on J_F by J_F^1. Show that $F^\times < J_F^1$ and that $J_F^1 \big/ F^\times$ (with the quotient topology) is compact.

b. Let $\eta : J_F \to \mathcal{I}_F$ be as before. Show that if $\mathcal{I}_F$ is given the discrete topology, then η is continuous.

c. Show that $\eta(J_F^1) = \mathcal{I}_F$.

d. Show that $\mathcal{I}_F \big/ \mathcal{P}_F$ is compact. (Since it is also a discrete group, it must be finite.) $\diamond$

Proposition 3.4. Let $\mathfrak{m}$ be a non-zero integral ideal of $\mathcal{O}_F$, and define

$$J_{F,\mathfrak{m}}^+ = \{a \in J_F : a_v > 0 \text{ for all real } v, \text{ and } a_v \equiv 1 \pmod{\mathfrak{p}_v^{\mathrm{ord}_v(\mathfrak{m})}} \text{ for all } \mathfrak{p}_v | \mathfrak{m}\},$$
$$\mathcal{E}_{F,\mathfrak{m}}^+ = J_{F,\mathfrak{m}}^+ \cap \mathcal{E}_F.$$

Then

$$J_F \Big/ F^\times \mathcal{E}_{F,\mathfrak{m}}^+ \cong \mathcal{R}_{F,\mathfrak{m}}^+.$$

Proof. Note that $J_{F,\mathfrak{m}}^+ \cap F^\times = \{\alpha \in F^\times : \alpha \gg 0, \alpha \overset{\times}{\equiv} 1 \pmod{\mathfrak{m}}\} = F_\mathfrak{m}^+$. We claim $J_{F,\mathfrak{m}}^+ F^\times = J_F$.

To prove the claim, let $(\ldots, a_v, \ldots) = \mathbf{a} \in J_F$, and let $\varepsilon > 0$. By the Approximation Theorem, and the density of $\iota_v(F)$ in F_v, there exists $\alpha \in F^\times$ such that $\|\iota_v(\alpha) - a_v\|_v < \varepsilon$ for all infinite real v, and also for all finite v with $\mathfrak{p}_v | \mathfrak{m}$. Thus, if ε is sufficiently small, then:

$$\text{sign}\,(\iota_v((\alpha))) = \text{sign}\,(a_v) \text{ for all infinite real } v, \text{ and}$$

$$\iota_v(\alpha)^{-1}a_v \equiv 1(\text{mod }\mathfrak{p}_v^{\text{ord}_v(\mathfrak{m})}) \text{ for all } v \text{ with } \mathfrak{p}_v|\mathfrak{m}.$$

We conclude that the idèle $\alpha^{-1}\mathbf{a}$ is in $J_{F,\mathfrak{m}}^+$, whence $\mathbf{a} = \alpha\mathbf{b}$ for some $\mathbf{b} \in J_{F,\mathfrak{m}}^+$. This implies $J_F = J_{F,\mathfrak{m}}^+ F^\times$, as claimed.

From the claim, we get

$$J_{F,\mathfrak{m}}^+ \Big/ F_\mathfrak{m}^+ = J_{F,\mathfrak{m}}^+ \Big/ J_{F,\mathfrak{m}}^+ \cap F^\times \cong J_{F,\mathfrak{m}}^+ F^\times \Big/ F^\times = J_F \Big/ F^\times,$$

whence

$$J_{F,\mathfrak{m}}^+ \Big/ \mathcal{E}_{F,\mathfrak{m}}^+ F_\mathfrak{m}^+ \cong J_F \Big/ F^\times \mathcal{E}_{F,\mathfrak{m}}^+.$$

As in the proof of Proposition 3.3, let

$$\eta_\mathfrak{m} : J_{F,\mathfrak{m}}^+ \to \mathcal{I}_F(\mathfrak{m})$$

be given by

$$\eta_\mathfrak{m} : \mathbf{a} = (\dots, a_v, \dots) \mapsto \langle \mathbf{a} \rangle = \prod_{v \text{ finite}} \mathfrak{p}_v^{\text{ord}_v a_v}.$$

Then clearly $\eta_\mathfrak{m} = \eta\big|_{J_{F,\mathfrak{m}}^+}$ and $\eta_\mathfrak{m}(J_{F,\mathfrak{m}}^+) = \mathcal{I}_F(\mathfrak{m})$. Also $\ker \eta_\mathfrak{m} = J_{F,\mathfrak{m}}^+ \cap \mathcal{E}_F = \mathcal{E}_{F,\mathfrak{m}}^+$ and we conclude

$$J_{F,\mathfrak{m}}^+ \Big/ \mathcal{E}_{F,\mathfrak{m}}^+ \cong \mathcal{I}_F(\mathfrak{m}).$$

Thus

$$J_{F,\mathfrak{m}}^+ \Big/ \mathcal{E}_{F,\mathfrak{m}}^+ F_\mathfrak{m}^+ \cong \mathcal{I}_F(\mathfrak{m}) \Big/ \eta_\mathfrak{m}(F_\mathfrak{m}^+)$$

$$= \mathcal{I}_F(\mathfrak{m}) \Big/ \mathcal{P}_{F,\mathfrak{m}}^+ = \mathcal{R}_{F,\mathfrak{m}}^+.$$

But then

$$J_F \Big/ F^\times \mathcal{E}_{F,\mathfrak{m}}^+ \cong J_{F,\mathfrak{m}}^+ \Big/ \mathcal{E}_{F,\mathfrak{m}}^+ F_\mathfrak{m}^+ \cong \mathcal{R}_{F,\mathfrak{m}}^+. \qquad \square$$

Corollary 3.5. The set of subgroups $\mathcal{H}$ of J_F, with $\mathcal{H} \supseteq F^\times \mathcal{E}_{F,\mathfrak{m}}^+$ for some $\mathfrak{m}$, corresponds to the set of open subgroups of J_F that contain $F^\times$.

Proof. We have that $\mathcal{E}^+_{F,\mathrm{m}}$ is an open subgroup of J_F; in fact

$$\mathcal{E}^+_{F,\mathrm{m}} \cong \prod_{\substack{v \text{ imaginary}}} \mathbb{C}^\times \prod_{\substack{v \text{ real}}} \mathbb{R}^\times_+ \prod_{\substack{v \text{ finite} \\ \mathfrak{p}_v | \mathrm{m}}} (1 + \mathfrak{p}^{\mathrm{ord}_v \mathrm{m}}_v) \prod_{\substack{v \text{ finite} \\ \mathfrak{p}_v \nmid \mathrm{m}}} \mathcal{U}_v,$$

which is open in $\mathcal{E}_F$, ($\mathcal{E}_F$ has the product topology), so is open in J_F. This gives that $F^\times \mathcal{E}^+_{F,\mathrm{m}}$ is open, whence any subgroup $\mathcal{H}$ of J_F, with $\mathcal{H} \supseteq F^\times \mathcal{E}^+_{F,\mathrm{m}}$ is open.

Conversely, if $\mathcal{H}$ is an open subgroup of J_F that contains $F^\times$, then we claim that $\mathcal{H} \supseteq \mathcal{E}^+_{F,\mathrm{m}}$ for some m. Now $1 \in \mathcal{H}$, so there is an open neighborhood, A, of 1 in $\mathcal{H}$. We may take A to be of the form: $A = \prod_v A_v$, where $A_v = \mathcal{U}_v$ for all but finitely many places v, and for the remaining places, A_v is an open neighborhood of 1 of the form

$$A_v = \begin{cases} 1 + \mathfrak{p}^{n_v}_v & \text{for } v \text{ finite} \\ \{x \in F_v : \|x - 1\|_v < \varepsilon\} & \text{for } v \text{ infinite.} \end{cases}$$

Now let $\mathcal{H}_o$ be the subgroup of J_F generated by A. We have $\mathcal{H}_o \subseteq \mathcal{H}$. Also,

$$\mathcal{H}_o = \prod_{v \in V_F} \langle A_v \rangle = \prod_{v \text{ infinite}} \langle A_v \rangle \prod_{v \text{ finite}} A_v$$

since $1 + \mathfrak{p}^{n_v}_v$ is already a group.

For v real, $\langle A_v \rangle = \mathbb{R}^\times_+$ or $\mathbb{R}^\times$. (This is clear, since if $\alpha \in \mathbb{R}^\times_+$, then $\alpha^{\frac{1}{n}} \in A_v$ for n sufficiently large, giving $\alpha \in \langle A_v \rangle$. Thus $\mathbb{R}^\times_+ \subseteq \langle A_v \rangle \subseteq \mathbb{R}^\times$.)

For v imaginary, $\langle A_v \rangle = \mathbb{C}^\times$. (This follows, since if $z \in \mathbb{C}^\times$, then $|z|^{\frac{1}{n}} e^{\frac{i\theta}{n}} \in A_v$ for n sufficiently large and $\theta = \arg z$, whence $z \in \langle A_v \rangle$.)

Putting everything together, we get

$$\mathcal{H}_o = \prod_{v \text{ imaginary}} \mathbb{C}^\times \prod_{v \text{ real}} \{\mathbb{R}^\times \text{ or } \mathbb{R}^\times_+\} \prod_{v \text{ finite}} A_v$$

where

$$A_v = \begin{cases} \mathcal{U}_v & \text{almost everywhere} \\ 1 + \mathfrak{p}^{n_v}_v & \text{else.} \end{cases}$$

Let $\mathrm{m} = \prod \mathfrak{p}^{n_v}_v$, where the product is over the finite places $v \in V_F$ for which $A_v = 1 + \mathfrak{p}^{n_v}_v$. Then $\mathcal{H}_o \supseteq \mathcal{E}^+_{F,\mathrm{m}}$, and we have $\mathcal{H} \supseteq \mathcal{H}_o \supseteq \mathcal{E}^+_{F,\mathrm{m}}$. $\quad\square$

We may now reformulate Takagi's class field theory in terms of idèles. The main ideas translate to the following claim.

There is an order reversing, bijective correspondence between the set of all finite abelian extensions K/F and the set of open subgroups $\mathcal{H}$ of J_F for which $\mathcal{H} \supseteq F^\times$. In this correspondence $\mathrm{Gal}\,(K/F) \cong J_F \big/ \mathcal{H}$.

Compare the result that $J_F \big/ F^\times \mathcal{E}_{F,\mathrm{m}}^+ \cong \mathcal{I}_F(\mathrm{m}) \big/ \mathcal{P}_{F,\mathrm{m}}^+ = \mathcal{R}_{F,\mathrm{m}}^+$. Subgroups $\mathcal{H}$ as in the claim will contain $F^\times \mathcal{E}_{F,\mathrm{m}}^+$ for appropriately chosen m. It remains for us to answer many questions about the precise nature of the "translation" between Takagi's ray classes and Chevalley's idèles. For example, if $\mathcal{H}$ is an open subgroup of J_F containing $F^\times \mathcal{E}_{F,\mathrm{m}}^+$, then we may consider the subgroup $\mathcal{H} \big/ F^\times \mathcal{E}_{F,\mathrm{m}}^+$ of the factor group $J_F \big/ F^\times \mathcal{E}_{F,\mathrm{m}}^+$. What is the corresponding subgroup of $\mathcal{R}_{F,\mathrm{m}}^+$?

Recall the Isomorphy Theorem asserts that if $\mathcal{P}_{F,\mathrm{m}}^+ \subseteq \tilde{\mathcal{H}} \subseteq \mathcal{I}_F(\mathrm{m})$ and $\tilde{K}$ is the class field for $\tilde{\mathcal{H}}$, then $\mathrm{Gal}\,(\tilde{K}/F) \cong \mathcal{I}_F(\mathrm{m}) \big/ \tilde{\mathcal{H}}$. If the subgroup $\tilde{\mathcal{H}} \big/ \mathcal{P}_{F,\mathrm{m}}^+$ of $\mathcal{R}_{F,\mathrm{m}}^+$ corresponds to the subgroup $\mathcal{H} \big/ F^\times \mathcal{E}_{F,\mathrm{m}}^+$ of $J_F \big/ F^\times \mathcal{E}_{F,\mathrm{m}}^+$, how are the class field $\tilde{K}$ of $\tilde{\mathcal{H}}$ and the extension K/F associated to $\mathcal{H}$ (in the claim) related?

We were also especially interested in the subgroups $\mathcal{P}_{F,\mathrm{m}}^+ N_{K/F}(\mathrm{m})$, where $N_{K/F}(\mathrm{m}) = \{\mathfrak{a} \in \mathcal{I}_F(\mathrm{m}) : \mathfrak{a} = N_{K/F}(\mathfrak{A}) \text{ for some } \mathfrak{A} \in \mathcal{I}_K\}$, (see the Universal Norm Index Inequality). It is perhaps not surprising that the open idèlic subgroups $\mathcal{H}$ containing $F^\times$ and the subgroups $\tilde{\mathcal{H}} = \mathcal{P}_{F,\mathrm{m}}^+ N_{K/F}(\mathrm{m})$ of $\mathcal{I}_F(\mathrm{m})$ are related.

We may make a more precise statement (to be proved later), if we introduce the notion of the norm of an idèle. Let K/F be an extension of number fields. Define $N_{K/F} : J_K \to J_F$ as follows. Let $(\ldots, a_w, \ldots) = \mathbf{a} \in J_K$, where the w are places of K. For a fixed $v \in V_F$, the set $\{w \in V_K : w|v\}$ is finite. We construct the norm of $\mathbf{a}$ as an idèle of F by computing each v-component in terms of the corresponding set $\{w \in V_K : w|v\}$. Specifically, we let $b_v = \prod_{w|v} N_{K_w/F_v}(a_w)$ and define $N_{K/F}(\mathbf{a}) = (\ldots, b_v, \ldots) \in J_F$.

Recall that if $\alpha \in K$, then for any fixed $v \in V_F$,

$$N_{K/F}(\alpha) = \prod_{w|v} N_{K_w/F_v}(\iota_w(\alpha)).$$

Hence if $\alpha \in K^\times$ is viewed as an idèle in J_K, then $N_{K/F}(\alpha)$ is the idèle in J_F arising from the usual norm of the element α. In other words, we have a commutative diagram:

$$
\begin{array}{ccc}
K^\times & \longrightarrow & J_K \\
\left\downarrow{\scriptstyle N_{K/F}}\right. & & \left\downarrow{\scriptstyle N_{K/F}}\right. \\
F^\times & \longrightarrow & J_F
\end{array}
$$

If $\mathcal{H}$ corresponds to K in the claim above, then it turns out that $\mathcal{H} = F^{\times} N_{K/F} J_K$. Moreover, soon we'll be able to show the following (it is Proposition 5.6).

Proposition. Let K/F be abelian Galois, and let

$$\mathcal{H} = F^{\times} N_{K/F} J_K$$

(so $F^{\times} \subseteq \mathcal{H} \subseteq J_F$). Then $\mathcal{H}$ is an open subgroup in J_F. Moreover, if $\mathfrak{m}$ is chosen so that

$$\mathcal{E}_{F,\mathfrak{m}}^{+} \subseteq \mathcal{H}$$

then the image of $\mathcal{H}$ under the isomorphism

$$J_F \Big/ F^{\times} \mathcal{E}_{F,\mathfrak{m}}^{+} \cong \mathcal{I}_F(\mathfrak{m}) \Big/ \mathcal{P}_{F,\mathfrak{m}}^{+}$$

is precisely $\mathcal{P}_{F,\mathfrak{m}}^{+} N_{K/F}(\mathfrak{m}) \Big/ \mathcal{P}_{F,\mathfrak{m}}^{+}$ and

J_F		$\mathcal{I}_F(\mathfrak{m})$
$\mathcal{H} = F^{\times} N_{K/F} J_K$	$\longleftrightarrow$	$\mathcal{P}_{F,\mathfrak{m}}^{+} N_{K/F}(\mathfrak{m})$
$F^{\times} \mathcal{E}_{F,\mathfrak{m}}^{+}$		$\mathcal{P}_{F,\mathfrak{m}}^{+}$

in particular, we have $[J_F : \mathcal{H}] = [\mathcal{I}_F(\mathfrak{m}) : \mathcal{P}_{F,\mathfrak{m}}^{+} N_{K/F}(\mathfrak{m})] \leq [K : F]$. $\qquad\square$

Compare the above proposition with the Universal Norm Index Inequality of Chapter 3. Before we can give a proof, we need to study (a very small amount of) cohomology of groups. Once we have done so, we shall return to idèles and prove the above proposition.

4 Cohomology of Finite Cyclic Groups and the Herbrand Quotient

Let G be a finite cyclic group, say $G = \langle \sigma \rangle$, and let A be a G-module, (so G acts on A and A is a module over the group ring $\mathbb{Z}[G]$). Let

$$s(G) = 1 + \sigma + \cdots + \sigma^{n-1},$$

where n is the order of G.

Consider the map $\sigma - 1$ on A. We have

$$\ker(\sigma - 1) = \{a \in A : \sigma(a) = a\} = A^G.$$

Note that $s(G)A \subseteq A^G$:

$$(\sigma - 1)(1 + \sigma + \cdots + \sigma^{n-1}) = (\sigma^n - 1) = 1 - 1 = 0$$

since $n = \#G$. Similarly

$$(\sigma - 1)A \subseteq \ker s(G).$$

We define

$$\mathfrak{Q}_G(A) = [A^G : s(G)A] \Big/ [\ker s(G) : (\sigma - 1)A]$$

when these indices are finite. The number $\mathfrak{Q}_G(A)$ is called the *Herbrand quotient* of A for the group G.

Example.

5. Let $G = \langle \sigma \rangle$ be cyclic of order n and let $A = \mathbb{Z}$, with G acting trivially on A. Then

$$A^G = \{a \in \mathbb{Z} : \sigma(a) = a\} = \mathbb{Z}$$

and

$$s(G)A = s(G)\mathbb{Z} = n\mathbb{Z}.$$

Also, $\ker s(G) = \{0\}$, and $(\sigma - 1)A = \{0\}$. We get

$$\mathfrak{Q}_G(A) = [A^G : s(G)A] \Big/ [\ker s(G) : (\sigma - 1)A] = [\mathbb{Z} : n\mathbb{Z}] = n.$$

Next we want to study some properties of the Herbrand quotient. For their proofs, we use a lemma about the *Tate cohomology groups*

$$H^0(A) = \ker{}_A(\sigma - 1) \Big/ s(G)A$$

$$H^1(A) = \ker{}_A s(G) \Big/ (\sigma - 1)A.$$

Note that $\#H^0(A) \Big/ \#H^1(A) = \mathfrak{Q}_G(A).$

Lemma 4.1. (Exact Hexagon Lemma). Suppose we have an exact sequence of $\mathbb{Z}[G]$-modules $0 \longrightarrow B \stackrel{f}{\longrightarrow} A \stackrel{g}{\longrightarrow} C \longrightarrow 0$. Then there are $\mathbb{Z}[G]$-homomorphisms $f_0, f_1, g_0, g_1, \delta_0, \delta_1$ such that

$$
\begin{array}{ccc}
H^0(B) & \stackrel{f_0}{\longrightarrow} & H^0(A) \\
{\scriptstyle \delta_1} \nearrow & & \searrow {\scriptstyle g_0} \\
H^1(C) & & H^0(C) \\
{\scriptstyle g_1} \nwarrow & & \swarrow {\scriptstyle \delta_0} \\
H^1(A) & \underset{f_1}{\longleftarrow} & H^1(B)
\end{array}
$$

is exact.

Proof. Define

$$
\begin{aligned}
&f_0 : H^0(B) \to H^0(A) \text{ by} \\
&\qquad b + s(G)B \mapsto f(b) + s(G)A \\
&g_0 : H^0(A) \to H^0(C) \text{ by} \\
&\qquad a + s(G)A \mapsto g(a) + s(G)C \\
&f_1 : H^1(B) \to H^1(A) \text{ by} \\
&\qquad b + (\sigma - 1)B \mapsto f(b) + (\sigma - 1)A \\
&g_1 : H^1(A) \to H^1(C) \text{ by} \\
&\qquad a + (\sigma - 1)A \mapsto g(a) + (\sigma - 1)C.
\end{aligned}
$$

All are clearly well-defined $\mathbb{Z}[G]$-homomorphisms.

Now define $\delta_0 : H^0(C) \to H^1(B)$ as follows. Let $c \in \ker_C(\sigma - 1)$. We must define $\delta_0(c + s(G)C)$. Now g is surjective, so there is an element $a_0 \in A$ such that $g(a_0) = c$. And since $c \in \ker_C(\sigma - 1)$, we have

$$
\begin{aligned}
(\sigma - 1)(g(a_0)) &= 0 \\
g((\sigma - 1)(a_0)) &= 0 \\
(\sigma - 1)(a_0) &\in \ker g = \operatorname{im} f.
\end{aligned}
$$

Hence there is an element $b_0 \in B$ such that $f(b_0) = (\sigma - 1)(a_0)$. Now $s(G)(\sigma - 1)$ is the zero map, so

$$
\begin{aligned}
0 &= s(G)(\sigma - 1)(a_0) \\
&= s(G)f(b_0) \\
&= f(s(G)(b_0)).
\end{aligned}
$$

But f is injective, so $s(G)(b_0) = 0$. This gives $b_0 \in \ker {}_B s(G)$, so we may let

$$\delta_0(c + s(G)C) = b_0 + (\sigma - 1)B \quad (\in H^1(B)).$$

We must show that δ_0 is well-defined. Suppose $c + s(G)C = c' + s(G)C$. Repeating the above for c', we obtain

$$a_0' \in A \text{ with } g(a_0') = c', \text{ and}$$
$$b_0' \in B \text{ with } f(b_0') = (\sigma - 1)(a_0').$$

It suffices to show that

$$b_0 - b_0' \in (\sigma - 1)B.$$

We have that $c - c' \in s(G)C$ and since $g : A \to C$ is surjective, there is some $a \in A$ with

$$c - c' = s(G)(g(a)) = g(s(G)(a)).$$

Also $c - c' = g(a_0) - g(a_0') = g(a_0 - a_0')$. Thus

$$g(a_0 - a_0' - s(G)(a)) = 0,$$

i.e.,

$$a_0 - a_0' - s(G)(a) \in \ker g = \operatorname{im} f.$$

But now there exists $b \in B$ with

$$f(b) = a_0 - a_0' - s(G)(a)$$
$$(\sigma - 1)(f(b)) = (\sigma - 1)(a_0 - a_0' - s(G)(a))$$
$$= f(b_0) - f(b_0')$$
$$f((\sigma - 1)(b)) = f(b_0 - b_0').$$

Since f is injective, we must have

$$(\sigma - 1)(b) = b_0 - b_0'$$
$$b_0 - b_0' \in (\sigma - 1)B.$$

We have shown δ_0 is well-defined. The proof that δ_0 is $\mathbb{Z}[G]$-linear is routine. We leave it as **Exercise 4.17.**

Now we must define $\delta_1 : H^1(C) \to H^0(B)$. Let $c \in \ker {}_C s(G)$. Since g is surjective, we know that there is some $a_1 \in A$ such that $g(a_1) = c$. Also, since $c \in \ker {}_C s(G)$, we have

$$s(G)(g(a_1)) = 0$$
$$g(s(G)(a_1)) = 0$$
$$s(G)(a_1) \in \ker g = \operatorname{im} f.$$

Hence there exists $b_1 \in B$ with $f(b_1) = s(G)(a_1)$. Now

$$0 = (\sigma - 1)s(G)(a_1)$$
$$= (\sigma - 1)(f(b_1))$$
$$= f((\sigma - 1)(b_1)).$$

Since f is injective, we must have

$$(\sigma - 1)(b_1) = 0$$
$$b_1 \in \ker {}_B(\sigma - 1).$$

Hence we may let $\delta_1(c + (\sigma - 1)C) = b_1 + s(G)B$.

Exercise 4.18. Show that δ_1 is well-defined, and that the "hexagon" is exact.   □

Proposition 4.2. If B is a G-submodule of A, and $C = {}^A\!/_B$, then

$$\mathfrak{Q}_G(A) = \mathfrak{Q}_G(B)\mathfrak{Q}_G(C).$$

(In particular, if any two of these exist, then so does the third.)

Proof. We have the canonical exact sequence

$$0 \longrightarrow B \xrightarrow{f} A \xrightarrow{g} C \longrightarrow 0$$

so we may apply Lemma 4.1 to get the the exact hexagon below.

$$
\begin{array}{ccc}
H^0(B) & \xrightarrow{f_0} & H^0(A) \\
{\scriptstyle \delta_1}\nearrow & & \searrow{\scriptstyle g_0} \\
H^1(C) & & H^0(C) \\
{\scriptstyle g_1}\nwarrow & & \swarrow{\scriptstyle \delta_o} \\
H^1(A) & \xleftarrow{f_1} & H^1(B)
\end{array}
$$

Now suppose $\mathfrak{Q}_G(B)$, $\mathfrak{Q}_G(A)$, $\mathfrak{Q}_G(C)$ are defined. (Note that if any two of them are defined, then so is the third.) Then

$$\#H^0(B) = \#(\ker f_0)\,\#(\operatorname{im} f_0)$$
$$\#H^0(C) = \#(\ker \delta_0)\,\#(\operatorname{im} \delta_0)$$
$$\#H^0(A) = \#(\ker g_0)\,\#(\operatorname{im} g_0),\ \text{etc.}$$

From the exact hexagon, we have

$$\#H^0(B)\,\#H^0(C)\,\#H^1(A)$$
$$= \#(\ker f_0)\,\#(\operatorname{im} f_0)\,\#(\ker \delta_0)\,\#(\operatorname{im} \delta_0)\,\#(\ker g_1)\,\#(\operatorname{im} g_1)$$
$$= \#(\operatorname{im} \delta_1)\,\underbrace{\#(\ker g_0)\,\#(\operatorname{im} g_0)}\,\underbrace{\#(\ker f_1)\,\#(\operatorname{im} f_1)}\,\#(\ker \delta_1)$$
$$= \#H^1(C)\,\#H^0(A)\,\#H^1(B),$$

whence

$$Q_G(B)\,Q_G(C) = \#H^0(B)\,\#H^0(C)\Big/{\#H^1(B)\,\#H^1(C)}$$
$$= \#H^0(A)\Big/{\#H^1(A)} = Q_G(A).$$

$\square$

Proposition 4.3. If A is a finite G-module, then $Q_G(A) = 1$.

Proof. Since A is a finite $\mathbb{Z}[G]$-module, we have

$$Q_G(A) = [\ker(\sigma - 1) : \operatorname{im} s(G)]\Big/[\ker s(G) : \operatorname{im}(\sigma - 1)]$$
$$= \frac{\#\ker(\sigma - 1)}{\#\operatorname{im} s(G)}\,\frac{\#\operatorname{im}(\sigma - 1)}{\#\ker s(G)} = \frac{\#A}{\#A} = 1.$$

(Note that since A is finite, all of the cardinalities above are finite.) $\square$

Corollary 4.4. if B is a G-submodule of A of finite index, then $Q_G(A) = Q_G(B)$.

Proof. Clear. $\square$

Proposition 4.5. (Shapiro's Lemma). Suppose $A = A_1 \oplus \cdots \oplus A_r$, where the A_j are subgroups of A (not submodules) and suppose that G transitively permutes $A_1, \ldots, A_r$. Let

$$G_j = \{\tau \in G : \tau(A_j) = A_j\}.$$

Then A_j is a G_j-module and $Q_G(A) = Q_{G_j}(A_j)$. (Note that since G is cyclic, each G_j is also cyclic so $Q_{G_j}(A_j)$ makes sense.)

Proof. We shall give the proof for $j = 1$ (clearly the proof for arbitrary j is the same).

STEP I. First, we show that the natural projection $\pi : A \to A_1$ induces an isomorphism $H_G^0(A) \cong H_{G_1}^0(A_1)$.

Proof. Let $G = \bigcup_{j=1}^{r} \sigma_j G_1$ be a coset decomposition of G, where σ_j satisfies $\sigma_j(A_1) = A_j$. Then

$$A^G = \ker_A(\sigma - 1) = \left\{ \sum_{j=1}^{r} \sigma_j(a_1) : a_1 \in A_1^{G_1} \right\}.$$

(For "$\supseteq$" suppose $a_1 \in A_1^{G_1}$. We have $\sigma \in \sigma_k G_1$ for some (unique) k, so $\sigma = \sigma_k \tau$ for some $\tau \in G_1$. Given $j \in \{1, \ldots, r\}$ there is a unique $i_j \in \{1, \ldots, r\}$ such that $\sigma_j \sigma_k^{-1} \in \sigma_{i_j} G_1$ or, equivalently, $\sigma_k \sigma_{i_j} \in \sigma_j G_1$, say $\sigma_k \sigma_{i_k} = \sigma_j \tau_j$, $(\tau_j \in G_1)$. Note the i_j are pairwise distinct, so

$$\sigma \left(\sum_{i=1}^{r} \sigma_i(a_1) \right) = \sigma \left(\sum_{j=1}^{r} \sigma_{i_j}(a_1) \right) = \sum_{j=1}^{r} \sigma_k \tau \sigma_{i_j}(a_1)$$

$$= \sum_{j=1}^{r} \sigma_k \sigma_{i_j}(a_1) = \sum_{j=1}^{r} \sigma_j \tau_j(a_1) = \sum_{j=1}^{r} \sigma_j(a_1).$$

For "$\subseteq$" suppose $a \in A^G$. We have $a = \sum_{j=1}^{r} a_j$ for a unique set of $a_j \in A_j$. Also, since $\sigma_j(A_1) = A_j$, we have $a_j = \sigma_j(\tilde{a}_j)$ for some $\tilde{a}_j \in A_1$. So $a = \sum_{j=1}^{r} \sigma_j(\tilde{a}_j) \in A^G$. For a fixed index i, apply σ_i^{-1}

$$a = \sigma_i^{-1}(a) = \sum_{j=1}^{r} \sigma_i^{-1} \sigma_j(\tilde{a}_j),$$

the first equality being true because $a \in A^G$. Thus the A_1-component of a is $\sigma_i^{-1} \sigma_i(\tilde{a}_i) = \tilde{a}_i$. But we know that the A_1-component of a is unique, so $a_1 = \tilde{a}_i$. This must hold for every i. Hence, $a = \sum_{j=1}^{r} \sigma_j(\tilde{a}_j) = \sum_{j=1}^{r} \sigma_j(a_1)$, as needed.)

Since $A^G = \{\sum_{j=1}^{r} \sigma_j(a_1) : a_1 \in A_1^{G_1}\}$ we see that an element of A^G is completely determined by its A_1-component. Thus $A_1^{G_1} \cong A^G$ via $\varphi : a_1 \mapsto \sum_{j=1}^{r} \sigma_j(a_1)$. Under φ, we also have

$$s(G_1)(a_1) \mapsto \sum_{j=1}^{r} \sigma_j s(G_1)(a_1) = s(G)(a_1) \in s(G)A.$$

Thus the map

$$\bar{\varphi} : A_1^{G_1} \Big/ s(G_1)A_1 \longrightarrow A^G \Big/ s(G)A$$

given by $a_1 + s(G_1)A_1 \mapsto \sum_{j=1}^{r} \sigma_j(a_1) + s(G)A$ is well-defined. It is also surjective, since φ was. Now for $a_1 + s(G_1)A_1$ to be in $\ker \bar{\varphi}$, we must have $\sum_{j=1}^{r} \sigma_j(a_1) = s(G)(b)$ for some $b \in A$. Write $b = b_1 + \cdots + b_r = \sigma_1(\bar{b}_1) + \cdots + \sigma_r(\bar{b}_r)$ where $\bar{b}_j \in A_1$. Then

$$s(G)(b_i) = \sum_{j=1}^{r} \sigma_i \sigma_j s(G_1)(\bar{b}_i).$$

But, since $s(G)(b_i) = \sigma_i^{-1} s(G)(b_i)$, we also have

$$s(G)(b_i) = \sum_{j=1}^{r} \sigma_j s(G_1)(\bar{b}_i).$$

If $a_1 + s(G_1)A_1 \in \ker \bar{\varphi}$, then

$$\sum_{j=1}^{r} \sigma_j(a_1) = \sum_{i=1}^{r} s(G)(b_i)$$

$$= \sum_{i=1}^{r} \sum_{j=1}^{r} \sigma_j s(G_1)(\bar{b}_i)$$

$$= \sum_{j=1}^{r} \sigma_j s(G_1) \left(\sum_{i=1}^{r} \bar{b}_i \right).$$

Compare A_1-components in the above to get

$$\sigma_1(a_1) = \sigma_1 s(G_1) \left(\sum_{i=1}^{r} \bar{b}_i \right)$$

$$a_1 = s(G_1) \left(\sum_{i=1}^{r} \bar{b}_i \right) \in s(G_1)A_1.$$

Thus $\ker \bar{\varphi}$ is trivial, and we have

$$A_1^{G_1} \big/ s(G_1)A_1 \cong A^G \big/ s(G)A.$$

STEP 2. Next we show $H_G^1(A) \cong H_{G_1}^1(A_1)$.

Proof. **Exercise 4.19.**

We now have

$$\mathcal{Q}_G(A) = \frac{\#H_G^0(A)}{\#H_G^1(A)} = \frac{\#H_{G_1}^0(A_1)}{\#H_{G_1}^1(A_1)} = \mathcal{Q}_{G_1}(A_1).   \square$$

Example.

6. Let $F/\mathbb{Q}$ be a cyclic extension of number fields, with $G = \text{Gal}\,(F/\mathbb{Q})$, and let
$A = \mathcal{O}_F$ (so G acts on A). What is $\mathcal{Q}_G(\mathcal{O}_F)$?

The Normal Basis Theorem gives the existence of a $\mathbb{Q}$-basis for F of the form
$\{a^\tau : \tau \in G\}$. (Here we are using the notation a^τ in place of $\tau(a)$.) We may
assume $a \in \mathcal{O}_F$ (if $a \notin \mathcal{O}_F$ some multiple of it is). Let $B = \mathbb{Z}a + \mathbb{Z}a^\sigma + \mathbb{Z}a^{\sigma^2} + \cdots + \mathbb{Z}a^{\sigma^{n-1}}$ where $G = \langle \sigma \rangle$ has order n. Now rank $B = n$, and $[\mathcal{O}_F : B]$ is
finite. Thus $\mathcal{Q}_G(\mathcal{O}_F) = \mathcal{Q}_G(B)$. Let $A_1 = \mathbb{Z}a$. Then $G_1 = \{\tau \in G : \tau(A_1) = A_1\} = \{1\}$. We have (by Shapiro's Lemma):

$$\mathcal{Q}_G(B) = \mathcal{Q}_{G_1}(A_1) = [A_1^{G_1} : s(G_1)A_1] \big/ [\ker s(G_1) : (\sigma_1 - 1)A_1]$$

$$= [A_1 : A_1] \big/ [\{0\} : \{0\}] = 1.$$

Hence $\mathcal{Q}_G(\mathcal{O}_F) = 1$.

5 Cyclic Galois Action on Idèles

Let K/F be a (not necessarily abelian) Galois extension of number fields. We return to our study of idèles by defining an action of $\mathrm{Gal}\,(K/F)$ on J_K.

Let $G = \mathrm{Gal}\,(K/F)$ and let $\mathbf{a} = (\cdots, a_w, \cdots) \in J_K$. Let $\sigma \in G$. For a place w of K, define the place σw by

$$\|\alpha\|_{\sigma w} = \|\sigma^{-1}(\alpha)\|_w,$$

or equivalently $\|\sigma(\alpha)\|_{\sigma w} = \|\alpha\|_w$. Note that $\tau(\sigma w) = (\tau\sigma)w$. It is clear that G transitively permutes the places of K, and $(K, \|\cdot\|_w)$ is isometric to $(K, \|\cdot\|_{\sigma w})$ via $\sigma : \alpha \mapsto \sigma(\alpha)$. Thus σ induces an isomorphism between the completions that we also denote by σ:

$$\sigma : K_w \xrightarrow{\;\cong\;} K_{\sigma w}.$$

We may now define for each $v \in V_F$

$$\sigma(\ldots, a_w, \ldots)_{w|v} = (\ldots, b_w, \ldots)_{w|v}$$

where

$$b_{\sigma w} = \sigma(a_w), \quad \text{i.e., } b_w = \sigma(a_{\sigma^{-1}w}).$$

This gives an action of σ on J_K.

Exercise 4.20. Show that this action is consistent with the usual action of G on $K^\times$, where we view $K^\times \subseteq J_K$ as before. $\Diamond$

Now we have that J_K is a G-module. What is J_K^G? We endeavor to find it: Suppose $\mathbf{a} \in J_K^G$, i.e., $\sigma(\mathbf{a}) = \mathbf{a}$ for all $\sigma \in G$. Then, for every place v of F, we have

$$\sigma(\ldots, a_w, \ldots)_{w|v} = (\ldots, a_w, \ldots)_{w|v}$$

whence

$$a_w = \sigma(a_{\sigma^{-1}w}) \qquad \text{for all } \sigma, \text{ and for all } w.$$

If we let $G_w = \{\sigma \in G : \sigma w = w\}$, then $a_w = \sigma(a_w)$ for every $\sigma \in G_w$. Note that G_w is the Galois group of K_w/F_v; in particular, if w is a finite place of K, then $G_w = Z(\mathfrak{P}_w/\mathfrak{p}_v)$, (the decomposition group). We have shown that $a_w \in F_v$ for every place v of F and for every place w of K above v. Now suppose w_1, w_2 both lie above v. Then there is some $\sigma \in G$ with $\sigma : w_2 \mapsto w_1$. Since $w_2 = \sigma^{-1}w_1$, we have $a_{w_1} = \sigma(a_{w_2})$. But $a_{w_1} \in F_v$ and $a_{w_2} \in F_v$. It follows that $a_{w_1} = a_{w_2}$.

We have shown that if $\sigma(\mathbf{a}) = \mathbf{a}$ for all $\sigma \in G$, then for every place v of F,

$$(\ldots, a_w, \ldots)_{w|v} = (b_v, \ldots, b_v)$$

for some $b_v \in F_v$. This gives

$$J_K^G = \{(\ldots, (b_v, \ldots, b_v), \ldots)\} = J_F,$$

where we have identified J_F with its image in J_K under the obvious embedding $J_F \hookrightarrow J_K$ sending $(\ldots, b_v, \ldots) \mapsto (\ldots, (b_v, \ldots, b_v), \ldots)$.

Now suppose G is cyclic. We want to study $s(G)J_K$. We shall use multiplicative notation for $s(G)$, so $s(G) : J_K \to J_K$ by $\mathbf{a} \mapsto \prod_{\sigma \in G} \sigma(\mathbf{a})$. This suggests that we consider norms. Is $\prod_{\sigma \in G} \sigma(\mathbf{a})$ the same as $N_{K/F}(\mathbf{a})$ (the *idèlic* norm)?

Recall for $\mathbf{a} = (\cdots, a_w, \ldots)_{w \in V_K}$, if for each $v \in V_F$, we set $b_v = \prod_{w|v} N_{K_w/F_v}(a_w)$, then $N_{K/F}(\mathbf{a}) = (\ldots, b_v, \ldots)$, an idèle of F.

Fix $v \in V_F$. Then

$$(\ldots, a_w, \ldots)_{w|v} = (a_{w_1}, 1, 1, \ldots 1)(1, a_{w_2}, 1, \ldots, 1) \cdots (1, \ldots, 1, a_{w_r}).$$

Now $\prod_{\sigma \in G} \sigma(a_{w_1}, 1, \ldots, 1)$ is invariant under the action of G. Hence, by our computation of J_K^G, we see that

$$\prod_{\sigma \in G} \sigma(a_{w_1}, 1, \ldots, 1) = (c_v, \ldots, c_v)$$

for some $c_v \in F_v$. On the other hand, if $\sigma \notin G_{w_1}$, then the first coordinate of $\sigma(a_{w_1}, 1, \ldots, 1)$ is 1. Thus the first coordinate of $\prod_{\sigma \in G} \sigma(a_{w_1}, 1, \ldots, 1)$ is

$$\prod_{\sigma \in G_{w_1}} \sigma(a_{w_1}) = N_{K_{w_1}/F_v}(a_{w_1}).$$

It follows that

$$\prod_{\sigma \in G} \sigma(a_{w_1}, 1, \ldots, 1) = (N_{K_{w_1}/F_v}(a_{w_1}), \ldots, N_{K_{w_1}/F_v}(a_{w_1})).$$

In the same way,

$$\prod_{\sigma \in G} \sigma(1, \ldots, 1, a_{w_j}, 1, \ldots, 1) = (N_{K_{w_j}/F_v}(a_{w_j}), \ldots, N_{F_{w_j}/F_v}(a_{w_j}))$$

for each j. We get

$$\prod_{\sigma \in G} \sigma(a_{w_1}, \ldots, a_{w_r}) = \left(\prod_{w|v} N_{K_w/F_v}(a_w), \ldots, \prod_{w|v} N_{K_w/F_v}(a_w) \right). \qquad (*)$$

Since $\prod_{w|v} N_{K_w/F_v}(a_w)$ is the vth coordinate of $N_{K/F}(\mathbf{a})$, if we embed $J_F \hookrightarrow J_K$ as before, we have

$$\prod_{\sigma \in G} \sigma(\mathbf{a}) = N_{K/F}(\mathbf{a})$$

as desired.

Now we may study the Herbrand quotient. Say $G = \langle \sigma \rangle$ is a finite cyclic group, with $G = \text{Gal}\,(K/F)$ as before. Recall, for a G-module A, we have

$$\mathcal{Q}_G(A) = [A^G : s(G)A] \Big/ [\ker s(G) : (\sigma - 1)A].$$

We want $[J_F : F^\times N_{K/F} J_K]$ to be of the form $[A^G : s(G)A]$. In order to make this statement more precise, (and to give a proof of it!), we need to know more about the Herbrand quotient on idèles and idèle class groups. We begin with the following lemma.

Lemma 5.1. Let $C_K = J_K \big/ K^\times$ be the group of idèle classes of K, and similarly let $C_F = J_F \big/ F^\times$. The embedding $J_F \hookrightarrow J_K$ induces an embedding $C_F \hookrightarrow C_K$. Furthermore, $C_K^G = C_F$.

Proof. Let $\mathbf{a} \in J_F$ and $\bar{\mathbf{a}} \in J_F \big/ F^\times$ be the image of $\mathbf{a}$ in C_F. Suppose when we consider $\mathbf{a}$ as an element of J_K, we find that it is in $K^\times$, i.e., $\bar{\mathbf{a}} = \bar{\mathbf{1}}$ in C_K. Now $\mathbf{a} \in J_F = J_K^G$ also, so $\mathbf{a} \in J_K^G \cap K^\times = (K^\times)^G = F^\times$. Thus $\bar{\mathbf{a}} = \bar{\mathbf{1}}$ in C_F. We have shown that there is an embedding $C_F \hookrightarrow C_K$ that arises from the embedding $J_F \hookrightarrow J_K$ as claimed. It remains to find C_K^G.

Suppose $\mathbf{b} \in J_K$ has image $\bar{\mathbf{b}}$ in C_K and that $\bar{\mathbf{b}} \in C_K^G$. Then for all $\sigma \in G$, $\sigma(\bar{\mathbf{b}}) = \bar{\mathbf{b}}$, i.e., for all $\sigma \in G$, $\overline{\sigma(\mathbf{b})\mathbf{b}^{-1}} = \bar{\mathbf{1}}$, i.e., for all $\sigma \in G$, $\frac{\sigma(\mathbf{b})}{\mathbf{b}} \in K^\times$. Let $\varphi_{\mathbf{b}} : G \to K^\times$ by $\varphi_{\mathbf{b}} : \sigma \mapsto \frac{\sigma(\mathbf{b})}{\mathbf{b}}$. Then, for $\tau, \sigma \in G$, we have

$$\tau(\varphi_{\mathbf{b}}(\sigma)) = \frac{\tau\sigma(\mathbf{b})}{\tau(\mathbf{b})}$$

$$= \varphi_{\mathbf{b}}(\tau\sigma)[\varphi_{\mathbf{b}}(\tau)]^{-1}.$$

(The above equation gives that $\varphi_{\mathbf{b}}$ is what is known as a *1-cocycle* or a *crossed homomorphism*.) By Hilbert's Theorem 90, (see below), there is some $\alpha \in K^\times$ such that

$$\varphi_{\mathbf{b}}(\sigma) = \frac{\sigma(\alpha)}{\alpha} \qquad \text{for all } \sigma \in G.$$

We have

$$\frac{\sigma(\alpha)}{\alpha} = \frac{\sigma(\mathbf{b})}{\mathbf{b}} \qquad \text{for all } \sigma \in G$$

$$\sigma(\alpha^{-1}\mathbf{b}) = \alpha^{-1}\mathbf{b} \qquad \text{for all } \sigma \in G$$

$$\alpha^{-1}\mathbf{b} \in J_K^G = J_F.$$

Now $\overline{\alpha^{-1}\mathbf{b}} = \bar{\mathbf{b}}$ in C_K, since $\alpha^{-1} \in K^\times$. But $\alpha^{-1}\mathbf{b} \in J_F$ by the above, so we must have $\overline{\alpha^{-1}\mathbf{b}} \in C_F$. The map $C_F \hookrightarrow C_K$ is injective. Thus $\bar{\mathbf{b}} \in C_F$, and we have shown $C_K^G \subseteq C_F$. The reverse inclusion is clear. □

We record the statement of Hilbert's Theorem 90 in its modern form here.

Theorem 5.2 (Hilbert Theorem 90). *Let $G = \mathrm{Gal}\,(K/F)$, and let $f : G \to K^\times$ satisfy $f(\tau\sigma) = \tau(f(\sigma))f(\tau)$ Then there is some $\alpha \in K^\times$ such that $f(\sigma) = \frac{\sigma(\alpha)}{\alpha}$ for all $\sigma \in G$.* □

The classical version of this theorem, which originated in the work of Gauss and Kummer, says that if we have $\beta \in K^\times$ and $N_{K/F}(\beta) = 1$, then $\beta = \frac{\sigma(\alpha)}{\alpha}$ for some $\alpha \in K^\times$, where $G = \mathrm{Gal}\,(K/F) = \langle\sigma\rangle$ is cyclic. Note that for a crossed homomorphism f, we have

$$N_{K/F}(f(\sigma)) = \prod_{\tau \in G} \tau(f(\sigma))$$

$$= \prod_{\tau \in G} \left(\frac{f(\tau\sigma)}{f(\tau)} \right)$$

$$= \frac{\prod_{\tau \in G} f(\tau\sigma)}{\prod_{\tau \in G} f(\tau)}$$

$$= 1.$$

Reviewing what we have done so far, we have

$$J_K^G = J_F$$

$$s(G)J_K = N_{K/F}J_K$$

$$C_F \hookrightarrow C_K$$

$$C_K^G = C_F.$$

Also, we have defined $N_{K/F}$ on C_K, and the diagram below commutes.

$$1 \longrightarrow K^\times \longrightarrow J_K \longrightarrow C_K \longrightarrow 1$$

$$N_{K/F} \downarrow \qquad N_{K/F} \downarrow \qquad N_{K/F} \downarrow$$

$$1 \longrightarrow F^\times \longrightarrow J_F \longrightarrow C_F \longrightarrow 1$$

Putting this together,

$$[C_K^G : s(G)C_K] = [C_F : N_{K/F}C_K]$$

$$= \left[J_F \Big/ F^\times : {}^{N_{K/F}J_K} \Big/ {}_{F^\times \cap N_{K/F}J_K} \right]$$

$$= [J_F : F^\times N_{K/F}J_K].$$

(As we'll see, $[J_F : F^\times N_{K/F}J_K] = [\mathcal{I}_F(\mathfrak{m}) : \mathcal{P}_{F,\mathfrak{m}}^+ N_{K/F}(\mathfrak{m})]$ for some $\mathfrak{m}$.) Given the above, it seems potentially useful to study $\mathcal{Q}_G(C_K)$ when $G = \mathrm{Gal}(K/F)$ is cyclic.

Our approach to this study begins by recalling that we have shown $J_K \Big/ {}_{K^\times \mathcal{E}_K} \cong \mathcal{C}_K$, the *ideal* class group of K. In particular, $K^\times \mathcal{E}_K$ has finite index in J_K, which implies that $K^\times \mathcal{E}_K \Big/ {}_{K^\times}$ has finite index in C_K, the *idèle* class group. This gives $\mathcal{Q}_G(C_K) = \mathcal{Q}_G\left(K^\times \mathcal{E}_K \Big/ {}_{K^\times} \right)$.

Now $\mathcal{E}_K \cap K^\times = \mathcal{U}_K = \mathcal{O}_K^\times$, so

$$K^\times \mathcal{E}_K \Big/ {}_{K^\times} \cong \mathcal{E}_K \Big/ {}_{\mathcal{E}_K \cap K^\times} = \mathcal{E}_K \Big/ {}_{\mathcal{U}_K}$$

whence

$$\mathcal{Q}_G(C_K) = \mathcal{Q}_G\left(\mathcal{E}_K \Big/ {}_{\mathcal{U}_K} \right).$$

On the other hand,

$$\mathcal{E}_K = \prod_{w \in V_K} \mathcal{U}_w = \prod_{v \in V_F} \left(\prod_{w|v} \mathcal{U}_w \right)$$

and $G = \mathrm{Gal}(K/F)$ permutes $\{\mathcal{U}_w : w|v\}$. This makes $\prod_{w|v} \mathcal{U}_w$ a G-module.

Exercise 4.21. If A and B are G-modules that have Herbrand quotients, show that

$$\mathcal{Q}_G(A \times B) = \mathcal{Q}_G(A)\mathcal{Q}_G(B). \qquad \Diamond$$

By Exercise 4.21, we may choose any finite set $\mathcal{S}$ of places $v \in V_F$, and write

$$\mathcal{Q}_G\left(\prod_{v \in V_F} \prod_{w|v} \mathcal{U}_w \right) = \left(\prod_{v \in \mathcal{S}} \mathcal{Q}_G\left(\prod_{w|v} \mathcal{U}_w \right) \right) \left(\mathcal{Q}_G\left(\prod_{v \notin \mathcal{S}} \prod_{w|v} \mathcal{U}_w \right) \right).$$

By Shapiro's Lemma, we know

$$\mathfrak{Q}_G\left(\prod_{w|v}\mathcal{U}_w\right) = \mathfrak{Q}_{Gw}(\mathcal{U}_w)$$

where $G_w = \{\sigma \in G : \sigma w = w\}$. In our study of $\mathfrak{Q}_G(C_K)$, we shall take

$$\mathbb{S} = \{v \in V_F : v \text{ is infinite or } v \text{ is finite and } \mathfrak{p}_v \text{ ramifies in } K/F\},$$

(see (iii) of Proposition 5.7 below).

From the above observations, it is apparent that it will be useful to study the Herbrand quotient on units in local fields. We continue to use multiplicative notation, so that from the Herbrand quotient, we obtain information about the norm index. The following lemma shows that for a cyclic extension of local fields this is governed by the ramification index for the extension.

Lemma 5.3. Let k_2/k_1 be an extension of local fields, (for us this means that for some p, $k_j/\mathbb{Q}_p$ is a finite extension), with $\mathrm{Gal}(k_2/k_1) = G$, a cyclic group. Let $\mathcal{U}_j$ denote the units of k_j, i.e., the elements of absolute value 1. Then $\mathfrak{Q}_G(\mathcal{U}_2) = 1$, and

$$[\mathcal{U}_2^G : s(G)\mathcal{U}_2] = [\mathcal{U}_1 : N_{k_2/k_1}\mathcal{U}_2] = e(k_2/k_1)$$

(whence also $[\ker s(G) : (\sigma - 1)\mathcal{U}_2] = e(k_2/k_1)$).

Proof. Consider the p-adic power series

$$\log X = \sum_{n=1}^{\infty}(-1)^{n-1}\frac{(X-1)^n}{n}$$

$$\exp X = \sum_{n=0}^{\infty}\frac{X^n}{n!}.$$

These converge (p-adically) on small discs about 1 and 0, respectively. If we take the radii of these discs to be less than $p^{-\frac{1}{p-1}}$, these functions satisfy the usual identities.

Let $D = p^N\mathcal{O}_2$, where $\mathcal{O}_2$ denotes the integers of k_2. (For large N, D is a small open subgroup of $\mathcal{O}_2$, preserved by the Galois group G.) Let $B = \exp(D)$ (this will be defined if N is sufficiently large).

Now $\log 1 = 0$, (and log is continuous), so taking $\varepsilon = |p^N|_p$, there exists $\delta > 0$ such that whenever $|x - 1|_p < \delta$ we have $|\log x|_p < \varepsilon$. Thus $|x - 1|_p < \delta$ implies $\log x \in D$, whence $x \in B$. We have shown that B contains an open neighborhood of 1:

$$B \supseteq \{x : |x - 1|_p < \delta\}.$$

Hence B is open.

The map $\exp : (D, +) \xrightarrow{\cong} (B, \cdot)$ is a group isomorphism. Also, for $\sigma \in G$, $\sigma(\exp x) = \exp(\sigma(x))$, so $\exp$ is in fact an isomorphism of G-modules.

Now D has finite index in $\mathcal{O}_2$. Also B has finite index in $\mathcal{U}_2$. (Since B is open, the cosets of B in $\mathcal{U}_2$ are all open, so that $\mathcal{U}_2$ is a disjoint union of open sets. But $\mathcal{U}_2$ is compact, so there exists a finite subcover of $\mathcal{U}_2$. Since the union is disjoint, we cannot eliminate any coset from the union and still cover $\mathcal{U}_2$. Thus the total number of cosets must be finite already.)

We conclude $\mathcal{Q}_G(\mathcal{U}_2) = \mathcal{Q}_G(B) = \mathcal{Q}_G(D) = \mathcal{Q}_G(\mathcal{O}_2)$. As in the number field case, (see the example following Shapiro's Lemma), $\mathcal{Q}_G(\mathcal{O}_2) = 1$. (Note: in the global situation we were discussing before the present lemma, we can now say that for a finite place w, $\mathcal{Q}_{G_w}(\mathcal{U}_w) = 1$.)

Since $\mathcal{Q}_G(\mathcal{O}_2) = 1 = \mathcal{Q}_G(\mathcal{U}_2)$, we have

$$[\mathcal{U}_1 : N_{k_2/k_1}\mathcal{U}_2] = [\mathcal{U}_2^G : s(G)\mathcal{U}_2] = [\ker_{\mathcal{U}_2} s(G) : (\sigma - 1)\mathcal{U}_2].$$

Let $A = \ker_{\mathcal{U}_2} s(G)$, (for $u \in A$, we have $s(G)(u) = 1$, i.e., $N_{k_2/k_1}(u) = 1$). By Hilbert's Theorem 90, if $u \in A$ then $u = \frac{\sigma(\alpha)}{\alpha}$ for some $\alpha \in k_2^\times$. Let π be a uniformizer in k_2 and write

$$\alpha = \pi^t \varepsilon, \text{ for some } t \in \mathbb{Z}, \varepsilon \in \mathcal{U}_2.$$

We get

$$u = \frac{\sigma(\alpha)}{\alpha} = \left(\frac{\sigma(\pi)}{\pi}\right)^t \left(\frac{\sigma(\varepsilon)}{\varepsilon}\right)$$

$$\equiv \left(\frac{\sigma(\pi)}{\pi}\right)^t (\mathrm{mod}\,(\sigma - 1)\mathcal{U}_2).$$

Hence $\frac{\sigma(\pi)}{\pi}$ generates $A \big/ (\sigma - 1)\mathcal{U}_2 = \ker_{\mathcal{U}_2} s(G) \big/ (\sigma - 1)\mathcal{U}_2$. To find the order of this group, we need only find the order of its generator, i.e., the smallest positive exponent t such that $\left(\frac{\sigma(\pi)}{\pi}\right)^t \in (\sigma - 1)\mathcal{U}_2$.

Say $\left(\frac{\sigma(\pi)}{\pi}\right)^t \in (\sigma - 1)\mathcal{U}_2$. Then $\left(\frac{\sigma(\pi)}{\pi}\right)^t = \frac{\sigma(\eta)}{\eta}$, for some $\eta \in \mathcal{U}_2$, so $\frac{\sigma(\pi^t)}{\sigma(\eta)} = \frac{\pi^t}{\eta}$, and we find that σ fixes $\frac{\pi^t}{\eta}$. Since $G = \langle \sigma \rangle = \mathrm{Gal}\,(k_2/k_1)$, we must have $\frac{\pi^t}{\eta} \in k_1$. But for any element $\gamma \in k_1^\times$, we have $e | \mathrm{ord}_\pi(\gamma)$, where $e = e(k_2/k_1)$ is the ramification index. Thus $t = \mathrm{ord}_\pi\left(\frac{\pi^t}{\eta}\right)$ is divisible by e.

Conversely, if we let ρ be a uniformizer in k_1, then $\pi^e = \delta\rho$ for some $\delta \in \mathcal{U}_2$, so $\frac{\sigma(\pi^e)}{\pi^e} = \frac{\sigma(\delta)}{\delta} \in (\sigma - 1)\mathcal{U}_2$. Thus $\left(\frac{\sigma(\pi)}{\pi}\right)^e \in (\sigma - 1)\mathcal{U}_2$.

We have shown that e is the order of $\frac{\sigma(\pi)}{\pi}$ in the group $\ker_{\mathcal{U}_2} s(G) \big/ (\sigma - 1)\mathcal{U}_2$. Since $\frac{\sigma(\pi)}{\pi}$ is a generator for this group, it follows that

$$e = [\ker_{\mathcal{U}_2} s(G) : (\sigma - 1)\mathcal{U}_2]$$
$$= [\mathcal{U}_1 : N_{k_2/k_1}\mathcal{U}_2]. \qquad \square$$

Note in particular, that if k_2/k_1 is an unramified extension of local fields, then $e = 1$ and Lemma 5.3 gives $\mathcal{U}_1 = N_{k_2/k_1}\mathcal{U}_2$. This shows that the norm is surjective on units in unramified local extensions. (If k_2/k_1 is unramified, then it is necessarily cyclic.)

What can be said in the case of a non-cyclic (hence ramified) abelian extension of local fields? Our strategy will be to decompose the extension into a tower of intermediate subfields, so that each stage in the tower is a (normal) extension with a cyclic Galois group.

Lemma 5.4. Let k_2/k_1 be an extension of local fields and suppose we have subgroups $B < A < \mathcal{U}_2$, with $[A : B] = d$. Then $N_{k_2/k_1}B \subseteq N_{k_2/k_1}A$ are subgroups of $\mathcal{U}_1$ and $[N_{k_2/k_1}A : N_{k_2/k_1}B]$ divides d.

Proof. That $N_{k_2/k_1}B \subseteq N_{k_2/k_1}A$ are subgroups of $\mathcal{U}_1$ is clear. Let $\varphi : A \to {}^A/_B$ be the canonical epimorphism, and define a map

$$f : {}^A/_B \to {}^{N_{k_2/k_1}A}/_{N_{k_2/k_1}B}$$

given by

$$f : aB \mapsto N_{k_2/k_1}(a)N_{k_2/k_1}B.$$

It is routine to verify that f is a well-defined epimorphism. We have $B = \ker\varphi \subseteq \ker(f \circ \varphi) \subseteq A$, whence $[A : \ker(f \circ \varphi)]$ divides $[A : B] = d$. The result now follows, since ${}^{N_{k_2/k_1}A}/_{N_{k_2/k_1}B} \cong {}^A/_{\ker(f \circ \varphi)}$. $\qquad \square$

Let $k_1 \subseteq k_2 \subseteq k_3$ be a tower of local fields, and suppose $[\mathcal{U}_1 : N_{k_2/k_1}\mathcal{U}_2] \le e(k_2/k_1)$, and $[\mathcal{U}_2 : N_{k_3/k_2}\mathcal{U}_3] \le e(k_3/k_2)$.

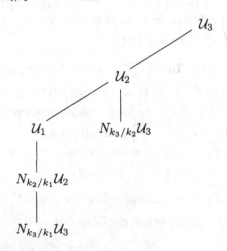

Since $N_{k_3/k_1} = N_{k_2/k_1} \circ N_{k_3/k_2}$, we have $N_{k_3/k_1}\mathcal{U}_3 \subseteq N_{k_2/k_1}\mathcal{U}_2$, and

$$[\mathcal{U}_1 : N_{k_3/k_1}\mathcal{U}_3] = [\mathcal{U}_1 : N_{k_2/k_1}\mathcal{U}_2][N_{k_2/k_1}\mathcal{U}_2 : N_{k_3/k_1}\mathcal{U}_3]$$
$$= [\mathcal{U}_1 : N_{k_2/k_1}\mathcal{U}_2][N_{k_2/k_1}\mathcal{U}_2 : N_{k_2/k_1}(N_{k_3/k_2}\mathcal{U}_3)].$$

By Lemma 5.4, $[N_{k_2/k_1}\mathcal{U}_2 : N_{k_2/k_1}(N_{k_3/k_2}\mathcal{U}_3)]$ divides $[\mathcal{U}_2 : N_{k_3/k_2}\mathcal{U}_3]$. It follows that

$$[\mathcal{U}_1 : N_{k_3/k_1}\mathcal{U}_3] \leq e(k_2/k_1)\, e(k_3/k_2) = e(k_3/k_1).$$

We have shown the following.

Corollary 5.5. If k_2/k_1 is an abelian extension of local fields, then

$$[\mathcal{U}_1 : N_{k_2/k_1}\mathcal{U}_2] \leq e(k_2/k_1). \qquad \square$$

Exercise 4.22. Let k_2/k_1 be an extension of local fields above $\mathbb{Q}_p$. Show that, with respect to the p-adic topology, $N_{k_2/k_1} : k_2 \to k_1$ is continuous. $\Diamond$

Returning to the (global) number field case, we may now prove the following proposition. A straightforward consequence of this proposition will be that for some $\mathfrak{m}$,

$$J_F \Big/ F^\times N_{K/F} J_K \cong I_F(\mathfrak{m}) \Big/ \mathcal{P}_{F,\mathfrak{m}}^+ N_{K/F}(\mathfrak{m}).$$

Proposition 5.6. For an abelian extension K/F of number fields with group G, let $\mathcal{H} = F^\times N_{K/F} J_K$. Then

i. $\mathcal{H}$ is open in J_F, so $\mathcal{H} \supseteq \mathcal{E}_{F,\mathfrak{m}}^+$ for some $\mathfrak{m}$,
ii. the image of $\mathcal{H}$ under the isomorphism

$$J_F \Big/ F^\times \mathcal{E}_{F,\mathfrak{m}}^+ \cong \mathcal{I}_F(\mathfrak{m}) \Big/ \mathcal{P}_{F,\mathfrak{m}}^+$$

(see Proposition 3.4) is precisely

$$\mathcal{P}_{F,\mathfrak{m}}^+ N_{K/F}(\mathfrak{m}) \Big/ \mathcal{P}_{F,\mathfrak{m}}^+.$$

Proof. i. Since $\mathcal{H} \supseteq N_{K/F} J_K$, we have

$$\mathcal{H} \supseteq N_{K/F}\mathcal{E}_K = \prod_v \left(\prod_{w|v} N_{K_w/F_v}\mathcal{U}_w \right).$$

Say v is finite. If $w|v$ is unramified, then $N_{K_w/F_v}\mathcal{U}_w = \mathcal{U}_v$. If $w|v$ is ramified, then $[\mathcal{U}_v : N_{K_w/F_v}\mathcal{U}_w] \leq e(w/v)$, the ramification index. Since N_{K_w/F_v} is continuous, we have that $N_{K_w/F_v}\mathcal{U}_w$ is compact. Thus $N_{K_w/F_v}\mathcal{U}_w$ is a closed subgroup of finite index in the compact group $\mathcal{U}_v$. This gives that $N_{K_w/F_v}\mathcal{U}_w$ is open (see Exercise 4.8).

If v is infinite, then $\mathcal{U}_w = \mathbb{C}^\times$ or $\mathbb{R}^\times$, and $N_{K_w/F_v}\mathcal{U}_w = \mathbb{C}^\times$, $\mathbb{R}^\times$ or $\mathbb{R}_+^\times$ is open.

Since $\mathcal{E}_F$ has the product topology, we conclude that $N_{K/F}\mathcal{E}_K$ is open in $\mathcal{E}_F$. But this implies that $\mathcal{H} = F^\times N_{K/F} J_K$ is open in J_F, (we may give an open neighborhood of $\mathbf{a} \in \mathcal{H}$ by noting that since $\mathbf{1} \in N_{K/F}\mathcal{E}_K$, we have $\mathbf{a} \in \mathbf{a}N_{K/F}\mathcal{E}_K \subseteq \mathcal{H}$, and $\mathbf{a}N_{K/F}\mathcal{E}_K$ is open since it is a basis set for the topology on J_F). Recall the result in Corollary 3.5. Since $\mathcal{H}$ is open and $F^\times \subseteq \mathcal{H}$, we know that there is some m for which $\mathcal{H} \supseteq \mathcal{E}_{F,\mathrm{m}}^+$.

ii. We have $F^\times \mathcal{E}_{F,\mathrm{m}}^+ \subseteq F^\times N_{K/F} J_K = \mathcal{H} \subseteq J_F = F^\times J_{F,\mathrm{m}}^+$ and (as we saw in the proof of Proposition 3.4)

$$J_F \Big/ F^\times = F^\times J_{F,\mathrm{m}}^+ \Big/ F^\times \cong J_{F,\mathrm{m}}^+ \Big/ J_{F,\mathrm{m}}^+ \cap F^\times = J_{F,\mathrm{m}}^+ \Big/ F_\mathrm{m}^+.$$

Now $F^\times N_{K/F} J_K = F^\times (F^\times N_{K/F} J_K \cap J_{F,\mathrm{m}}^+)$, so

$$F^\times N_{K/F} J_K \Big/ F^\times \cong F^\times N_{K/F} J_K \cap J_{F,\mathrm{m}}^+ \Big/ F^\times \cap F^\times N_{K/F} J_K \cap J_{F,\mathrm{m}}^+$$

$$\cong F^\times N_{K/F} J_K \cap J_{F,\mathrm{m}}^+ \Big/ F_\mathrm{m}^+.$$

The map $\eta_\mathrm{m} : J_{F,\mathrm{m}}^+ \longrightarrow \mathcal{I}_F(\mathrm{m})$ given by

$$(\ldots, a_v, \ldots) \mapsto \prod_{v \text{ finite}} \mathfrak{p}_v^{\mathrm{ord}_v(a_v)}$$

is surjective with kernel $\mathcal{E}_{F,\mathrm{m}}^+$.

Consider the restriction of η_m to $F^\times N_{K/F} J_K \cap J_{F,\mathrm{m}}^+$. Since $\mathcal{E}_{F,\mathrm{m}}^+ \subseteq F^\times N_{K/F} J_K$, the kernel of the restriction is still $\mathcal{E}_{F,\mathrm{m}}^+$. We leave it as **Exercise 4.23** to show that its image is $\mathcal{P}_{F,\mathrm{m}}^+ \mathcal{N}_{K/F}(\mathrm{m})$. Thus

$$F^\times N_{K/F} J_K \cap J_{F,\mathrm{m}}^+ \Big/ \mathcal{E}_{F,\mathrm{m}}^+ \cong \mathcal{P}_{F,\mathrm{m}}^+ \mathcal{N}_{K/F}(\mathrm{m})$$

and

$$F^\times N_{K/F} J_K \Big/ F^\times \mathcal{E}_{F,\mathrm{m}}^+ \cong F^\times N_{K/F} J_K \cap J_{F,\mathrm{m}}^+ \Big/ F_\mathrm{m}^+ \mathcal{E}_{F,\mathrm{m}}^+$$

$$\cong \mathcal{P}_{F,\mathrm{m}}^+ \mathcal{N}_{K/F}(\mathrm{m}) \Big/ \mathcal{P}_{F,\mathrm{m}}^+. \qquad \square$$

In the above proposition, we may say a bit more about the ideals m that satisfy $\mathcal{H} \supseteq \mathcal{E}_{F,\mathrm{m}}^+$. Recall that we proved (in Corollary 3.5) that such an ideal must exist for an open subgroup $\mathcal{H}$ of J_F containing $F^\times$ by first finding an open neighborhood of 1 in $\mathcal{H}$ of the form $A = \prod A_v$, where $A_v = \mathcal{U}_v$ for all but finitely many places and

$$A_v = \begin{cases} 1 + \mathfrak{p}_v^{n_v} & \text{for } v \text{ finite} \\ \{x \in F_v : \|x - 1\|_v < \varepsilon\} & \text{for } v \text{ infinite} \end{cases}$$

for the remaining places. We then set $\mathfrak{m} = \prod \mathfrak{p}_v^{n_v}$, where the product was over the finite places v of F for which $A_v = 1 + \mathfrak{p}_v^{n_v}$. Since we know that the local norm is surjective for the unramified finite places, it is clear that we may find an open neighborhood of 1 in $\mathcal{H} = F^\times N_{K/F} J_K$ for which $A_v = \mathcal{U}_v$ for *all* unramified v. Thus we may find an ideal $\mathfrak{m}$ that satisfies the propostion and that is divisible only by primes that ramify in K/F.

Example.

7. Let $p > 2$ be a prime and let $K = \mathbb{Q}(\zeta_p)$, $F = \mathbb{Q}$, so $\mathcal{H} = \mathbb{Q}^\times N_{K/\mathbb{Q}} J_K$. We want to find $\mathfrak{m} = m\mathbb{Z}$ so that $\mathcal{H} \supseteq \mathcal{E}_{\mathbb{Q},\mathfrak{m}}^+$. Now

$$\mathcal{E}_{\mathbb{Q},\mathfrak{m}}^+ = \mathbb{R}_+^\times \times \prod_{q \mid m} \left(1 + q^{\operatorname{ord}_q m} \mathbb{Z}_q\right) \times \prod_{q \nmid m} \mathbb{Z}_q^\times.$$

For a prime $q \neq p$, we know that q is unramified in $K/\mathbb{Q}$. If w is a place of K above q, then the local norm map is surjective on units, i.e., $N_{K_w/\mathbb{Q}_q} \mathcal{U}_w = \mathbb{Z}_q^\times$. On the other hand, the prime p is totally ramified in $K/\mathbb{Q}$, so if w is a place of K above p, then $[\mathbb{Z}_p^\times : N_{K_w/\mathbb{Q}_p} \mathcal{U}_w] = p - 1$. Since $\mathbb{Z}_p^\times \cong \mu_{p-1} \times (1 + p\mathbb{Z}_p)$ has only one subgroup of index $p - 1$, we must have $N_{K_w/\mathbb{Q}_p} \mathcal{U}_w = 1 + p\mathbb{Z}_p$. Thus

$$N_{K/\mathbb{Q}} J_K = \mathbb{R}_+^\times \times (1 + p\mathbb{Z}_p) \times \prod_{q \neq p} \mathbb{Z}_q^\times.$$

This allows us to take $\mathfrak{m} = p\mathbb{Z}$ for $\mathbb{Q}(\zeta_p)/\mathbb{Q}$.

Exercise 4.24. Let E/F be an arbitrary extension of number fields (not necessarily Galois). Let $\mathcal{H} = F^\times N_{E/F} J_E$. Use an argument similar to the proof of part (i) of Proposition 5.6 to show that $\mathcal{H}$ is an open subgroup of J_F. $\diamond$

For an arbitrary abelian extension of number fields K/F, let $\mathfrak{m}$ be divisible only by ramified primes and such that $F^\times N_{K/F} J_K \supseteq \mathcal{E}_{F,\mathfrak{m}}^+$. Let us examine the proof of Proposition 5.6 more closely. It gives us an isomorphism

$$\left. J_F \middle/ F^\times N_{K/F} J_K \right. \cong \left. I_F(\mathfrak{m}) \middle/ P_{F,\mathfrak{m}}^+ \mathcal{N}_{K/F}(\mathfrak{m}), \right.$$

which we see arises from the isomorphism $\left. J_F \middle/ F^\times \right. \cong \left. J_{F,\mathfrak{m}}^+ \middle/ F_{\mathfrak{m}}^+ \right.$. Recall that we first encountered this isomorphism during the proof of Proposition 3.4:

$$J_F \Big/ {F^\times \mathcal{E}_{F,\mathfrak{m}}^+} \cong \mathcal{R}_{F,\mathfrak{m}}^+.$$

In that proof, we argued that given $\mathbf{a} \in J_F$, there is some $\alpha \in F^\times$ such that $\alpha \mathbf{a} \in J_{F,\mathfrak{m}}^+$. It was from this that we deduced the isomorphism $J_F \Big/ {F^\times} \cong J_{F,\mathfrak{m}}^+ \Big/ {F_\mathfrak{m}^+}$.

   Explicitly, consider the map

$$\varphi : J_F \longrightarrow J_{F,\mathfrak{m}}^+ \Big/ {F_\mathfrak{m}^+} \longrightarrow I_F(\mathfrak{m}) \Big/ {\mathcal{P}_{F,\mathfrak{m}}^+}$$

given by

$$\varphi : \mathbf{a} \longrightarrow \alpha \mathbf{a} F_\mathfrak{m}^+ \longrightarrow \langle \alpha \mathbf{a} \rangle \mathcal{P}_{F,\mathfrak{m}}^+$$

where $\alpha \in F^\times$ is chosen so that $\alpha \mathbf{a} \in J_{F,\mathfrak{m}}^+$. Specifically, we are taking $\alpha \in F^\times$ to satisfy $\iota_v(\alpha)a_v > 0$ for all real places v of F, $\iota_v(\alpha)a_v \in \mathcal{O}_v$ and $\iota_v(\alpha)a_v \equiv 1$ (mod $\mathfrak{p}_v^{\mathrm{ord}_v \mathfrak{m}}$) for all finite places v. (Recall that for an idèle $\mathbf{b} \in J_{F,\mathfrak{m}}^+$ we have $\langle \mathbf{b} \rangle = \eta_\mathfrak{m}(\mathbf{b})$ as before.)

Exercise 4.25. Show that the map φ is well-defined. What is its kernel?   ◊

   Our results on cyclic extensions of local fields also allow us to prove the following proposition on Herbrand quotients of local units arising from a cyclic extension of number fields, which in turn will allow us to evaluate the Herbrand quotient for the idèle class group C_K.

Proposition 5.7. Let K/F be a Galois extension of number fields, with cyclic Galois group $G = \langle \sigma \rangle$. Let v be a place of F and let w be a place of K above v. Then

   i. $\mathcal{Q}_{G_w}(\mathcal{U}_w) = 1$ if w is finite, if w is real, or if v is imaginary.
   ii. $\mathcal{Q}_{G_w}(\mathcal{U}_w) = 2$ if w is imaginary but v is real.
   iii. $\mathcal{Q}_G(\prod_{v \notin S} \prod_{w|v} \mathcal{U}_w) = 1$, where

$$S = \{v \in V_F : \ v \text{ is infinite, or } v \text{ ramifies in } K/F\}.$$

Proof.

   i. First assume that w is real (so v is real). Then $G_w = \mathrm{Gal}(K_w/F_v) = \{1\}$ and $\mathcal{U}_w = \mathbb{R}^\times$. Thus $\mathcal{Q}_{G_w}(\mathcal{U}_w) = 1$.
      Next assume that v is imaginary. Again $G_w = \{1\}$ so $\mathcal{Q}_{G_w}(\mathcal{U}_w) = 1$.
      If w is finite, then Lemma 5.3 gives that $\mathcal{Q}_{G_w}(\mathcal{U}_w) = 1$.
   ii. Suppose that w is imaginary and v is real. Then G_w is generated by complex conjugation. We have $\mathcal{U}_w^{G_w} = (\mathbb{C}^\times)^G = \mathbb{R}^\times$, and $s(G_w)\mathcal{U}_w = s(G_w)\mathbb{C}^\times = \{z\bar{z} : z \in \mathbb{C}^\times\} = \mathbb{R}_+^\times$. Also

$$\ker{}_{\mathcal{U}_w} s(G_w) = \{z \in \mathbb{C}^\times : z\bar{z} = 1\},$$

the "unit circle group," and if τ denotes complex conjugation,

$$(\tau - 1)\mathcal{U}_w = (\tau - 1)\mathbb{C}^\times = \{\frac{z}{\bar{z}} : z \in \mathbb{C}^\times\}$$

is also the unit circle group. It follows that

$$[\mathcal{U}_w^{G_w} : s(G_w)\mathcal{U}_w] = [\mathbb{R}^\times : \mathbb{R}_+^\times] = 2$$
$$[\ker_{\mathcal{U}_w} s(G_w) : (\tau - 1)\mathcal{U}_w] = 1,$$

whence $\mathcal{Q}_{G_w}(\mathcal{U}_w) = 2$.

iii. Let

$$\mathcal{S} = \{v \in V_F : v \text{ is infinite or } v \text{ is ramified in } K/F\}.$$

Note that $\mathcal{S}$ is a finite set. Let

$$A = \prod_{v \notin \mathcal{S}} A_v, \quad \text{where } A_v = \prod_{w|v} \mathcal{U}_w.$$

By Shapiro's Lemma, $\mathcal{Q}_G(A_v) = \mathcal{Q}_{G_w}(\mathcal{U}_w)$ and this equals 1 by part (i).

Now $A_v^G = \{(u, \ldots, u) : u \in \mathcal{U}_v\} \cong \mathcal{U}_v$ (see (*) in the proof that $s(G)(\mathbf{a}) = N_{K/F}(\mathbf{a})$ for $\mathbf{a} \in J_K$). Also

$$s(G)A_v = \prod_{w|v} N_{K_w/F_v}\mathcal{U}_w \cong \mathcal{U}_v,$$

since

$$s(G)(\ldots, a_w, \ldots)_{w|v} = \left(\ldots, \prod_{w|v} N_{K_w/F_v}(a_w), \ldots\right)_{w|v}.$$

As before,

$$\prod_{w|v} N_{K_w/F_v}\mathcal{U}_w = \mathcal{U}_v$$

since $v \notin \mathcal{S}$ implies v is unramified so that the local norm N_{K_w/F_v} is surjective on units by Lemma 5.3.

We have $[A_v^G : s(G)A_v] = 1$. But also $\mathcal{Q}_G(A_v) = 1$. Thus

$$[\ker_{A_v} s(G) : (\sigma - 1)A_v] = 1.$$

We get

$$A^G = \prod_{v \notin \mathcal{S}} A_v^G = \prod_{v \notin \mathcal{S}} \mathcal{U}_v = \prod_{v \notin \mathcal{S}} s(G) A_v = s(G) A.$$

It follows that $[A^G : s(G)A] = 1$. Since $\ker_{A_v} s(G) = (\sigma - 1)A_v$, we obtain

$$\ker_A s(G) = \prod_{v \notin \mathcal{S}} \ker_{A_v} s(G) = \prod_{v \notin \mathcal{S}} (\sigma - 1)A_v = (\sigma - 1)A.$$

It follows that $[\ker_A s(G) : (\sigma - 1)A] = 1$ and

$$\mathcal{Q}_G(A) = \mathcal{Q}_G\left(\prod_{v \notin \mathcal{S}} \prod_{w|v} \mathcal{U}_w\right) = 1. \qquad \square$$

Observe that Proposition 5.7 allows us to compute the Herbrand quotient of $\mathcal{E}_K$:

$$\mathcal{Q}_G(\mathcal{E}_K) = 2^a,$$

where $a = \#\{v \in V_F : v \text{ is real on } F, \text{ but extends to imaginary places on } K\}$.

We want to find the Herbrand quotient of $\mathcal{U}_K$, where K/F is a cyclic extension of number fields. Once we have done so, we can combine it with the above information about $\mathcal{Q}_G(\mathcal{E}_K)$ to find the Herbrand quotient of the idèle class group C_K. This in turn will lead to a proof of a result on the norm index in the case of a cyclic global extension.

We shall need two preliminary lemmas.

Lemma 5.8. Let $\mathcal{S}$ be a finite set and let $V = \bigoplus_{w \in \mathcal{S}} \mathbb{R}X_w$ be a real vector space. For an element $\sum_{w \in \mathcal{S}} a_w X_w$ of V, define

$$\left\| \sum_{w \in \mathcal{S}} a_w X_w \right\|_0 = \max_w \{|a_w| : w \in \mathcal{S}\},$$

(the sup-norm on V). If $\{X'_w : w \in \mathcal{S}\}$ is given so that $\|X'_w - X_w\|_0 < \frac{1}{\dim_{\mathbb{R}} V}$ for each w, then $\{X'_w : w \in \mathcal{S}\}$ is also a basis for V.

Proof. Suppose not. Then there is a set of scalars $\{b_w : w \in \mathcal{S}\} \subseteq \mathbb{R}$ such that $\sum_{w \in \mathcal{S}} b_w X'_w = \mathbf{0}$ and not all of the b_w are 0. Without loss of generality we may assume $\max_w |b_w| = 1$. Then

$$\mathbf{0} = \sum_{w \in \mathcal{S}} b_w X'_w = \sum_{w \in \mathcal{S}} b_w (X'_w - X_w) + \sum_{w \in \mathcal{S}} b_w X_w$$

whence

$$\sum_{w \in \mathcal{S}} b_w X_w = -\sum_{w \in \mathcal{S}} b_w (X'_w - X_w).$$

Taking norms, we obtain

$$
\begin{aligned}
1 &= \max_w |b_w| \\
&\leq \sum_{w \in \mathcal{S}} |b_w| \, \|X'_w - X_w\|_0 \\
&\leq \sum_{w \in \mathcal{S}} \|X'_w - X_w\|_0 \\
&< \sum_{w \in \mathcal{S}} \frac{1}{\dim_{\mathbb{R}} V} \\
&= \sum_{w \in \mathcal{S}} \frac{1}{\#\mathcal{S}} \\
&= 1,
\end{aligned}
$$

a contradiction. $\qquad\square$

Next, we prove the following, which appears in the Artin-Tate class field theory notes of 1961, [AT].

Lemma 5.9. Let G be a finite group acting on a finite set $\mathcal{S}$. Let $V = \bigoplus_{w \in \mathcal{S}} \mathbb{R}X_w$ be a vector space. Then G acts on V via

$$\sigma\left(\sum_{w \in \mathcal{S}} a_w X_w\right) = \sum_{w \in \mathcal{S}} a_w X_{\sigma w}.$$

Note that the action of G preserves sup-norms: $\|\sigma(X)\|_0 = \|X\|_0$ for all $X \in V$. Let $\mathcal{L} \subseteq V$ be a lattice preserved by G. Then there is a basis $\{Y_w\}_{w \in \mathcal{S}}$ of V contained in $\mathcal{L}$ such that $\sigma(Y_w) = Y_{\sigma w}$ for all $\sigma \in G$ and for all $w \in \mathcal{S}$.

Proof. We have a lattice $\mathcal{L}$ in V, so $\mathbb{R}\mathcal{L} = V$. Our finite group G acts on the finite set $\mathcal{S}$; let r denote the number of orbits of G in $\mathcal{S}$, and let $w_1, \ldots, w_r$ be a complete set of representatives for the orbits, (one w_j from each orbit). For each $j = 1, \ldots, r$, choose $X'_{w_j} \in \mathbb{Q}\mathcal{L}$ such that

$$\|X'_{w_j} - X_{w_j}\|_0 < \frac{1}{\dim_{\mathbb{R}} V}$$

and define

$$X''_{w_j} = \frac{1}{\#G_{w_j}}\left(\sum_{\sigma \in G_{w_j}} \sigma(X'_{w_j})\right).$$

Then

$$\|X''_{w_j} - X_{w_j}\|_0 = \left\|\frac{1}{\#G_{w_j}}\left(\sum_{\sigma \in G_{w_j}} \sigma(X'_{w_j})\right) - X_{w_j}\right\|_0$$

$$= \left\|\frac{1}{\#G_{w_j}}\sum_{\sigma \in G_{w_j}}\left(\sigma(X'_{w_j}) - \sigma(X_{w_j})\right)\right\|_0$$

$$\text{(since } \sigma(X_{w_j}) = X_{w_j} \text{ for all } \sigma \in G_{w_j})$$

$$\leq \frac{1}{\#G_{w_j}}\sum_{\sigma \in G_{w_j}} \|\sigma(X'_{w_j} - X_{w_j})\|_0$$

$$= \frac{1}{\#G_{w_j}}\sum_{\sigma \in G_{w_j}} \|X'_{w_j} - X_{w_j}\|_0$$

$$\text{(since } \|\sigma(X)\|_0 = \|X\|_0 \text{ for all } \sigma, \text{ for all } X)$$

$$= \|X'_{w_j} - X_{w_j}\|_0$$

$$< \frac{1}{\dim_{\mathbb{R}} V}.$$

Thus, for any $w \in \mathcal{S}$, one can write $w = \sigma w_j$ for some j and some σ, and define

$$X''_w = \sigma(X''_{w_j}).$$

(This is independent of σ, so long as $\sigma w_j = w$, so is well-defined. This is why we needed to define the X''. We couldn't have used the X'_{w_j}: The above would not be independent of σ.)

By Lemma 5.8, we have that $\{X''_w : w \in \mathcal{S}\}$ is a basis for V. Note also that $X''_w \in \mathbb{Q}\mathcal{L}$ for all w. Hence, there exists a sufficiently large integer N so that $NX''_w \in \mathcal{L}$ for all $w \in \mathcal{S}$.

We may now take $Y_w = NX''_w$ for each $w \in \mathcal{S}$. Then $\{Y_w : w \in \mathcal{S}\}$ is a basis for V. It remains only to show that $\sigma(Y_w) = Y_{\sigma w}$ for all σ, for all w. We leave this as

Exercise 4.26. after which our proof of Lemma 5.9 will be complete. $\square$

Proposition 5.10. Let K/F be a Galois extension of number fields, with cyclic Galois group $G = \langle \sigma \rangle$. Then

$$\mathcal{Q}_G(\mathcal{U}_K) = \frac{2^a}{[K:F]}$$

where $a = \#\{v \in V_F : v \text{ is real on } F, \text{ but extends to imaginary places on } K\}$.

Proof. Let $a = \#\{v \in V_F : v \text{ is real, but becomes imaginary in } V_K\}$. Let $\mathcal{U}_K = \mathcal{O}_K^\times$.
By Dirichlet's Unit Theorem, $\text{rank}_{\mathbb{Z}} \mathcal{U}_K = r_1 + r_2 - 1$. Let

$$\ell : \mathcal{U}_K \to \mathbb{R}^{r_1 + r_2}$$

be given by

$$\ell(\varepsilon) = (\ldots, \log \|\varepsilon\|_w, \ldots)_{w|\infty}.$$

Then $\ell(\mathcal{U}_K)$ is a discrete subgroup of $V = \mathbb{R}^{r_1 + r_2}$ (i.e., $\ell(\mathcal{U}_K)$ is a lattice in a subspace of V. In fact, it is a lattice in $V_0 = \{(\ldots, x_w, \ldots) \in V : \sum_w x_w = 0\}$. Note that $\dim_{\mathbb{R}} V_0 = r_1 + r_2 - 1$.)

Let $\mathcal{S} = \{w \in V_K : w \text{ is infinite}\}$, and for $w \in \mathcal{S}$, let $X_w = (0, \ldots, 0, 1, 0, \ldots, 0)$ where the 1 is in the wth component. Then $V = \underset{w \in \mathcal{S}}{\to} \oplus \mathbb{R} X_w$. Let $G = \text{Gal}(K/F)$ act via $\sigma(X_w) = X_{\sigma w}$ and let

$$\mathcal{L} = \ell(\mathcal{U}_K) \oplus \mathbb{Z}\ell_0, \qquad \text{where } \ell_0 = \sum_w X_w = (1, \ldots, 1).$$

Note that $\mathcal{L}$ is a lattice in $V = V_0 \oplus \mathbb{R}\ell_0$. We shall compute $\mathcal{Q}_G(\mathcal{U}_K)$ by relating it to $\mathcal{Q}_G(\mathcal{L})$.

Now $\sigma(\ell_0) = \sum_{w \in \mathcal{S}} \sigma(X_w) = \sum_{w \in \mathcal{S}} X_{\sigma w} = \ell_0$, and for $\varepsilon \in \mathcal{U}_K$, we have

$$\begin{aligned}
\ell(\sigma(\varepsilon)) &= (\ldots, \log \|\sigma(\varepsilon)\|_w, \ldots) \\
&= \sum_{w \in \mathcal{S}} \log \|\sigma(\varepsilon)\|_w X_w \\
&= \sum_{w \in \mathcal{S}} \log \|\sigma(\varepsilon)\|_{\sigma w} X_{\sigma w} \\
&= \sum_{w \in \mathcal{S}} \log \|\varepsilon\|_w X_{\sigma w} \\
&= \sigma \left(\sum_{w \in \mathcal{S}} \log \|\varepsilon\|_w X_w \right) \\
&= \sigma(\ldots, \log \|\varepsilon\|_w, \ldots) \\
&= \sigma(\ell(\varepsilon)).
\end{aligned}$$

Thus $\sigma(\ell(\mathcal{U}_K)) = \ell(\sigma(\mathcal{U}_K)) = \ell(\mathcal{U}_K)$. Since we also have $\sigma(\ell_0) = \ell_0$, it follows that $\sigma(\mathcal{L}) = \mathcal{L}$.

Recall that $\ker \ell = \mathcal{W}_K$, the (finite) group of roots of unity in K, so we have an exact sequence

$$0 \longrightarrow \mathcal{W}_K \longrightarrow \mathcal{U}_K \longrightarrow \ell(\mathcal{U}_K) \longrightarrow 0,$$

and $\mathcal{U}_K / \mathcal{W}_K \cong \ell(\mathcal{U}_K)$. We get

$$\mathcal{Q}_G(\mathcal{U}_K) = \mathcal{Q}_G(\mathcal{W}_K)\mathcal{Q}_G(\ell(\mathcal{U}_K)) = \mathcal{Q}_G(\ell(\mathcal{U}_K)).$$

Now G acts trivially on $\mathbb{Z}\ell_0 \cong \mathbb{Z}$, so $\mathcal{Q}_G(\mathbb{Z}\ell_0) = \#G$. Thus, as a G-module, the lattice $\mathcal{L} = \ell(\mathcal{U}_K) \oplus \mathbb{Z}\ell_0$ has Herbrand quotient

$$\begin{aligned} \mathcal{Q}_G(\mathcal{L}) &= \mathcal{Q}_G(\ell(\mathcal{U}_K))\mathcal{Q}_G(\mathbb{Z}\ell_0) \\ &= \mathcal{Q}_G(\mathcal{U}_K)[K : F]. \end{aligned}$$

Now we may apply Lemma 5.9. There is a basis $\{Y_w\}$ of V with $Y_w \in \mathcal{L}$ and $\tau(Y_w) = Y_{\tau w}$ for all $w \in \mathcal{S}$, for all $\tau \in G$.

Let $\mathcal{L}' = \underset{w \in \mathcal{S}}{\longrightarrow} \oplus \mathbb{Z}Y_w$, a sublattice of $\mathcal{L}$: $\mathrm{rank}_z\, \mathcal{L}' = \#\mathcal{S} = \mathrm{rank}_z\, \mathcal{L}$ and $\mathcal{L}' \subseteq \mathcal{L}$.

Hence $\mathcal{L}/\mathcal{L}'$ is finite and $\mathcal{Q}_G(\mathcal{L}) = \mathcal{Q}_G(\mathcal{L}')$. We may reorder the w to get

$$\mathcal{L}' = \bigoplus_{\substack{v \in V_F \\ v|\infty}} \left(\bigoplus_{w|v} \mathbb{Z}Y_w \right).$$

Each $\underset{w|v}{\oplus}\mathbb{Z}Y_w$ is a G-module, and G permutes its summands transitively, so we may apply Shapiro's Lemma to get

$$\mathcal{Q}_G(\bigoplus_{w|v} \mathbb{Z}Y_w) = \mathcal{Q}_{G_w}(\mathbb{Z}Y_w).$$

Since $\mathbb{Z}Y_w \cong \mathbb{Z}$ with trivial G_w-action we get $\mathcal{Q}_{G_w}(\mathbb{Z}Y_w) = \#G_w$. Hence

$$\mathcal{Q}_G(\mathcal{L}') = \prod_{\substack{v|\infty \\ v \in V_F}} \#G_w$$

where we take one arbitrary $w|v$ for each factor of the product. Now

$$\#G_w = [K_w : F_v] = \begin{cases} 1 & \text{if } w, v \text{ are both real or both imaginary} \\ 2 & \text{otherwise.} \end{cases}$$

Thus

$$\mathcal{Q}_G(\mathcal{L}') = 2^a$$

where $a = \#\{v \in V_F : v \text{ is real but extends to an imaginary } w \in V_K\}$. We get

$$\begin{aligned} \mathcal{Q}_G(\mathcal{U}_K) &= \frac{\mathcal{Q}_G(\mathcal{L})}{[K : F]} \\ &= \frac{\mathcal{Q}_G(\mathcal{L}')}{[K : F]} \\ &= \frac{2^a}{[K : F]}. \end{aligned}$$

□

Given Proposition 5.10, finally we have enough information to find the Herbrand quotient of the idèle class group and to use it to prove the Global Cyclic Norm Index Equality.

Corollary 5.11. $\mathcal{Q}_G(C_K) = [K : F]$.

Proof. $\mathcal{Q}_G(C_K) = \mathcal{Q}_G\left(\mathcal{E}_K \big/ \mathcal{U}_K\right) = 2^a \left(\frac{2^a}{[K:F]}\right)^{-1}$. $\square$

Theorem 5.12 (Global Cyclic Norm Index Equality). *If K/F is a cyclic extension of number fields and $\mathfrak{m}$ is an integral ideal of $\mathcal{O}_F$ that is divisible by a sufficiently high power of every ramified prime in K/F, then*

$$[\mathcal{I}_F(\mathfrak{m}) : \mathcal{P}^+_{F,\mathfrak{m}}\mathcal{N}_{K/F}(\mathfrak{m})] = [K : F].$$

Proof. Since $\mathcal{Q}_G(C_K) = [K : F]$, we have

$$[C_K^G : s(G)C_K] = [K : F][\ker {}_{C_K}s(G) : (\sigma - 1)C_K].$$

But also $[C_K^G : s(G)C_K] = [C_F : N_{K/F}C_K] = [\mathcal{I}_F(\mathfrak{m}) : \mathcal{P}^+_{F,\mathfrak{m}}\mathcal{N}_{K/F}(\mathfrak{m})]$, whenever $\mathfrak{m}$ satisfies $F^\times N_{K/F} J_K \supseteq \mathcal{E}^+_{F,\mathfrak{m}}$. Thus $[K : F]$ divides $[\mathcal{I}_F(\mathfrak{m}) : \mathcal{P}^+_{F,\mathfrak{m}}\mathcal{N}_{K/F}(\mathfrak{m})]$ and we have obtained the "Global Cyclic Norm Index Inequality"

$$[K : F] \leq [\mathcal{I}_F(\mathfrak{m}) : \mathcal{P}^+_{F,\mathfrak{m}}\mathcal{N}_{K/F}(\mathfrak{m})].$$

Meanwhile, on the other hand we have already shown the Universal Norm Index Inequality, which gives $[\mathcal{I}_F(\mathfrak{m}) : \mathcal{P}^+_{F,\mathfrak{m}}\mathcal{N}_{K/F}(\mathfrak{m})] \leq [K : F]$. $\square$

Note that the preceeding proof also gives us $(\sigma - 1)C_K = \ker {}_{C_K}s(G)$ when $G = \langle \sigma \rangle$ is cyclic.

The hypothesis that K/F is cyclic may be weakened in Theorem 5.12, but we obtain the resulting stronger statement (for arbitrary abelian extensions) as a consequence of Artin Reciprocity, which we haven't yet discussed (see Chapter 5). Indeed, Takagi took the equation $[\mathcal{I}_F(\mathfrak{m}) : \mathcal{H}] = [K : F]$ to be the defining property for K to be a class field for $\mathcal{H}$.

The term "Second Fundamental Inequality" has previously been used for the inequality $[K : F] \leq [\mathcal{I}_F(\mathfrak{m}) : \mathcal{P}^+_{F,\mathfrak{m}}\mathcal{N}_{K/F}(\mathfrak{m})]$ because historically its proof came later (Takagi, [T], 1920) than that of the Universal Norm Index Inequality, which was proved by Weber, [We2], (using Dirichlet L-series) in the late 19th century. This terminology was the convention originally. Many authors however, (including Artin and Tate, [AT]), reverse the order in which the two inequalities are presented, and consequently refer to the above as the First Inequality and the Universal Norm Index Inequality as the Second Inequality. It is perhaps best to avoid this confusion by using the names "Global Cyclic Norm Index" and "Universal Norm Index" for these inequalities.

Exercise 4.27. Let K/F be a cyclic extension of number fields and suppose $D < J_F$ satifies

i. $D \subseteq N_{K/F} J_K$
ii. $F^\times D$ is dense in J_F.

Show that $K = F$. Does this result generalize to arbitrary abelian extensions K/F? $\Diamond$

We must emphasize that the techniques we have used in this chapter (idèles, cohomology) were not available to Takagi. They came onto the scene only in the 1930s. The original proof of the Global Cyclic Norm Index Inequality, for example, uses what Hasse called a "far-reaching generalization" of Gauss' theory of genera of quadratic forms, [CF]. See *Disquisitiones Arithmeticae*, 1801, for Gauss' results.

To conclude this chapter, we discuss a result about norms that is a nice application of what we have done so far. Let

$$A = \ker {}_{J_K}s(G)$$
$$B = \ker {}_{C_K}s(G)$$

and let $\theta : A \to B$ be the restriction of the natural homomorphism $J_K \to C_K$. We have the following commutative diagram with exact rows and columns.

$$
\begin{array}{ccccccccc}
 & & & & 1 & & 1 & & \\
 & & & & \downarrow & & \downarrow & & \\
 & & & & A & \xrightarrow{\theta} & B & & \\
 & & & & \downarrow & & \downarrow & & \\
1 & \longrightarrow & K^\times & \xrightarrow{\iota} & J_K & \longrightarrow & C_K & \longrightarrow & 1 \\
 & & N_{K/F}\downarrow & & N_{K/F}\downarrow {=}s(G) & & N_{K/F}\downarrow {=}s(G) & & \\
1 & \longrightarrow & F^\times & \xrightarrow{\iota} & J_F & \longrightarrow & C_F & \longrightarrow & 1 \\
 & & \downarrow & & \downarrow & & & & \\
 & & {F^\times}\big/{N_{K/F}K^\times} & \xrightarrow{\tilde{\iota}} & {J_F}\big/{N_{K/F}J_K} & & & & \\
 & & \downarrow & & \downarrow & & & & \\
 & & 1 & & 1 & & & & \\
\end{array}
$$

The Snake Lemma gives the existence of $\delta : B \longrightarrow {F^\times}\big/{N_{K/F}K^\times}$ so that

$$A \xrightarrow{\theta} B \xrightarrow{\delta} {F^\times}\big/{N_{K/F}K^\times} \xrightarrow{\tilde{\iota}} {J_F}\big/{N_{K/F}J_K}$$

is exact.

Since $s(G)(\sigma - 1) = 0$, we have $(\sigma - 1)J_K \subseteq A = \ker_{J_K} s(G)$. Thus $(\sigma - 1)C_K \subseteq \theta(A)$, whence $B = \theta(A)$ by Theorem 5.12. Now exactness gives $\ker \delta = \theta(A)$, so $\delta(B) = \{1\}$. It follows that the map $\bar{\iota} : \left. F^{\times} \middle/ N_{K/F} K^{\times} \right. \longrightarrow \left. J_F \middle/ N_{K/F} J_K \right.$ has trivial kernel, i.e., $\bar{\iota}$ is an injection. From this, we conclude that if $\alpha \in F^{\times} \subseteq J_F$ is in $N_{K/F} J_K$, then $\alpha \in N_{K/F} K^{\times}$, whence $\alpha = N_{K/F}(\beta)$ for some $\beta \in K^{\times}$.

Exercise 4.28. Let K/F be a cyclic extension. Suppose $\alpha \in F^{\times}$, and view $F^{\times} \subseteq J_F$ as usual. Show that $\alpha \in N_{K/F} J_K$ if and only if $\alpha \in N_{K_w/F_v} K_w^{\times}$ for all places v of F and all places w of K with $w | v$. $\Diamond$

Given Exercise 4.28, we have

HASSE'S NORM PRINCIPLE. For any cyclic extension K/F, an element $\alpha \in F^{\times}$ is a norm from K if and only if α is a local norm in each completion.

The hypothesis that K/F is cyclic is necessary here. Hasse's norm principle fails in general for K/F non-cyclic abelian. Indeed, the simplest example occurs for the simplest non-cyclic abelian Galois group: In the extension $\mathbb{Q}(\sqrt{13}, \sqrt{17})/\mathbb{Q}$, there are many examples of rational numbers that are local norms for every completion but are not global norms. For more about the defect in the non-cyclic abelian case one must look at higher cohomology. See the article by Tate in *Algebraic Number Theory*, (Cassels and Fröhlich, [CF]) for details.

Hasse's norm prinicple is useful in proving the Hasse-Minkowski Theorem on quadratic forms: A non-degenerate quadratic form over a number field F that represents zero in F_v for all v must represent zero in F. Exercise 4.4 in Cassels and Fröhlich, [CF], outlines the steps of the proof.

Chapter 5
Artin Reciprocity

Artin Reciprocity answers explicitly the problem of finding a proof for the Isomorphy Theorem. Takagi ([T], 1920) proved the isomorphism by considering cyclic groups and their orders, but he did not give a canonical map for it. Indeed, mathematicians working at the time did not seem to be concerned with finding such a map.

Artin ([A2], 1927), using some ideas from Chebotarev's work on a conjecture of Frobenius, was able to give explicitly a map that was the desired isomorphism. This was a more complete answer than the original problem had required and it allowed Artin to prove a result on L-functions, which had been his motivation at the time. Together with some other work of Artin, it also leads to a proof of a conjecture of Hilbert that says that every ideal of a number field generates a principal ideal in its Hilbert class field. The first proof of this, the *Principal Ideal Theorem* or *Principal Divisor Theorem*, was given by Furtwängler ([Fur2], 1930). A simpler proof, using an idea of Artin ([A3], 1930) that reduces the problem to a question in group theory was given subsequently (Iyanaga, [Iy], 1934). We discuss the Hilbert class field and the Principal Ideal Theorem in Chapter 6.

Artin Reciprocity is regarded as the central result in class field theory, even though its proof came later than the proofs of nearly all of the other results. Once it was known, many of the other main theorems of class field theory were seen to follow from it. It was also seen to imply all previously known reciprocity laws.

In this chapter, we define the Artin symbol, and give a proof of Artin Reciprocity. The proof we give is Artin's, although not his original proof from 1927. We also show how Artin Reciprocity can be used to obtain a proof of Quadratic Reciprocity. The other main theorems of class field theory will be treated in Chapter 6.

1 The Conductor of an Abelian Extension of Number Fields and the Artin Symbol

Recall that for an ideal $\mathfrak{m}$ of $\mathcal{O}_F$, we set

$$J_{F,\mathfrak{m}}^+ = \{\mathbf{a} \in J_F : a_v > 0 \text{ for all real } v, \ a_v \equiv 1 \pmod{\mathfrak{p}_v^{\mathrm{ord}_v(\mathfrak{m})}} \text{ for all finite } v\}$$

$$\mathcal{E}_{F,\mathfrak{m}}^+ = J_{F,\mathfrak{m}}^+ \cap \mathcal{E}_F$$

N. Childress, *Class Field Theory*, Universitext, DOI 10.1007/978-0-387-72490-4_5,
© Springer Science+Business Media, LLC 2009

$$F_{\mathfrak{m}}^+ = J_{F,\mathfrak{m}}^+ \cap F^\times$$

$$\mathcal{I}_F(\mathfrak{m}) = \{\mathfrak{a} \in \mathcal{I}_F : \mathrm{ord}_\mathfrak{p}\mathfrak{a} = 0 \text{ for all } \mathfrak{p}|\mathfrak{m}\}$$

$$\mathcal{P}_{F,\mathfrak{m}}^+ = \{\langle \alpha \rangle \in \mathcal{P}_F : \alpha \gg 0,\ \alpha \overset{\times}{\equiv} 1 \pmod{\mathfrak{m}}\}$$

$$\mathcal{N}_{K/F}(\mathfrak{m}) = \{\mathfrak{a} \in \mathcal{I}_F(\mathfrak{m}) : \mathfrak{a} = N_{K/F}(\mathfrak{A}) \text{ for some } \mathfrak{A} \in \mathcal{I}_K\}$$

and showed

$$J_F \Big/ {}_{F^\times \mathcal{E}_{F,\mathfrak{m}}^+} \cong J_{F,\mathfrak{m}}^+ \Big/ {}_{\mathcal{E}_{F,\mathfrak{m}}^+ F_{\mathfrak{m}}^+} \cong \mathcal{I}_F(\mathfrak{m}) \Big/ {}_{\mathcal{P}_{F,\mathfrak{m}}^+}$$

via the homomorphism $\eta_\mathfrak{m} : J_{F,\mathfrak{m}}^+ \longrightarrow \mathcal{I}_F(\mathfrak{m})$ given by

$$\mathbf{a} = (\dots, a_v, \dots) \mapsto \langle \mathbf{a} \rangle = \prod_{v \text{ finite}} \mathfrak{p}_v^{\mathrm{ord}_v(a_v)}.$$

Given an abelian Galois extension of number fields K/F, we have also shown that the subgroup $\mathcal{H} = F^\times N_{K/F} J_K$ of J_F contains $\mathcal{E}_{F,\mathfrak{m}}^+$ for some (integral) ideal $\mathfrak{m} \in \mathcal{I}_F$ divisible by sufficiently high powers of the primes that ramify in K/F. The ideal $\mathfrak{m}$ is not unique however. Suppose $\mathfrak{n}$ is another ideal of $\mathcal{O}_F$ for which $\mathcal{E}_{F,\mathfrak{n}}^+ \subseteq \mathcal{H}$. Let $(\mathfrak{m}, \mathfrak{n})$ denote the g.c.d. of $\mathfrak{m}$ and $\mathfrak{n}$.

Exercise 5.1. Show that $\mathcal{E}_{F,\mathfrak{m}}^+ \mathcal{E}_{F,\mathfrak{n}}^+ = \mathcal{E}_{F,(\mathfrak{m},\mathfrak{n})}^+$.   ◇

By Exercise 5.1, we have that there is a minimal ideal $\mathfrak{f}$ of $\mathcal{O}_F$ such that $\mathcal{E}_{F,\mathfrak{f}}^+ \subseteq \mathcal{H}$. This ideal $\mathfrak{f}$ is called the *conductor* of $\mathcal{H}$ (or of K/F), denoted $\mathfrak{f} = \mathfrak{f}(K/F)$. By minimality here, we mean precisely that if $\mathcal{E}_{F,\mathfrak{m}}^+ \subseteq \mathcal{H}$ then $\mathfrak{f}|\mathfrak{m}$.

Exercise 5.2. Let K/F be an abelian extension of number fields, and let $\mathfrak{f} = \mathfrak{f}(K/F)$.

a. Show: if $\mathfrak{m}$ is an ideal of $\mathcal{O}_F$ such that $\mathcal{E}_{F,\mathfrak{m}}^+ \subseteq F^\times N_{K/F} J_K$, then $\mathcal{I}_F(\mathfrak{m}) \subseteq \mathcal{I}_F(\mathfrak{f})$.

b. Prove or disprove and salvage: If $\mathcal{E}_{F,\mathfrak{m}}^+ \subseteq F^\times N_{K/F} J_K$, then $\mathcal{P}_{F,\mathfrak{m}}^+ \mathcal{N}_{K/F}(\mathfrak{m}) = \mathcal{P}_{F,\mathfrak{f}}^+ \mathcal{N}_{K/F}(\mathfrak{f}) \cap \mathcal{I}_F(\mathfrak{m})$.

c. Suppose $\mathcal{I}_F(\mathfrak{m}) \subseteq \mathcal{I}_F(\mathfrak{f})$. Show that there is a natural embedding

$$\mathcal{I}_F(\mathfrak{m}) \Big/ {}_{\mathcal{P}_{F,\mathfrak{m}}^+ \mathcal{N}_{K/F}(\mathfrak{m})} \hookrightarrow \mathcal{I}_F(\mathfrak{f}) \Big/ {}_{\mathcal{P}_{F,\mathfrak{f}}^+ \mathcal{N}_{K/F}(\mathfrak{f})}$$

   induced by inclusion.

d. Under what circumstances is the embedding of part c an isomorphism?   ◇

Exercise 5.3. Suppose $F \subseteq E \subseteq K$ are number fields and K/F is abelian. How are $\mathfrak{f}(K/F)$ and $\mathfrak{f}(E/F)$ related?   ◇

Which primes of $\mathcal{O}_F$ can divide the conductor $\mathfrak{f}(K/F)$? Recall from Chapter 4, we can find an ideal $\mathfrak{m}$ that is divisible only by the primes that ramify in K/F and for which $\mathcal{E}_{F,\mathfrak{m}}^+ \subseteq F^\times N_{K/F} J_K$ (see the discussion following the proof of Proposition 4.5.6). For any $\mathfrak{m}$, as in the proof of Corollary 4.3.5, we have

$$\mathcal{E}_{F,\mathfrak{m}}^+ = \prod_{v \text{ imaginary}} \mathbb{C}^\times \prod_{v \text{ real}} \mathbb{R}_+^\times \prod_{\mathfrak{p}_v | \mathfrak{m}} (1 + \mathfrak{p}_v^{\text{ord}_v \mathfrak{m}}) \prod_{\mathfrak{p}_v \nmid \mathfrak{m}} \mathcal{U}_v.$$

Since $N_{K/F}\mathcal{E}_K$ is open in $\mathcal{E}_F$, we can choose $\mathfrak{m}$ so that

$$\mathcal{E}_{F,\mathfrak{m}}^+ \subseteq N_{K/F}\mathcal{E}_K.$$

(Such an $\mathfrak{m}$ is said to be *admissible* for K/F.) But for an unramified prime $\mathfrak{p}_v$, and any place w of K above v, we know that N_{K_w/F_v} is surjective on units. Thus

$$N_{K/F}\mathcal{E}_K = \Big(\prod_{\substack{v \text{ finite} \\ \text{unramified}}} \mathcal{U}_v \Big) \Big(\prod_{\substack{v \text{ finite} \\ \text{ramified}}} \prod_{w|v} N_{K_w/F_v}\mathcal{U}_w \Big) \Big(\prod_{v \text{ infinite}} \prod_{w|v} N_{K_w/F_v} K_w^\times \Big).$$

It follows that we only need $\text{ord}_v \mathfrak{m} > 0$ when $N_{K_w/F_v}\mathcal{U}_w \neq \mathcal{U}_v$, which cannot happen unless v is ramified. By the minimality of $\mathfrak{f}$, we therefore have that if $\mathfrak{p}_v$ is unramified then $\mathfrak{p}_v \nmid \mathfrak{f}$. The conductor *cannot* be divisible by any unramified prime.

If $\mathfrak{p}_v$ is a ramified prime, then we have not yet determined whether $\mathfrak{p}_v$ *must* divide $\mathfrak{f}$. We shall return to this question in Chapter 6 after we have developed a bit more of the theory.

Exercise 5.4. Suppose K/F is a *cyclic* extension of number fields and $\mathfrak{m}$ is an admissible ideal for K/F. What can you conclude from Lemma 4.5.3 about divisibility of $\mathfrak{m}$ by ramified primes? ◇

For some authors, the conductor is a divisor, and includes factors involving the infinite real primes of F that extend to imaginary primes of K (the *ramified infinite primes*). This makes sense, as it is precisely for these infinite places that $N_{K_w/F_v} K_w^\times = N_{\mathbb{C}/\mathbb{R}}\mathbb{C}^\times = \mathbb{R}_+^\times \neq \mathbb{R}^\times$. For us, however, the conductor is an ideal of $\mathcal{O}_F$.

We may incorporate the notion of conductor into some of our results from Chapter 4. To summarize, we have shown (via the Global Cyclic Norm Index Equality)

$$[K : F] = [C_F : N_{K/F} C_K]$$
$$= [J_F : F^\times N_{K/F} J_K]$$
$$= [\mathcal{I}_F(\mathfrak{f}) : \mathcal{P}_{F,\mathfrak{f}}^+ N_{K/F}(\mathfrak{f})]$$

for K/F cyclic and $\mathfrak{f} = \mathfrak{f}(K/F)$. This is actually still true if K/F is only assumed to be abelian, but we have not yet shown it. (It does not generalize to non-abelian Galois extensions.)

Let us return briefly to the question of the primes of $\mathcal{O}_F$ that split completely in an abelian extension K/F. Recall we set

$$S_{K/F} = \{\text{primes } \mathfrak{p} \text{ of } F \text{ that split completely in } K/F\}.$$

We have seen in Chapter 3 that $S_{K/F}$ is central to Weber's notion of class field and that the Dirichlet density $\delta_F(S_{K/F}) = \frac{1}{[K:F]}$. Now consider the set

$$\mathcal{T}_{K/F} = \{\text{primes } \mathfrak{p} \in \mathcal{P}_{F,\mathfrak{f}}^+ \mathcal{N}_{K/F}(\mathfrak{f})\}.$$

If $\mathfrak{p}$ splits completely in K/F, then $(\mathfrak{p}, \mathfrak{f}) = 1$ and for $\mathfrak{P}|\mathfrak{p}$, we have $N_{K/F}\mathfrak{P} = \mathfrak{p}$. Thus $S_{K/F} \subseteq \mathcal{T}_{K/F}$. Also, (see Proposition 3.2.3), if we can show $L_{\mathfrak{f}}(1, \chi) \neq 0$ for all characters $\chi \neq \chi_0$ of $\mathcal{I}_F(\mathfrak{f})$ that are trivial on $\mathcal{P}_{F,\mathfrak{f}}^+ \mathcal{N}_{K/F}(\mathfrak{f})$, and if we have the generalization to abelian extensions of the Global Cyclic Norm Index Equality mentioned above, then

$$\delta_F(\mathcal{T}_{K/F}) = \frac{1}{[\mathcal{I}_F(\mathfrak{f}) : \mathcal{P}_{F,\mathfrak{f}}^+ \mathcal{N}_{K/F}(\mathfrak{f})]} = \frac{1}{[K : F]} = \delta_F(S_{K/F}),$$

so $S_{K/F} \approx \mathcal{T}_{K/F}$. Recall, according to Weber's definition of class field, it would then follow that K is the class field over F of $\mathcal{P}_{F,\mathfrak{f}}^+ \mathcal{N}_{K/F}(\mathfrak{f})$, and the Completeness Theorem would be established.

After we have proved Artin Reciprocity, we shall be able to say more about $S_{K/F}$ and $\mathcal{T}_{K/F}$. Artin Reciprocity gives an explicit isomorphism between the Galois group of an abelian extension K/F and the group $\mathcal{I}_F(\mathfrak{m})\big/ \mathcal{P}_{F,\mathfrak{m}}^+ \mathcal{N}_{K/F}(\mathfrak{m})$ for suitable $\mathfrak{m}$, yielding (among other things) the generalization of the Global Cyclic Norm Index Equality to arbitrary abelian extensions. Moreover, the nature of the isomorphism is such that information about $S_{K/F}$ is readily obtained. This isomorphism is defined using the Artin symbol, which we introduce next.

Let K/F be a Galois extension of number fields with abelian Galois group G. Let $\mathfrak{p}$ be a prime ideal of $\mathcal{O}_F$ that is unramified in K/F. Then the decomposition group $G_\mathfrak{p} = Z(\mathfrak{p})$ must be cyclic (inertia is trivial) with a canonical generator $\sigma_\mathfrak{p} = \left(\frac{\mathfrak{p}}{K/F}\right)$, the *Artin automorphism*.

Let $\mathfrak{m}$ be an ideal of $\mathcal{O}_F$ that is divisible by all the primes that ramify in the extension K/F and no others. The map $\mathfrak{p} \mapsto \sigma_\mathfrak{p}$ induces a homomorphism $\mathcal{A} = \mathcal{A}_{K/F} : \mathcal{I}_F(\mathfrak{m}) \to G$ given by $\mathfrak{a} \mapsto \sigma_\mathfrak{a} = \left(\frac{\mathfrak{a}}{K/F}\right)$ where, for $\mathfrak{a} = \prod_\mathfrak{p} \mathfrak{p}^{n_\mathfrak{p}} \in \mathcal{I}_F(\mathfrak{m})$, we set

$$\sigma_\mathfrak{a} = \prod_\mathfrak{p} \sigma_\mathfrak{p}^{n_\mathfrak{p}} := \left(\frac{\mathfrak{a}}{K/F}\right).$$

($\sigma_{\mathfrak{a}}$ does not depend on the choice of m.) The map $\mathcal{A}$ is called the *Artin map* and $\left(\frac{\mathfrak{a}}{K/F}\right)$ is the *Artin symbol*. Note that since m is divisible by all the ramifying primes, $\sigma_{\mathfrak{p}}$ is defined for all $\mathfrak{p} \nmid \mathfrak{m}$.

Recall that if $\tau : K \to K'$ is an isomorphism, and $\tau(F) = F'$, then

$$\tau\left(\frac{\mathfrak{p}}{K/F}\right)\tau^{-1} = \left(\frac{\mathfrak{p}'}{K'/F'}\right)$$

where $\tau(\mathfrak{p}) = \mathfrak{p}'$, (a prime of $\mathcal{O}_{F'}$).

Next we prove a result on the Artin symbol in towers of number fields.

Proposition 1.1 (Consistency Property). Let $F \subseteq L \subseteq K$, $F \subseteq E \subseteq K$ be number fields and suppose K/F is abelian. Let $\mathfrak{p}$ be a prime ideal of $\mathcal{O}_F$ that is unramified in K/F and let $\mathfrak{P}_K$ be a prime ideal of $\mathcal{O}_K$ that divides $\mathfrak{p}$. Let $\mathfrak{P}_L = \mathfrak{P}_K \cap L$, $\mathfrak{P}_E = \mathfrak{P}_K \cap E$ (prime ideals of $\mathcal{O}_E$, $\mathcal{O}_L$, respectively, that divide $\mathfrak{p}$).

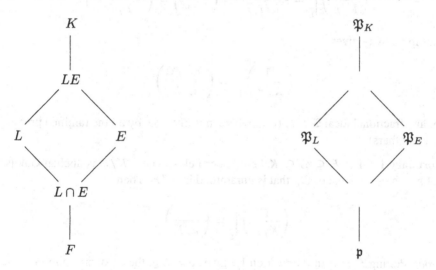

Then

$$\left(\frac{\mathfrak{P}_E}{K/E}\right)\bigg|_L = \left(\frac{\mathfrak{p}}{L/F}\right)^f$$

where $f = f(\mathfrak{P}_E/\mathfrak{p})$ is the residue field degree.

Proof. Let $\sigma_{\mathfrak{p}} = \left(\frac{\mathfrak{p}}{L/F}\right)$. Recall, for $\alpha \in \mathcal{O}_L$, we have

$$\sigma_{\mathfrak{p}}(\alpha) \equiv \alpha^{N\mathfrak{p}} \pmod{\mathfrak{P}_L},$$

and this congruence completely characterizes $\sigma_{\mathfrak{p}}$.

Let $\sigma_{\mathfrak{P}_E} = \left(\frac{\mathfrak{P}_E}{K/E}\right)$. Then $\sigma_{\mathfrak{P}_E}(\alpha) \equiv \alpha^{N\mathfrak{P}_E} \pmod{\mathfrak{P}_K}$ for all $\alpha \in \mathcal{O}_K$. If $\alpha \in \mathcal{O}_L$, then $\sigma_{\mathfrak{P}_E}(\alpha) \equiv \alpha^{N\mathfrak{P}_E} \pmod{\mathfrak{P}_K \cap L}$. Of course $\mathfrak{P}_K \cap L = \mathfrak{P}_L$, so $\sigma_{\mathfrak{P}_E}\big|_L (\alpha) \equiv \alpha^{N\mathfrak{P}_E} \pmod{\mathfrak{P}_L}$.

Now $N\mathfrak{P}_E = (N\mathfrak{p})^f$, so

$$\sigma_{\mathfrak{p}}^f(\alpha) \equiv \alpha^{N\mathfrak{p}^f} \equiv \alpha^{N\mathfrak{P}_E} \pmod{\mathfrak{P}_L}.$$

Hence $\sigma_{\mathfrak{p}}^f(\alpha) \equiv \sigma_{\mathfrak{P}_E}\big|_L (\alpha) \pmod{\mathfrak{P}_L}$. It follows that $\sigma_{\mathfrak{p}}^f = \sigma_{\mathfrak{P}_E}\big|_L$ as desired. (They are equal on the residue field, and the map to the residue field has kernel equal to the inertia group, which is trivial here.) $\qquad\square$

By the Consistency Property, we have

$$\left(\frac{\mathfrak{P}_E}{K/E}\right)\bigg|_L = \left(\frac{\mathfrak{p}}{L/F}\right)^f = \left(\frac{\mathfrak{p}^f}{L/F}\right) = \left(\frac{N_{E/F}\mathfrak{P}_E}{L/F}\right).$$

Multiplicativity gives

$$\left(\frac{\mathfrak{A}}{K/E}\right)\bigg|_L = \left(\frac{N_{E/F}\mathfrak{A}}{L/F}\right)$$

for any fractional ideal $\mathfrak{A} \in \mathcal{I}_E(\mathfrak{m})$, (where $\mathfrak{m}$ is divisible by all the ramified primes, and no others).

Corollary 1.2. Let $F \subseteq L \subseteq K$ be number fields, where K/F is abelian Galois. Let $\mathfrak{p}$ be a prime ideal of $\mathcal{O}_F$ that is unramified in K/F. Then

$$\left(\frac{\mathfrak{p}}{K/F}\right)\bigg|_L = \left(\frac{\mathfrak{p}}{L/F}\right).$$

Proof. Putting $E = F$ in Proposition 1.1 gives $\mathfrak{p} = \mathfrak{P}_E$; the corollary follows. $\qquad\square$

Corollary 1.3. Let $F \subseteq E \subseteq K$ be number fields, where K/F is abelian Galois. Let $\mathfrak{p}$ be a prime ideal of $\mathcal{O}_F$ that is unramified in K/F and let $\mathfrak{P}_E$ be a prime of $\mathcal{O}_E$ above $\mathfrak{p}$. Then

$$\left(\frac{\mathfrak{P}_E}{K/E}\right) = \left(\frac{N_{E/F}\mathfrak{P}_E}{K/F}\right).$$

Proof. Putting $L = K$ in Proposition 1.1 gives the result. $\qquad\square$

Corollary 1.4. Let K/F be an abelian Galois extension of number fields. Let $\mathfrak{m}$ be an ideal of $\mathcal{O}_F$ that is divisible by all the primes that ramify in K/F. Then

$$\mathcal{N}_{K/F}(\mathfrak{m}) \subseteq \ker(\mathcal{A} : \mathcal{I}_F(\mathfrak{m}) \to G).$$

(See (iii) of Artin Reciprocity in the next section.)

Proof. Putting $L = K = E$ in Proposition 1.1 gives

$$\left(\frac{N_{K/F}\mathfrak{P}_K}{K/F}\right) = \left(\frac{\mathfrak{P}_K}{K/K}\right)\Bigg|_{\kappa} = 1.$$

If $\mathfrak{A}$ is any ideal of K that is prime to the ramifying primes of K/F, then by factoring $\mathfrak{A}$, we have

$$\left(\frac{N_{K/F}\mathfrak{A}}{K/F}\right) = 1.$$ □

Exercise 5.5. Let $K = \mathbb{Q}(\sqrt{5}, i)$. Then $\text{Gal}(K/\mathbb{Q}) = \{1, \sigma, \tau, \sigma\tau\}$, where σ is complex conjugation, while τ fixes i and sends $\sqrt{5} \mapsto -\sqrt{5}$. Suppose $p\mathbb{Z}$ is an unramified prime in $K/\mathbb{Q}$.

a. Compute $\left(\frac{p\mathbb{Z}}{K/\mathbb{Q}}\right)$, i.e., give conditions (in terms of congruences) on p that determine whether the Artin symbol is 1, σ, τ, or $\sigma\tau$. (HINT: If you can find some cyclotomic field that contains K, then Example 1 of Chapter 1 may be of use.)

b. Give necessary and sufficient conditions (in terms of congruences) for the prime $p\mathbb{Z}$ to split completely in $K/\mathbb{Q}$. Compare your answer with part a and Theorem 1.1.8.

c. Suppose $p\mathbb{Z}$ is inert in $\mathbb{Q}(i)/\mathbb{Q}$. What can you say about $\left(\frac{p\mathbb{Z}[i]}{K/\mathbb{Q}(i)}\right)$?

d. Suppose $p\mathbb{Z}$ splits in $\mathbb{Q}(i)/\mathbb{Q}$, say $p\mathbb{Z}[i] = \mathfrak{p}\mathfrak{p}'$. What can you say about $\left(\frac{\mathfrak{p}}{K/\mathbb{Q}(i)}\right)$?

2 Artin Reciprocity

Theorem 2.1 (Artin Reciprocity). *Let K/F be an abelian extension of number fields, and assume $\mathfrak{m}$ is an ideal of $\mathcal{O}_F$, divisible by all the ramifying primes. Let $G = \text{Gal}(K/F)$. Then*

i. $\mathcal{A} : \mathcal{I}_F(\mathfrak{m}) \longrightarrow G$ *is surjective,*

ii. *the ideal $\mathfrak{m}$ of $\mathcal{O}_F$ can be chosen so that it is divisible only by the ramified primes and satisfies $\mathcal{P}_{F,\mathfrak{m}}^+ \subseteq \ker(\mathcal{A})$; thus we have an epimorphism $\mathcal{I}_F(\mathfrak{m})/\mathcal{P}_{F,\mathfrak{m}}^+ \longrightarrow$ G (it is surjective by (i)),*

iii. $N_{K/F}(\mathfrak{m}) \subseteq \ker(\mathcal{A})$.

Choosing $\mathfrak{m}$ *as in (ii), we have a well-defined homomorphism*

$$\mathcal{I}_F(\mathfrak{m}) \Big/ \mathcal{P}_{F,\mathfrak{m}}^+ \mathcal{N}_{K/F}(\mathfrak{m}) \to G$$

(still surjective). Since $\# \left(\mathcal{I}_F(\mathfrak{m}) \Big/ \mathcal{P}_{F,\mathfrak{m}}^+ \mathcal{N}_{K/F}(\mathfrak{m}) \right) \leq [K : F] = \#G$ *by the Universal Norm Index Inequality, in fact we have*

$$\mathcal{I}_F(\mathfrak{m}) \Big/ \mathcal{P}_{F,\mathfrak{m}}^+ \mathcal{N}_{K/F}(\mathfrak{m}) \cong G.$$

Note that this isomorphism is given explicitly by the Artin map. (Compare this to the Isomorphy Theorem.)

Proof. It will take some effort to give the proof. However, we have shown (iii) already, as a corollary to the Consistency Property in the previous section.

Next we prove (i), (i.e., $\mathcal{A} : \mathcal{I}_F(\mathfrak{m}) \to G$ is surjective). Let $H = \mathcal{A}(\mathcal{I}_F(\mathfrak{m})) \subseteq G$, and let E be the fixed field of H. Note that

$$\left(\frac{\mathfrak{a}}{E/F} \right) = \left(\frac{\mathfrak{a}}{K/F} \right) \Big|_E$$

for all $\mathfrak{a}$ prime to $\mathfrak{m}$, by Corollary 1.2.

By our choice of H, we have $\left(\frac{\mathfrak{a}}{K/F} \right) \in H$, and since E is the fixed field of H, we have

$$\left(\frac{\mathfrak{a}}{K/F} \right) \Big|_E = 1 = \left(\frac{\mathfrak{a}}{E/F} \right).$$

In particular, if $\mathfrak{a} = \mathfrak{p}$ is any prime with $\mathfrak{p} \nmid \mathfrak{m}$, then $\left(\frac{\mathfrak{p}}{E/F} \right) = 1$ generates the decomposition group $Z_{E/F}(\mathfrak{p})$. Therefore $\mathfrak{p}$ splits completely in E/F.

Let $\mathcal{S}_{E/F}$ be as before, and let $\mathcal{S}_F = \{$all primes $\mathfrak{p}$ of $F\}$. Then, by the above $\mathcal{S}_F \setminus \mathcal{S}_{E/F}$ is finite, whence

$$\frac{1}{[E : F]} = \delta_F(\mathcal{S}_{E/F}) = \delta_F(\mathcal{S}_F) = 1.$$

Thus $F = E$ is the fixed field of H. It follows that $H = G$.

Finally, we prove (ii), i.e., there exists $\mathfrak{m}$ such that $\mathcal{P}_{F,\mathfrak{m}}^+ \subseteq \ker(\mathcal{A} : \mathcal{I}_F(\mathfrak{m}) \to G)$. Moreover, we may take $\mathfrak{m}$ to be divisible only by primes that ramify in K/F.

We begin with the special case $F = \mathbb{Q}$, and $K = \mathbb{Q}(\zeta_m)$ where ζ_m is a primitive m^{th} root of unity. Let $p\mathbb{Z}$ be a prime of $\mathbb{Z}$, where $(p, m) = 1$. Then $\left(\frac{p\mathbb{Z}}{K/F} \right) = \sigma_p$, where $\sigma_p : \zeta_m \mapsto \zeta_m^p$. Let $a \in \mathbb{Z}_+$, say $a = p_1^{e_1} \cdots p_r^{e_r}$, and suppose the p_j are all prime to m. Then

$$\left(\frac{a\mathbb{Z}}{K/F}\right) = \prod_{j=1}^{r} \sigma_{p_j}^{e_j} = \sigma_a$$

where $\sigma_a : \zeta_m \mapsto \zeta_m^a$. Now say $a = \frac{b}{c}$, (where $b, c \in \mathbb{Z}_+$). Then $ca = b$, so

$$\left(\frac{ca\mathbb{Z}}{K/F}\right) = \sigma_{ca} = \sigma_c\sigma_a = \sigma_b.$$

Thus, $\sigma_a = \sigma_b\,\sigma_c^{-1}$. Choose $d \in \mathbb{Z}_+$ with $dc \equiv 1 \pmod{m}$. Then $\sigma_c^{-1} = \sigma_d$, and $\sigma_a = \sigma_{bd}$. When do we have $\sigma_a = 1$, i.e., when is $a\mathbb{Z} \in \ker(\mathcal{I}_\mathbb{Q}(m\mathbb{Z}) \to G)$? Our choice of d gives $\sigma_a = 1$ if and only if $\sigma_{bd} = 1$, so if and only if $bd \equiv 1 \pmod{m}$. But this happens if and only if $a \overset{\times}{\equiv} 1 \pmod{m\mathbb{Z}}$, (since a and bd are congruent modulo $p^{\operatorname{ord}_p m}$ in $\mathbb{Q}_p$ for any prime p dividing m). Hence $\ker(\mathcal{I}_\mathbb{Q}(m\mathbb{Z}) \to G) = \mathcal{P}^+_{\mathbb{Q},m\mathbb{Z}}$, and we have shown that (ii) is true for $\mathbb{Q}(\zeta_m)/\mathbb{Q}$.

Next suppose F is an arbitrary number field and $K = F(\zeta_m)$.

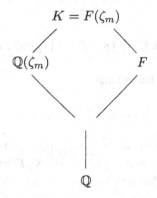

$$K = F(\zeta_m)$$
$$\mathbb{Q}(\zeta_m) \qquad F$$
$$\mathbb{Q}$$

We have

$$\left(\frac{\mathfrak{a}}{K/F}\right)\bigg|_{\mathbb{Q}(\zeta_m)} = \left(\frac{N\mathfrak{a}}{\mathbb{Q}(\zeta_m)/\mathbb{Q}}\right)$$

by the Consistency Property. Note that an automorphism $\sigma \in \operatorname{Gal}(K/F)$ is trivial if and only if $\sigma\big|_{\mathbb{Q}(\zeta_m)} = 1$.

Let $\mathfrak{m} = m\mathcal{O}_F$, and suppose $\mathfrak{a} \in \mathcal{P}^+_{F,\mathfrak{m}}$. Then $\mathfrak{a} = \langle\alpha\rangle$, where $\alpha \overset{\times}{\equiv} 1 \pmod{\mathfrak{m}}$, and $\alpha \gg 0$. Thus

$$\left(\frac{\mathfrak{a}}{K/F}\right)\bigg|_{\mathbb{Q}(\zeta_m)} = \left(\frac{\langle\alpha\rangle}{K/F}\right)\bigg|_{\mathbb{Q}(\zeta_m)} = \left(\frac{N\langle\alpha\rangle}{\mathbb{Q}(\zeta_m)/\mathbb{Q}}\right) = \left(\frac{\langle N\alpha\rangle}{\mathbb{Q}(\zeta_m)/\mathbb{Q}}\right).$$

Since $\alpha \gg 0$, we must have $N\alpha > 0$. Moreover, since $\alpha \overset{\times}{\equiv} 1 \pmod{\mathfrak{m}}$, we must have $N\alpha \overset{\times}{\equiv} 1 \pmod{\mathfrak{m}}$. But we already know from our discussion of $\mathbb{Q}(\zeta_m)/\mathbb{Q}$ that $\left(\frac{\langle N\alpha\rangle}{\mathbb{Q}(\zeta_m)/\mathbb{Q}}\right) = 1$ if and only if $N\alpha \overset{\times}{\equiv} 1 \pmod{m}$. Thus

$$\left(\frac{\mathfrak{a}}{K/F}\right)\bigg|_{\mathbb{Q}(\zeta_m)} = \left(\frac{\langle N\alpha\rangle}{\mathbb{Q}(\zeta_m)/\mathbb{Q}}\right) = 1$$

and (as we have already observed), this is only possible if $\left(\frac{\mathfrak{a}}{K/F}\right) = 1$ i.e., if $\mathfrak{a} \in \ker(\mathcal{A} : \mathcal{I}_F(\mathfrak{m}) \to G)$. It follows that $\mathcal{P}_{F,\mathfrak{m}}^+ \subseteq \ker(\mathcal{A})$ and (ii) is true for $F(\zeta_m)/F$, where F is any number field.

Exercise 5.6. Show that (ii) is true for E/F where F is any number field and $E \subseteq F(\zeta_m)$. $\Diamond$

Next let K/F be an arbitrary cyclic extension of number fields. We have seen that $[K : F] = [\mathcal{I}_F(\mathfrak{m}) : \mathcal{P}_{F,\mathfrak{m}}^+ \mathcal{N}_{K/F}(\mathfrak{m})]$ for $\mathfrak{m}$ "sufficiently large," i.e., for $\mathfrak{m}$ divisible by all the ramifying primes and such that $\mathcal{E}_{F,\mathfrak{m}}^+ \subseteq F^\times N_{K/F} J_K$. Since $\mathcal{A} : \mathcal{I}_F(\mathfrak{m}) \to G$ is surjective, we find:

$$[\mathcal{I}_F(\mathfrak{m}) : \ker\mathcal{A}] = \#G = [K : F] = [\mathcal{I}_F(\mathfrak{m}) : \mathcal{P}_{F,\mathfrak{m}}^+ \mathcal{N}_{K/F}(\mathfrak{m})].$$

We'll show (Proposition 2.2 below) that

$$\ker\mathcal{A} \subseteq \mathcal{P}_{F,\mathfrak{m}}^+ \mathcal{N}_{K/F}(\mathfrak{m}).$$

Given this, we conclude $\ker\mathcal{A} = \mathcal{P}_{F,\mathfrak{m}}^+ \mathcal{N}_{K/F}(\mathfrak{m}) \supseteq \mathcal{P}_{F,\mathfrak{m}}^+$. Proposition 2.2 thus suffices to complete the proof that (ii) is true for K/F arbitrary cyclic.

Finally, let K/F be an arbitrary abelian extension. By the above, for each cyclic E/F there is some $\mathfrak{m}_E$ such that $\mathcal{P}_{F,\mathfrak{m}_E}^+$ is contained in the kernel of the Artin map for E/F. Let $\mathfrak{m} = \prod_{\substack{E/F \text{ cyclic} \\ E \subseteq K}} \mathfrak{m}_E$. Then $\mathcal{P}_{F,\mathfrak{m}}^+ \subseteq \mathcal{P}_{F,\mathfrak{m}_E}^+$ for all such E. This means that $\mathcal{P}_{F,\mathfrak{m}}^+$ is contained in the kernel of the Artin map for every such E/F.

Suppose $\mathfrak{a} \in \mathcal{P}_{F,\mathfrak{m}}^+$. Then $\left(\frac{\mathfrak{a}}{E/F}\right) = 1$ for every cyclic subextension E/F. Let $\sigma = \left(\frac{\mathfrak{a}}{K/F}\right)$; so $\sigma\big|_E = 1$ for every E. If $\sigma \neq 1$, then there exists a non-trivial character $\chi : \langle\sigma\rangle \to \mathbb{C}^\times$ (so $\chi(\sigma) \neq 1$). Extend χ to a character of G and let $H = \ker\chi$. Then G/H is cyclic, since the image of χ is a finite subgroup of $\mathbb{C}^\times$. (In fact, $H = \langle\chi\rangle^\perp$ so $\langle\chi\rangle \cong H^\perp \cong \widehat{G/H} \cong G/H$.)

We have that G/H is cyclic. If we take E to be the fixed field of H, then we have that $\mathrm{Gal}(E/F) \cong G/H$ is cyclic. By the above, $\sigma\big|_E = 1$, which means we must have $\sigma \in H$ and $\chi(\sigma) = 1$, a contradiction.

Hence $\sigma = 1$ on all of K, i.e., $\left(\frac{\mathfrak{a}}{K/F}\right) = 1$. Thus $\mathfrak{a}$ must be in the kernel of the Artin map for K/F. It follows that $\mathcal{P}_{F,\mathfrak{m}}^+$ is contained in the kernel of the Artin map for K/F. □

For the proof of Artin Reciprocity to be complete, it only remains to show the following proposition.

Proposition 2.2. If K/F is a cyclic extension of number fields with Galois group G, and $\mathfrak{m}$ is an ideal of $\mathcal{O}_F$ sufficiently large so that it is divisible by all the ramifying primes in K/F and so that $\mathcal{E}_{F,\mathfrak{m}}^+ \subseteq F^\times N_{K/F} J_K$, then the kernel of the Artin map satisfies $\ker\left(\mathcal{A} : \mathcal{I}_F(\mathfrak{m}) \to G\right) \subseteq \mathcal{P}_{F,\mathfrak{m}}^+ N_{K/F}(\mathfrak{m})$.

Proof. Let $\mathfrak{a} \in \mathcal{I}_F(\mathfrak{m})$ with $\left(\frac{\mathfrak{a}}{K/F}\right) = 1$. We'll show

$$\mathfrak{a} = \langle \alpha \rangle N_{K/F}(\mathfrak{A})$$

where $\alpha \gg 0$, $\alpha \overset{\times}{\equiv} 1 \pmod{\mathfrak{m}}$, $\mathfrak{A} \in \mathcal{I}_K(\mathfrak{m}\mathcal{O}_K)$. This will require several lemmas; the proofs of the first few are due to Van der Waerden (1934). Also see the work of Birkhoff-Vandiver, Chevalley, Iyanaga and Takagi.

Lemma 2.3. Let $r > 1$, $a > 1$ be integers, and let q be a prime number. There is a prime number p such that the order of $a \bmod p$ in $\left(\mathbb{Z}/p\mathbb{Z}\right)^\times$ is q^r.

Proof. Let $t = \frac{a^{q^r}-1}{a^{q^{r-1}}-1}$, and $u = a^{q^{r-1}} - 1$. We have

$$a^{q^r} = (1+u)^q = 1 + qu + \binom{q}{2}u^2 + \cdots + u^q, \quad \text{whence}$$

$$t = \frac{a^{q^r} - 1}{u} = \frac{(1+u)^q - 1}{u}$$

$$= q + \binom{q}{2}u + \cdots + u^{q-1}$$

$$\equiv q \pmod{u}.$$

Let p be a prime dividing t. If $p|u$, then $p|q$, and we get $p = q$. Thus, unless t is a power of q, we may choose $p|t$, $p \neq q$ and therefore have $p \nmid u$. We get

$$a^{q^r} \equiv 1 \pmod{p} \quad \text{but} \quad a^{q^{r-1}} \not\equiv 1 \pmod{p}.$$

This will give that the order of $a \bmod p$ in $\left(\mathbb{Z}/p\mathbb{Z}\right)^\times$ is q^r as claimed. It remains only to consider the case when t is a power of q.

Suppose $t = q^e$, for some $e \in \mathbb{Z}_+$. Note in fact $e > 1$, since the binomial expansion forces $t > q$. So $q^2|t$. Using the binomial expansion, we find $q|u$. Since $q|u$, we have

$$0 \equiv t = q + \binom{q}{2} u + \cdots + u^{q-1}$$
$$\equiv q + u^{q-1} \pmod{q^2}$$

If $q > 2$, then $u^{q-1} \equiv 0 \pmod{q^2}$, and we get $0 \equiv q \pmod{q^2}$, a contradiction. Thus $q = 2$.

If $q = 2$, then we have $2|u$ so that $2|a^{2^{r-1}} - 1$ whence a is odd. Since $r > 1$, we get

$$u^{q-1} = u \equiv (a^{2^{r-2}})^2 - 1 \equiv 0 \pmod{8}.$$

But $q + u^{q-1} \equiv 0 \pmod{q^2}$, which (for $q = 2$) gives

$$q + u^{q-1} = 2 + u \equiv 0 \pmod{4},$$

a contradiction.

Our proof is complete, since t cannot be a power of q. □

Corollary 2.4. Let $a > 1$ be an integer. Given q^r as before, there are infinitely many primes p such that q^r divides the order of a mod p in $\left(\mathbb{Z}/p\mathbb{Z}\right)^{\times}$.

Proof. Apply Lemma 2.3 to q^{r+k}, for all $k \in \mathbb{Z}_+$. □

Lemma 2.5. Let S be a finite set of primes, and let $a > 1$, $n > 1$ be integers. There is an integer d, prime to all the elements of S, such that n divides the order of a mod d in $\left(\mathbb{Z}/d\mathbb{Z}\right)^{\times}$.

Proof. Write $n = q_1^{r_1} \cdots q_s^{r_s}$. The above corollary implies that for any j there is a prime $p_j \notin S$ such that the order of a mod p_j in $\left(\mathbb{Z}/p_j\mathbb{Z}\right)^{\times}$ is divisible by $q_j^{r_j}$. Hence n divides the order of a mod d in $\left(\mathbb{Z}/d\mathbb{Z}\right)^{\times}$ where $d = p_1 \cdots p_s$. □

Now fix $n, a \in \mathbb{Z}$, with $n, a > 1$, and a finite set of primes S. Find d as in Lemma 2.5, i.e., n dividing the order of a mod d in $\left(\mathbb{Z}/d\mathbb{Z}\right)^{\times}$.

Let $S' = S \cup \{\text{primes } p : p|d\}$, and let n' be the order of a mod d in $\left(\mathbb{Z}/d\mathbb{Z}\right)^{\times}$, (so $n|n'$). Apply Lemma 2.5 to S', n', to get $d' \in \mathbb{Z}$, prime to all the elements of S' (i.e., to S and to d), and such that n' divides the order of a mod d' in $\left(\mathbb{Z}/d'\mathbb{Z}\right)^{\times}$. Let $m = dd'$.

Lemma 2.6. Given integers $n > 1$, $a > 1$, and a finite set S of primes, there is a positive integer m such that

   i. m is prime to all the elements of S
   ii. n divides the order of a mod m in $\left(\mathbb{Z}/m\mathbb{Z}\right)^{\times}$

iii. there exists $b \in \mathbb{Z}$ such that n divides the order of b mod m in $\left(\mathbb{Z}/m\mathbb{Z}\right)^{\times}$ but a and b are independent mod m, (i.e., $\langle a \bmod m \rangle \cap \langle b \bmod m \rangle = 1$).

Proof. Let $m, d, d', n', \mathcal{S}'$ as before. Then (i) and (ii) are clear. For (iii), take

$$b \equiv a \quad (\bmod \ d)$$
$$b \equiv 1 \quad (\bmod \ d').$$

(The Chinese Remainder Theorem implies that b exists.) Then

$$\text{order of } b \bmod m \text{ in } \left(\mathbb{Z}/m\mathbb{Z}\right)^{\times} = \text{order of } a \bmod d \text{ in } \left(\mathbb{Z}/d\mathbb{Z}\right)^{\times}$$
$$\equiv 0 \quad (\bmod \ n).$$

Suppose a and b are not independent modulo m. Then there exist integers i, j such that $1 \not\equiv a^i \equiv b^j \pmod{m}$. Since $b \equiv 1 \pmod{d'}$, we have

$$a^i \equiv b^j \equiv 1 \pmod{d'}.$$

Since n' divides the order of a mod d' in $\left(\mathbb{Z}/d'\mathbb{Z}\right)^{\times}$, we get $n' | i$. Now n' is the order of a mod d in $\left(\mathbb{Z}/d\mathbb{Z}\right)^{\times}$, so $a^i \equiv 1 \pmod{d}$. But $m = dd'$, whence $a^i \equiv 1 \pmod{m}$, a contradiction. Thus a and b must be independent modulo m as claimed. □

Lemma 2.7. Let F be a number field, $\mathcal{S}$ a finite set of primes in $\mathbb{Z}$, $\mathfrak{p}$ a prime of $\mathcal{O}_F$. Then for any integer $n > 1$, there exists $m \in \mathbb{Z}$, prime to $\mathcal{S}$ and to $\mathfrak{p}$, such that if ζ_m is a primitive m^{th} root of unity, then

i. $\text{Gal}\,(F(\zeta_m)/F) \cong \left(\mathbb{Z}/m\mathbb{Z}\right)^{\times}$.

ii. $\left(\frac{\mathfrak{p}}{F(\zeta_m)/F}\right)$ has order divisible by n in $\text{Gal}\,(F(\zeta_m)/F)$.

iii. there is some $\tau \in \text{Gal}\,(F(\zeta_m)/F)$ of order divisible by n, such that τ is independent to $\left(\frac{\mathfrak{p}}{F(\zeta_m)/F}\right)$. [Note: independence implies $\langle \tau \rangle \cap Z(\mathfrak{p}) = 1$, since $\left(\frac{\mathfrak{p}}{F(\zeta_m)/F}\right)$ generates the decomposition group $Z(\mathfrak{p})$.]

Proof. Let

$$\mathcal{S}' = \{p : p\mathbb{Z} \text{ ramifies in } F/\mathbb{Q}\} \cup \{\mathfrak{p} \cap \mathbb{Z}\} \cup \mathcal{S}.$$

Apply Lemma 2.6 to $\mathcal{S}'$, $a = N\mathfrak{p}$, and $n \in \mathbb{Z}$, $n > 1$. There is some integer m, prime to all the elements of $\mathcal{S}'$, such that the order of a mod m in $\left(\mathbb{Z}/m\mathbb{Z}\right)^{\times}$ is divisible by n. In addition, there is an integer b, also having order divisible by n, which is independent modulo m to a.

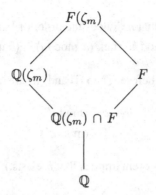

Now, since m is prime to the elements of $\mathcal{S}'$, the primes dividing m are unramified in $F/\mathbb{Q}$. But primes that do *not* divide m cannot ramify in $\mathbb{Q}(\zeta_m)$. Thus $F \cap \mathbb{Q}(\zeta_m)$ is everywhere unramified over $\mathbb{Q}$. By Minkowski theory, $\mathbb{Q}$ doesn't have a non-trivial extension that is everywhere unramified. We must have $F \cap \mathbb{Q}(\zeta_m) = \mathbb{Q}$, whence

$$\mathrm{Gal}\,(F(\zeta_m)/F) \cong \mathrm{Gal}\,(\mathbb{Q}(\zeta_m)/\mathbb{Q}) \cong \left(\mathbb{Z}/m\mathbb{Z}\right)^\times .$$

Under this isomorphism,

$$\left(\frac{\mathfrak{p}}{F(\zeta_m)/F}\right) \mapsto \left(\frac{\mathfrak{p}}{F(\zeta_m)/F}\right)\Bigg|_{\mathbb{Q}(\zeta_m)} = \left(\frac{N\mathfrak{p}}{\mathbb{Q}(\zeta_m)/\mathbb{Q}}\right) \mapsto a \bmod m.$$

Since n divides the order of $a \bmod m$, we have shown (i) and (ii).

Meanwhile, there is some integer b that has order divisible by n in $\left(\mathbb{Z}/m\mathbb{Z}\right)^\times$, and that is independent modulo m to a. If we take $\tau \in \mathrm{Gal}\,(F(\zeta_m)/F)$ corresponding to $b \bmod m$, then (iii) follows also.   □

Lemma 2.8 (Artin's Lemma). Let K/F be a cyclic extension of number fields of degree n, $\mathcal{S}$ a finite set of primes of $\mathbb{Z}$, $\mathfrak{p}$ a prime of $\mathcal{O}_F$. Then there is some $m \in \mathbb{Z}_+$, prime to the elements of $\mathcal{S}$ and to $\mathfrak{p}$, and an extension E/F such that

i. $K \cap E = F$
ii. $K(\zeta_m) = E(\zeta_m)$, i.e., $KE \subseteq K(\zeta_m) = E(\zeta_m)$
iii. $K \cap F(\zeta_m) = F$
iv. $\mathfrak{p}$ splits completely in E/F.

Proof. Enlarge $\mathcal{S}$ to contain all the primes that ramify in $K/\mathbb{Q}$. Using $n = [K : F]$ in Lemma 2.7, we obtain $m \in \mathbb{Z}$, prime to the enlarged $\mathcal{S}$ and to $\mathfrak{p}$, such that $\mathrm{Gal}\,(F(\zeta_m)/F) \cong \left(\mathbb{Z}/m\mathbb{Z}\right)^\times$, with the order of $\left(\frac{\mathfrak{p}}{F(\zeta_m)/F}\right)$ divisible by n, and τ independent to $\left(\frac{\mathfrak{p}}{F(\zeta_m)/F}\right)$, etc. Thus $F \cap \mathbb{Q}(\zeta_m) = \mathbb{Q}$, $K \cap \mathbb{Q}(\zeta_m) = \mathbb{Q}$, $K \cap F(\zeta_m) = F$, and (iii) is proved.

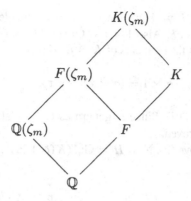

Let $G = \operatorname{Gal}(K/F) = \langle \sigma \rangle$. Now

$$\operatorname{Gal}(K(\zeta_m)/F) \cong \operatorname{Gal}(K/F) \times \operatorname{Gal}(F(\zeta_m)/F)$$

$$\cong \langle \sigma \rangle \times \left(\mathbb{Z}/m\mathbb{Z} \right)^{\times}$$

We shall identify $\operatorname{Gal}(K(\zeta_m)/F)$ and $\operatorname{Gal}(K/F) \times \operatorname{Gal}(F(\zeta_m)/F)$. Note that, under this identification,

$$\left(\frac{\mathfrak{p}}{K/F} \right) \times \left(\frac{\mathfrak{p}}{F(\zeta_m)/F} \right) = \left(\frac{\mathfrak{p}}{K(\zeta_m)/F} \right).$$

Let τ satisfy the conditions from the previous lemma, (so $\tau \in \operatorname{Gal}(F(\zeta_m)/F)$). Let H be the subgroup of $\operatorname{Gal}(K(\zeta_m)/F)$ generated by $\sigma \times \tau$ and by $\left(\frac{\mathfrak{p}}{K/F} \right) \times \left(\frac{\mathfrak{p}}{F(\zeta_m)/F} \right)$. Let E be the fixed field of H.

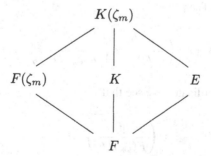

Since $\left(\frac{\mathfrak{p}}{K(\zeta_m)/F} \right)$ generates the decomposition group $Z_{K(\zeta_m)/F}(\mathfrak{p})$, we have

$$Z_{K(\zeta_m)/F}(\mathfrak{p}) \subseteq H.$$

Thus E is a subfield of the decomposition field for $\mathfrak{p}$ in $K(\zeta_m)/F$. This shows that $\mathfrak{p}$ splits completely in E/F, i.e., (iv) is proved.

Now, since $\sigma \times \tau \in H$, we have that $\sigma \times \tau$ fixes the elements of E and hence in particular, those of $K \cap E$. Also $1 \times \tau \in \mathrm{Gal}\,(K/F) \times \mathrm{Gal}\,(F(\zeta_m)/F)$ fixes the elements of K, so it fixes the elements of $K \cap E$. Hence

$$\sigma \times 1 = (\sigma \times \tau)(1 \times \tau)^{-1}$$

fixes the elements of $K \cap E$. Since σ generates $G = \mathrm{Gal}\,(K/F)$, we must have $K \cap E = F$, and (i) is proved.

It remains only to prove (ii). Now $H = \mathrm{Gal}\,(K(\zeta_m)/E)$. A typical element of H is

$$(\sigma \times \tau)^i \left(\left(\frac{\mathfrak{p}}{K/F} \right) \times \left(\frac{\mathfrak{p}}{F(\zeta_m)/F} \right) \right)^j.$$

Also $\mathrm{Gal}\,(K(\zeta_m)/F(\zeta_m)) \cong G \times 1$ has typical element $\sigma^a \times 1$. Note that $H \cap (G \times 1)$ has fixed field $E(\zeta_m)$.

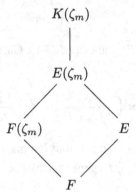

Say $b \in H \cap (G \times 1)$. Then

$$b = (\sigma \times \tau)^i \left(\left(\frac{\mathfrak{p}}{K/F} \right) \times \left(\frac{\mathfrak{p}}{F(\zeta_m)/F} \right) \right)^j = \sigma^a \times 1.$$

Comparing second coordinates, we see that

$$\tau^i \left(\frac{\mathfrak{p}}{F(\zeta_m)/F} \right)^j = 1.$$

By the previous lemma, n divides the order of τ and also divides the order of $\left(\frac{\mathfrak{p}}{F(\zeta_m)/F} \right)$. Thus $n|i$ and $n|j$, (since τ and $\left(\frac{\mathfrak{p}}{F(\zeta_m)/F} \right)$ are independent).

Now we consider first coordinates and find

$$\sigma^i \left(\frac{\mathfrak{p}}{K/F} \right)^j = \sigma^a.$$

Since $\#G = n$ and $n \mid j$, $n \mid i$, we must have $\sigma^a = 1$ whence $b = \sigma^a \times 1 = 1$. Thus $H \cap (G \times 1) = \{1\} = \mathrm{Gal}\,(K(\zeta_m)/E(\zeta_m))$. It follows that $K(\zeta_m) = E(\zeta_m)$ completing the proof of (ii).   □

We are now ready to prove the proposition that was all that was needed to complete the proof of Artin Reciprocity. Recall Proposition 2.2 asserts that when K/F is cyclic and $\mathfrak{m}$ is sufficiently large, $\ker(\mathcal{A} : \mathcal{I}_F(\mathfrak{m}) \to G) \subseteq \mathcal{P}^+_{F,\mathfrak{m}} \mathcal{N}_{K/F}(\mathfrak{m})$.

Proof. Let K/F be cyclic, and suppose $G = \mathrm{Gal}\,(K/F) = \langle \sigma \rangle$ with $\#G = n = [K : F]$. Let $\mathfrak{a} \in \ker \mathcal{A}_{K/F}$, i.e., suppose $\left(\frac{\mathfrak{a}}{K/F}\right) = 1$, and let $\mathfrak{m}$ be a multiple of $\mathfrak{f} = \mathfrak{f}(K/F)$, chosen divisible by all the ramified primes and no others. We may factor

$$\mathfrak{a} = \prod_{i=1}^{r} \mathfrak{p}_i^{\gamma_i}.$$

Let d_i be defined by

$$\left(\frac{\mathfrak{p}_i^{\gamma_i}}{K/F}\right) = \sigma^{d_i}.$$

Since $\left(\frac{\mathfrak{a}}{K/F}\right) = 1$, we find $\sigma^{d_1 + \cdots + d_r} = 1$, whence $n \mid d_1 + \cdots + d_r$.

Apply Artin's Lemma to $\mathfrak{p}_1, \ldots, \mathfrak{p}_r$ in succession to get integers $m_1, \ldots, m_r$ and fields $E_1, \ldots, E_r$ as in the following diagram.

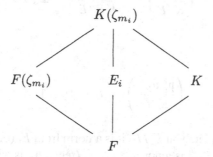

Let $G_i = \mathrm{Gal}\,(F(\zeta_{m_i})/F)$. We may assume that $m_i, \ldots, m_r$ are pairwise relatively prime (enlarge S each time we apply Artin's Lemma), and that each is prime to all the primes that ramify in $K/\mathbb{Q}$ and to $\mathfrak{p}_1, \ldots, \mathfrak{p}_r$ (again by enlarging S). Thus $\mathfrak{a} \in \mathcal{I}_F(m_1 \cdots m_r \mathfrak{m})$.

Working in $L = K(\zeta_{m_1}, \ldots, \zeta_{m_r})$, put $E = E_1 \cdots E_r$. Then $K \cap E = F = K \cap E_i$ and $\mathrm{Gal}\,(L/F) \cong G \times G_1 \times \cdots \times G_r$. Since $K \cap E = F$, we have

$$G = \mathrm{Gal}\,(K/F) \cong \mathrm{Gal}\,(KE_i/E_i) \cong \mathrm{Gal}\,(KE/E),$$

the isomorphisms being given by restriction.

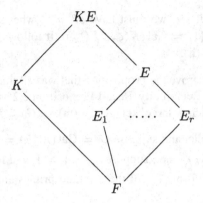

Let $\mathfrak{B}_E \in \mathcal{I}_E(m_1 \cdots m_r \mathfrak{m} \mathcal{O}_E)$ be such that

$$\left(\frac{\mathfrak{B}_E}{KE/E}\right)\Big|_K = \sigma$$

(the Artin map is surjective so some such $\mathfrak{B}_E$ exists) and let $\mathfrak{b}_F = N_{E/F} \mathfrak{B}_E$. (Note that $\mathfrak{b}_F \in \mathcal{I}_F(m_1 \cdots m_r \mathfrak{m})$.) Then

$$\left(\frac{\mathfrak{b}_F}{K/F}\right) = \left(\frac{\mathfrak{B}_E}{KE/E}\right)\Big|_K = \sigma$$

and for each i,

$$\left(\frac{\mathfrak{p}_i^{\gamma_i} \mathfrak{b}_F^{-d_i}}{K/F}\right) = \sigma^{d_i} \sigma^{-d_i} = 1.$$

Since $\mathfrak{p}_i$ splits completely in E_i/F, it is a norm from E_i (of any prime of E_i that lies above it). Thus $\mathfrak{p}_i^{\gamma_i} \mathfrak{b}_F^{-d_i}$ is a norm from E_i (recall $\mathfrak{b}_F$ is a norm from $E \supseteq E_i$). Hence $\mathfrak{p}_i^{\gamma_i} \mathfrak{b}_F^{-d_i} = N_{E_i/F} \mathfrak{A}_{E_i}$, for some fractional ideal $\mathfrak{A}_{E_i}$ of E_i. (Note that the $\mathfrak{A}_{E_i}$ are prime to $m_i \mathfrak{m}$.) By the Consistency Property, we have

$$\left(\frac{\mathfrak{A}_{E_i}}{KE_i/E_i}\right)\Big|_K = \left(\frac{\mathfrak{p}_i^{\gamma_i} \mathfrak{b}_F^{-d_i}}{K/F}\right) = 1.$$

Since $\mathrm{Gal}(KE_i/E_i) \cong \mathrm{Gal}(K/F)$ via restriction, we must have $\left(\frac{\mathfrak{A}_{E_i}}{KE_i/E_i}\right) = 1$. Thus $\mathfrak{A}_{E_i} \in \ker(\mathcal{I}_{E_i}(m_i \mathfrak{m} \mathcal{O}_{E_i}) \to \mathrm{Gal}(KE_i/E_i))$.

For each i we have $KE_i \subseteq E_i(\zeta_{m_i})$.

$$K(\zeta_{m_i}) = E_i(\zeta_{m_i})$$

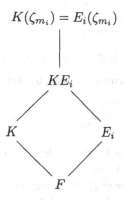

We have already shown that Artin Reciprocity is true for the extension $E_i(\zeta_{m_i})/E_i$. By Exercise 5.6, it is also true for the subextension KE_i/E_i; we use this fact to get

$$\mathfrak{A}_{E_i} = \langle \alpha_{E_i} \rangle N_{KE_i/E_i} \mathfrak{A}_{KE_i}$$

where $\alpha_{E_i} \overset{\times}{\equiv} 1 (\mathrm{mod}\, m_i \mathfrak{m} \mathcal{O}_{E_i})$, $\alpha_{E_i} \gg 0$ and $\mathfrak{A}_{KE_i} \in \mathcal{I}_{KE_i}(m_i \mathfrak{m} \mathcal{O}_{KE_i})$. Thus

$$
\begin{aligned}
\mathfrak{p}_i^{\gamma_i} \mathfrak{b}_F^{-d_i} &= N_{E_i/F} \mathfrak{A}_{E_i} \\
&= N_{E_i/F} \big(\langle \alpha_{E_i} \rangle N_{KE_i/E_i} \mathfrak{A}_{KE_i} \big) \\
&= \langle N_{E_i/F}(\alpha_{E_i}) \rangle N_{KE_i/F} \mathfrak{A}_{KE_i} \\
&\in \mathcal{P}_{F,m_i\mathfrak{m}}^+ \mathcal{N}_{K/F}(m_i\mathfrak{m})
\end{aligned}
$$

because

$$
\begin{aligned}
N_{E_i/F}(\alpha_{E_i}) &\overset{\times}{\equiv} 1 (\mathrm{mod}\, m_i \mathfrak{m}) \\
N_{E_i/F}(\alpha_{E_i}) &\gg 0 \\
N_{KE_i/F} \mathfrak{A}_{KE_i} &\in \mathcal{N}_{K/F}(m_i \mathfrak{m}).
\end{aligned}
$$

We have $\mathfrak{p}_i^{\gamma_i} \mathfrak{b}_F^{-d_i} \in \mathcal{P}_{F,m_i\mathfrak{m}}^+ \mathcal{N}_{K/F}(m_i\mathfrak{m})$ for all i, so $\prod_{i=1}^r \mathfrak{p}_i^{\gamma_i} \mathfrak{b}_F^{-d_i} \in \mathcal{P}_{F,\mathfrak{m}}^+ \mathcal{N}_{K/F}(\mathfrak{m})$, i.e., $\mathfrak{a}\mathfrak{b}_F^{-d_1-d_2-\cdots-d_r} \in \mathcal{P}_{F,\mathfrak{m}}^+ \mathcal{N}_{K/F}(\mathfrak{m})$. Now $n | (-d_1 - \cdots - d_r)$ (where $n = [K:F]$). Say $-d_1 - \cdots - d_r = -dn$. Then

$$\mathfrak{a}(\mathfrak{b}_F^{-d})^n = \mathfrak{a} N_{K/F}(\mathfrak{b}_F^{-d} \mathcal{O}_K) \in \mathcal{P}_{F,\mathfrak{m}}^+ \mathcal{N}_{K/F}(\mathfrak{m}).$$

But $N_{K/F}(\mathfrak{b}_F^{-d} \mathcal{O}_K) \in \mathcal{P}_{F,\mathfrak{m}}^+ \mathcal{N}_{K/F}(\mathfrak{m})$ already, since $\mathfrak{b}_F \in \mathcal{I}_F(\mathfrak{m})$. It follows that $\mathfrak{a} \in \mathcal{P}_{F,\mathfrak{m}}^+ \mathcal{N}_{K/F}(\mathfrak{m})$, and our proof is complete. $\qquad \square$

Exercise 5.7. Let K/F be an abelian extension of number fields. Show that if $\mathfrak{m}$ is any admissible ideal for K/F then

$$\mathcal{I}_F(\mathfrak{m})\Big/\mathcal{P}_{F,\mathfrak{m}}^+ \mathcal{N}_{K/F}(\mathfrak{m}) \cong \mathrm{Gal}(K/F)$$

and in particular, the conductor $\mathfrak{f}(K/F)$ satisfies Artin Reciprocity.   ◊

If K/F is abelian, Artin Reciprocity gives us (for an appropriately chosen $\mathfrak{m}$)

$$\mathcal{I}_F(\mathfrak{m})\Big/\mathcal{P}_{F,\mathfrak{m}}^+ \mathcal{N}_{K/F}(\mathfrak{m}) \cong \mathrm{Gal}(K/F).$$

In the case of a cyclic extension K/F, we already knew that the orders were equal by the norm index inequalities. From the above isomorphism, we see that the orders are equal for all abelian extensions K/F, as was claimed in Chapter 4.

Now that we have completed the proof of Artin Reciprocity, we may also revisit our consideration of the primes that split completely in an abelian extension of number fields. As before, suppose $\mathfrak{m}$ satisfies Artin reciprocity and let

$$\mathcal{S}_{K/F} = \{\text{prime ideals } \mathfrak{p} \text{ of } \mathcal{O}_F : \mathfrak{p} \text{ splits completely in } K/F\}$$
$$\mathcal{T}_{K/F} = \{\text{prime ideals } \mathfrak{p} \text{ of } \mathcal{O}_F : \mathfrak{p} \in \mathcal{P}_{F,\mathfrak{m}}^+ \mathcal{N}_{K/F}(\mathfrak{m})\}.$$

Recall we have seen that if $L_\mathfrak{m}(1, \chi) \neq 0$ for all characters $\chi \neq \chi_0$ of $\mathcal{R}_{F,\mathfrak{m}}^+$ that are trivial on $\mathcal{P}_{F,\mathfrak{m}}^+ \mathcal{N}_{K/F}(\mathfrak{m})$, then Dirichlet density can be used to show that $\mathcal{S}_{K/F} \approx \mathcal{T}_{K/F}$, i.e., K is the class field over F of $\mathcal{P}_{F,\mathfrak{m}}^+ \mathcal{N}_{K/F}(\mathfrak{m})$, thus proving the Completeness Theorem. Also, we have seen that $\mathcal{S}_{K/F} \subseteq \mathcal{T}_{K/F}$. Using Artin Reciprocity, we may obtain the Completeness Theorem without the result on the Weber L-functions. Compare the following corollary to Theorem 1.1.8.

Corollary 2.9. Let K/F be an abelian extension of number fields, say with $[K : F] = n$, and let $\mathfrak{p}$ be a prime of $\mathcal{O}_F$, unramified in K/F. Suppose $\mathfrak{m}$ is divisible by all the ramified primes and no others, and suppose $\mathfrak{m}$ satisfies Artin Reciprocity. Let f be the smallest positive integer such that $\mathfrak{p}^f \in \mathcal{P}_{F,\mathfrak{m}}^+ \mathcal{N}_{K/F}(\mathfrak{m})$. Then, in $\mathcal{O}_L$, we have a factorization $\mathfrak{p}\mathcal{O}_L = \mathfrak{P}_1 \cdots \mathfrak{P}_g$, where each $\mathfrak{P}_i$ is a prime of $\mathcal{O}_L$ with residue degree f over $\mathfrak{p}$, and where $g = n/f$. In particular, $\mathcal{S}_{K/F} = \mathcal{T}_{K/F}$.

Proof. Let $\mathcal{A} : \mathcal{I}_F(\mathfrak{m}) \to G$ as in Artin Reciprocity, so $\ker(\mathcal{A}) = \mathcal{P}_{F,\mathfrak{m}}^+ \mathcal{N}_{K/F}(\mathfrak{m})$. From our choice of the integer f, it follows that f is the smallest positive integer such that $\mathfrak{p}^f \in \ker(\mathcal{A})$. Hence f is the order of $\left(\frac{\mathfrak{p}}{K/F}\right)$ in $\mathrm{Gal}(K/F)$.

Since $\mathfrak{p}$ is unramified and the extension is Galois, we have a factorization $\mathfrak{p}\mathcal{O}_L = \mathfrak{P}_1 \cdots \mathfrak{P}_g$ where the primes $\mathfrak{P}_i$ are distinct and have equal residue degrees. This common residue degree is the order of the decomposition group, which is cyclic and generated by $\left(\frac{\mathfrak{p}}{K/F}\right) = \mathcal{A}(\mathfrak{p})$. Hence the residue degree is equal to f as claimed.

For the assertion about $\mathcal{S}_{K/F}$, note we have

$$\mathfrak{p} \in \mathcal{S}_{K/F} \iff Z(\mathfrak{p}) = 1$$
$$\iff \mathfrak{p} \in \ker(\mathcal{A}) = \mathcal{P}_{F,\mathfrak{m}}^+ \mathcal{N}_{K/F}(\mathfrak{m})$$
$$\iff \mathfrak{p} \in \mathcal{T}_{K/F} \qquad\qquad\qquad \square$$

Exercise 5.8. Now that we have $\mathcal{S}_{K/F} = \mathcal{T}_{K/F}$, we may apply Theorem 3.2.4 to deduce that $L_{\mathfrak{m}}(1, \chi) \neq 0$ for all characters $\chi \neq \chi_0$ of $\mathcal{R}_{F,\mathfrak{m}}^+$ that are trivial on $\mathcal{P}_{F,\mathfrak{m}}^+ \mathcal{N}_{K/F}(\mathfrak{m})$. Is this sufficient to finish the proof of the generalization of the Theorem on Primes in Arithmetic Progressions? Explain.   ◊

In the following exercises, we continue to revisit some of the ideas from Chapter 3. To answer them, it may be necessary to assume that for any number field F, any ideal $\mathfrak{m}$ of $\mathcal{O}_F$ and any non-trivial character χ of $\mathcal{R}_{F,\mathfrak{m}}^+$, the Weber L-function satisfies $L_{\mathfrak{m}}(1, \chi) \neq 0$. (As we saw in Chapter 3, this follows from the Existence Theorem, which we shall prove in Chapter 6.)

Exercise 5.9. Let K/F be an abelian extension of number fields with $\mathrm{Gal}\,(K/F) = G$. Let $\sigma \in G$ and define

$$\mathcal{S}_\sigma = \{\text{primes } \mathfrak{p} \text{ of } \mathcal{O}_F : \mathfrak{p} \text{ is unramified in } K/F \text{ and } \left(\frac{\mathfrak{p}}{K/F}\right) = \sigma\}.$$

Show that $\delta_F(\mathcal{S}_\sigma) = \frac{1}{[K:F]}$.   ◊

Exercise 5.10. Let K/F be a (possibly non-abelian) Galois extension of number fields with $\mathrm{Gal}\,(K/F) = G$. Because the extension is not necessarily abelian, instead of the Artin automorphism associated to a prime $\mathfrak{p}$ of $\mathcal{O}_F$, we must consider Frobenius elements at the primes above $\mathfrak{p}$. Let $\sigma \in G$ and let $[\sigma]_G = \{\tau\sigma\tau^{-1} : \tau \in G\}$ be the conjugacy class of σ in G. Let $E = K^{\langle\sigma\rangle}$ and put $n = [K : F]$, $d = [K : E]$, $c = \#[\sigma]_G$. Define

$$\mathcal{S}_\sigma = \{\text{unramified primes } \mathfrak{p} \text{ of } \mathcal{O}_F : \left(\frac{\mathfrak{P}}{K/F}\right) \in [\sigma]_G \text{ for } \mathfrak{P}|\mathfrak{p}\mathcal{O}_K\}$$

$$\mathcal{S}_{E,\sigma} = \{\text{primes } \mathfrak{Q} \text{ of } \mathcal{O}_E : \mathfrak{p} = \mathfrak{Q} \cap \mathcal{O}_F \text{ is unramified in } K/F, \; f(\mathfrak{Q}/\mathfrak{p}) = 1$$
$$\text{and } \left(\frac{\mathfrak{Q}}{K/E}\right) = \sigma\}$$

$$\mathcal{S}_{K,\sigma} = \{\text{primes } \mathfrak{P} \text{ of } \mathcal{O}_K : e(\mathfrak{P}/\mathfrak{P} \cap \mathcal{O}_F) = 1 \text{ and } \left(\frac{\mathfrak{P}}{K/F}\right) = \sigma\}.$$

a. Show that $\delta_E(\mathcal{S}_{E,\sigma}) = \frac{1}{d}$.

b. Show that $\mathfrak{P} \mapsto \mathfrak{P} \cap \mathcal{O}_E$ gives a bijection $\mathcal{S}_{K,\sigma} \to \mathcal{S}_{E,\sigma}$.

c. Show that $\mathfrak{P} \mapsto \mathfrak{P} \cap \mathcal{O}_F$ sends $\frac{n}{cd}$ primes of $\mathcal{S}_{K,\sigma}$ to each prime of $\mathcal{S}_\sigma$.

d. Prove the CHEBOTAREV DENSITY THEOREM: $\delta_F(\mathcal{S}_\sigma) = \frac{c}{n}$.   ◊

Exercise 5.11. Let F be a number field, and let $f(X) \in F[X]$ be a polynomial. Suppose $f(X)$ splits into linear factors modulo $\mathfrak{p}$ for all but finitely many prime ideals $\mathfrak{p}$ of $\mathcal{O}_F$. Use the Chebotarev Density Theorem, applied to the splitting field of $f(X)$, to show that $f(X)$ splits in $F[X]$.   ◊

Exercise 5.12. Let $f(X), g(X) \in \mathbb{Z}[X]$ be irreducible and denote their splitting fields (over $\mathbb{Q}$) by K_f, K_g, respectively. Let

$$\mathrm{Spl}(f) = \{\text{primes } p \in \mathbb{Z} : f(X) \text{ factors completely modulo } p\}$$
$$\mathrm{Spl}(g) = \{\text{primes } p \in \mathbb{Z} : g(X) \text{ factors completely modulo } p\}.$$

a. What is the relationship between $\mathrm{Spl}(f)$ and $\mathcal{S}_{K_f/\mathbb{Q}}$?

b. Prove the INCLUSION THEOREM: $K_f \supseteq K_g$ if and only if $\mathrm{Spl}(f)$ is "almost" a subset of $\mathrm{Spl}(g)$, i.e., with finitely many exceptions, $\mathrm{Spl}(f)$ is a subset of $\mathrm{Spl}(g)$.   ◊

The statement and proof of Artin Reciprocity were formulated in terms of ideals. In order to understand how all this fits together with the results of Chapter 4, we need to find an interpretation in terms of idèles. If $\mathfrak{m}$ is chosen so that $\mathcal{E}_{F,\mathfrak{m}}^+ \subseteq F^\times N_{K/F} J_K$, then we have

$$J_F \xrightarrow[\substack{\text{canonical}\\\text{surjection}}]{} {}^{J_F}\!\big/_{F^\times N_{K/F} J_K} \xrightarrow{\cong} {}^{\mathcal{I}_F(\mathfrak{m})}\!\big/_{\mathcal{P}_{F,\mathfrak{m}}^+ N_{K/F}(\mathfrak{m})} \xrightarrow[\text{Artin map}]{\cong} \mathrm{Gal}\,(K/F).$$

Let

$$\rho_{K/F} : J_F \longrightarrow \mathrm{Gal}\,(K/F)$$

be this composition of functions; it is a surjective homomorphism of groups with kernel $F^\times N_{K/F} J_K$. We say K is the class field over F of $F^\times N_{K/F} J_K$ and we call $\rho_{K/F}$ the *idèlic Artin map*. For $\mathbf{a} \in J_F$, we sometimes denote

$$\rho_{K/F}(\mathbf{a}) = \left(\frac{\mathbf{a}}{K/F}\right).$$

Can we give $\rho_{K/F}$ explicitly? Recall from Chapter 4, the isomorphism

$$^{J_F}\!\big/_{F^\times N_{K/F} J_K} \cong {}^{\mathcal{I}_F(\mathfrak{m})}\!\big/_{\mathcal{P}_{F,\mathfrak{m}}^+ N_{K/F}(\mathfrak{m})}$$

arises via the isomorphism $^{J_F}\!\big/_{F^\times} \cong {}^{J_{F,\mathfrak{m}}^+}\!\big/_{F_\mathfrak{m}^+}$ so arises via the map

$$J_F \longrightarrow {}^{J_{F,\mathfrak{m}}^+}\!\big/_{F_\mathfrak{m}^+} \longrightarrow {}^{\mathcal{I}_F(\mathfrak{m})}\!\big/_{\mathcal{P}_{F,\mathfrak{m}}^+}$$

given by

$$\mathbf{a} \mapsto \mathbf{b} F_\mathfrak{m}^+ \mapsto \langle \mathbf{b} \rangle \mathcal{P}_{F,\mathfrak{m}}^+$$

where $\mathbf{b} = \alpha\mathbf{a}$, with $\alpha \in F^\times$ chosen so that $\alpha\mathbf{a} \in J_{F,\mathfrak{m}}^+$. This means we must choose $\alpha \in F^\times$ so that

i. for any finite place v of F, $\iota_v(\alpha) \in \mathcal{O}_v$, and $\iota_v(\alpha)a_v \equiv 1 \pmod{\mathfrak{p}_v^{\mathrm{ord}_v\mathfrak{m}}}$
ii. for any real place v of F, $\iota_v(\alpha)a_v > 0$.

By Exercise 4.25, we know that the map $\mathbf{a} \mapsto \mathbf{b}F_\mathfrak{m}^+ \mapsto \langle\mathbf{b}\rangle\mathcal{P}_{F,\mathfrak{m}}^+$ is well-defined.

Exercise 5.13. Show that the idèlic Artin map is given by

$$\rho_{K/F}(\mathbf{a}) = \left(\frac{\langle\alpha\mathbf{a}\rangle}{K/F}\right) \in \mathrm{Gal}\,(K/F)$$

where α is as in (i) and (ii) above, and

$$\langle\alpha\mathbf{a}\rangle = \prod_{v \text{ finite}} \mathfrak{p}_v^{\mathrm{ord}_v(\iota_v(\alpha)a_v)}$$

as usual.   ◊

Exercise 5.14. For K/F abelian, we know that the idèlic Artin map $J_F \to \mathrm{Gal}\,(K/F)$ is surjective with kernel $F^\times N_{K/F}J_K$, (Artin Reciprocity). Alternatively, show how we may write this in terms of the idèle class group $C_F = {}^{J_F}\!/_{F^\times}$. In particular, show that the Artin map gives rise to a homomorphism $C_F \to \mathrm{Gal}\,(K/F)$, which is surjective with kernel $N_{K/F}C_K$. Thus,

$$J_F\big/_{F^\times N_{K/F}J_K} \cong C_F\big/_{N_{K/F}C_K} \cong \mathrm{Gal}\,(K/F).$$

(We shall not pursue it now, but those familiar with inverse limits may also wish to consider what happens if we take inverse limits here. Doing so yields a map $C_F \to \mathrm{Gal}\,(F^{\mathrm{ab}}/F)$ where F^{ab} is the maximal abelian extension of F. This map is also surjective. See the discussion on the norm residue symbol in Chapter 6 for more about this.)   ◊

Exercise 5.15. Let $F \subseteq L \subseteq K$, $F \subseteq E \subseteq K$ be number fields and suppose K/F is abelian. If $\mathbf{a} \in J_E$, do we have

$$\left(\frac{\mathbf{a}}{K/E}\right)\bigg|_L = \left(\frac{N_{E/F}(\mathbf{a})}{L/F}\right)$$

as we did with the classical Artin maps on fractional ideals?   ◊

3 An Example: Quadratic Reciprocity

Throughout this section, let $F = \mathbb{Q}$, $K = \mathbb{Q}(\sqrt{(-1)^{\frac{p-1}{2}} p})$, where $p > 2$ is a prime. Then p is the only finite prime that ramifies in $K/\mathbb{Q}$, so the ideal m in the statement of Artin Reciprocity is a positive power of $p\mathbb{Z}$. Also, ∞ ramifies in $K/\mathbb{Q}$ if and only if $\frac{p-1}{2}$ is odd. Consider the Artin map

$$\rho_{K/\mathbb{Q}} : J_{\mathbb{Q}} \longrightarrow \mathrm{Gal}\,(K/\mathbb{Q}) = \{\pm 1\}.$$

We have (by Artin Reciprocity), that $\ker \rho_{K/\mathbb{Q}} = \mathbb{Q}^{\times} N_{K/\mathbb{Q}} J_K$. We want to compute the image under $\rho_{K/\mathbb{Q}}$ of several particular idèles.

Let $\mathbf{c}_{\infty} = (-1, 1, 1, \ldots) \in J_{\mathbb{Q}}$, (where the -1 is in the component at ∞). Take $\alpha = 1 - p^{\mathrm{ord}_p \mathrm{m}}$ so that $\alpha \mathbf{c}_{\infty} \in J_{\mathbb{Q}_{\mathrm{m}}}^{+}$. Then

$$\rho_{K/\mathbb{Q}}(\mathbf{c}_{\infty}) = \left(\frac{\langle \alpha \mathbf{c}_{\infty} \rangle}{K/\mathbb{Q}} \right).$$

Now $\rho_{K/\mathbb{Q}}(\mathbf{c}_{\infty}) = 1$ if and only if $\mathbf{c}_{\infty} \in \ker \rho_{K/\mathbb{Q}}$. This is easily seen to be true if and only if $-1 \in N_{K_{\infty}/\mathbb{Q}_{\infty}}(K_{\infty}^{\times}) = N_{K_{\infty}/\mathbb{R}}(K_{\infty}^{\times})$, which occurs if and only if $K_{\infty} = \mathbb{R}$, i.e., if and only if $(-1)^{\frac{p-1}{2}} = 1$. We have shown that

$$\rho_{K/\mathbb{Q}}(\mathbf{c}_{\infty}) = (-1)^{\frac{p-1}{2}} = \left(\frac{-1}{p} \right),$$

the Legendre symbol.

Now let ℓ be a prime of $\mathbb{Q}$ and let $\mathbf{c}_{\ell} = (1, \ldots 1, -1, 1, \ldots) \in J_{\mathbb{Q}}$, (where the -1 is in the component at v_{ℓ}). If $\ell \neq p, \infty$, then $\mathbf{c}_{\ell} \in \mathbb{Q}_{\mathrm{m}}^{+}$ so we may take $\alpha = 1$. We get

$$\rho_{K/\mathbb{Q}}(\mathbf{c}_{\ell}) = \left(\frac{\langle \mathbf{c}_{\ell} \rangle}{K/\mathbb{Q}} \right) = \prod_{v} \left(\frac{\mathfrak{p}_v}{K/\mathbb{Q}} \right)^{\mathrm{ord}_v(c_v)} = 1.$$

Artin Reciprocity says that $-1 \in \mathbb{Q}^{\times} \subseteq \ker \rho_{K/\mathbb{Q}}$. Since, as idèles, $\iota(-1) = \prod \mathbf{c}_{\ell}$, where ℓ ranges over all the primes (including ∞) of $\mathbb{Q}$, we must have

$$1 = \rho_{K/\mathbb{Q}}(\iota(-1)) = \prod_{\ell} \rho_{K/\mathbb{Q}}(\mathbf{c}_{\ell})$$

$$= \rho_{K/\mathbb{Q}}(\mathbf{c}_{\infty}) \rho_{K/\mathbb{Q}}(\mathbf{c}_p).$$

We conclude that

$$\rho_{K/\mathbb{Q}}(\mathbf{c}_p) = \left(\frac{-1}{p} \right).$$

Let q be a prime of $\mathbb{Z}$ and let $\mathbf{b}_\ell = (1, \ldots, 1, q, 1, \ldots) \in J_\mathbb{Q}$, where the q is in the component at v_ℓ. For $\ell \neq p$, $\ell \neq q$, we have $\mathbf{b}_\ell \in J_{\mathbb{Q}_m}^+$ and

$$\rho_{K/\mathbb{Q}}(\mathbf{b}_\ell) = \prod_v \left(\frac{\mathfrak{p}_v}{K/\mathbb{Q}}\right)^{\mathrm{ord}_v(b_v)} = 1.$$

If $\ell = p \neq q$, then we have $\alpha \mathbf{b}_p \in J_{\mathbb{Q}_m}^+$ for some positive $\alpha \in \mathbb{Z}$ with $\alpha q \equiv 1$ (mod $p^{\mathrm{ord}_p m}$). Now $\rho_{K/\mathbb{Q}}(\mathbf{b}_p) = 1$ if and only if $\mathbf{b}_p \in \ker \rho_{K/\mathbb{Q}}$, which, by Artin Reciprocity, happens if and only if $\mathbf{b}_p \in \mathbb{Q}^\times N_{K/\mathbb{Q}} J_K$, i.e., if and only if q is a norm from K_p to $\mathbb{Q}_p$. Now $K_p/\mathbb{Q}_p$ is totally ramified of degree 2, so the norms in $\mathcal{U}_p = \mathbb{Z}_p^\times$ have index $e = 2$ in the group $\mathbb{Z}_p^\times$. (Note that $q \in \mathbb{Z}_p^\times$.) We know

$$\mathbb{Z}_p^\times \cong \mu_{p-1} \times (1 + p\mathbb{Z}_p),$$

where μ_{p-1} denotes the $(p-1)^{\mathrm{th}}$ roots of unity in $\mathbb{Z}_p^\times$. Since p is odd, the only subgroup of index 2 is $\mu_{p-1}^2 \times (1 + p\mathbb{Z}_p) = (\mathbb{Z}_p^\times)^2$. Thus, $\rho_{K/\mathbb{Q}}(\mathbf{b}_p) = 1$ if and only if $q \in (\mathbb{Z}_p^\times)^2$, which happens if and only if q is a square modulo p, i.e., if and only if $\left(\frac{q}{p}\right) = 1$. We have shown $\rho_{K/\mathbb{Q}}(\mathbf{b}_p) = \left(\frac{q}{p}\right)$.

If $\ell = q \neq p$, then we have $\alpha \mathbf{b}_q \in J_{\mathbb{Q}_m}^+$ for $\alpha = 1$, and

$$\rho_{K/\mathbb{Q}}(\mathbf{b}_q) = \prod_v \left(\frac{\mathfrak{p}_v}{K/\mathbb{Q}}\right)^{\mathrm{ord}_v(\alpha(\mathbf{b}_q)_v)} = \left(\frac{q\mathbb{Z}}{K/\mathbb{Q}}\right).$$

Thus

$$\rho_{K/\mathbb{Q}}(\mathbf{b}_q) = 1 \Longleftrightarrow \left(\frac{q\mathbb{Z}}{K/\mathbb{Q}}\right) = 1$$

$$\Longleftrightarrow q\mathbb{Z} \text{ splits completely in } K/\mathbb{Q}$$
$$\text{(by Corollary 2.9)}$$

$$\Longleftrightarrow (-1)^{\frac{p-1}{2}} p \text{ is a square in } \mathbb{Z}_q$$

$$\Longleftrightarrow \left(\frac{-1}{q}\right)^{\frac{p-1}{2}} \left(\frac{p}{q}\right) = 1 \text{ if } q \neq 2$$

$$\text{or } (-1)^{\frac{p-1}{2}} p \equiv 1 \pmod 8 \text{ if } q = 2.$$

Artin Reciprocity says that $q \in \mathbb{Q}^\times \subseteq \ker \rho_{K/\mathbb{Q}}$. For $q \neq p, 2$, we have

$$1 = \rho_{K/\mathbb{Q}}(\iota(q)) = \prod_{\ell} \rho_{K/\mathbb{Q}}(\mathbf{b}_\ell)$$

$$= \rho_{K/\mathbb{Q}}(\mathbf{b}_q)\rho_{K/\mathbb{Q}}(\mathbf{b}_p)$$

$$= (-1)^{\frac{q-1}{2}\cdot\frac{p-1}{2}} \left(\frac{p}{q}\right)\left(\frac{q}{p}\right).$$

Meanwhile for $q = 2$, we have

$$1 = \left(\frac{2}{p}\right)(-1)^{\frac{p^2-1}{8}}$$

since $(-1)^{\frac{p-1}{2}} p \equiv 1 \pmod 8$ if and only if $p \equiv \pm 1 \pmod 8$.

The above example shows that Quadratic Reciprocity follows from Artin Reciprocity! Indeed, after Artin Reciprocity was proved, all previously known reciprocity laws were seen to follow from it. For more on this, see the exercises in Cassels and Fröhlich, [CF].

4 Some Preliminary Results about the Artin Map on Local Fields

We want to study the idèlic Artin map on local fields. First, we consider the image of the local units. Let K/F be an abelian extension of number fields, and consider the map $J_F \twoheadrightarrow \mathrm{Gal}(K/F)$ given by $\mathbf{a} \mapsto \left(\frac{\mathbf{a}}{K/F}\right)$. Let $\mathfrak{m}$ be an ideal of $\mathcal{O}_F$ that is divisible only by the ramified primes in K/F, and such that $\mathcal{E}_{F,\mathfrak{m}}^+ \subseteq F^\times N_{K/F} J_K$.

Let v be any place of F. Recall if v is infinite, then we take $\mathcal{U}_v = F_v^\times$ by convention. We may view the local units $\mathcal{U}_v$ as a subgroup of J_F as follows. Map $u \in \mathcal{U}_v$ to the idèle

$$\varphi_v(u) = (1, \ldots, 1, u, 1, \ldots) \in J_F,$$

where the u is in the component corresponding to v. Via the map φ_v, we see readily that

$$\mathcal{U}_v \cong (1, \ldots, 1, \mathcal{U}_v, 1, \ldots) < J_F.$$

Example.

1. Let $F = \mathbb{Q}$, $K = \mathbb{Q}(\zeta_{15})$, $v = 3$, $u = 5$. For this extension, we know that the conductor is divisible only by the primes that ramify, i.e., only by 3 and 5. Eventually, we shall be able to prove that the conductor $\mathfrak{m}$ of this extension is exactly $15\mathbb{Z}$. We have $\mathcal{U}_v = \mathbb{Z}_3^\times$. Of course, $5 \in \mathbb{Z}_3^\times$, so what is $\rho_{K/\mathbb{Q}}(\varphi_3(5))$? We must find $\alpha \in \mathbb{Q}^\times$ such that

$$\alpha\varphi_3(5) = (\alpha, \ldots, \alpha, 5\alpha, \alpha, \ldots) \in J_{\mathbb{Q},\mathfrak{m}}^+$$

so we need $\alpha > 0$, $\alpha \equiv 1 \pmod{\mathfrak{p}_v^{\mathrm{ord}_v \mathfrak{m}}}$ for finite $v \neq 3$, and $5\alpha \equiv 1 \pmod{3^{\mathrm{ord}_v \mathfrak{m}}}$. Using the future result (Chapter 6) that $\mathfrak{m} = 15\mathbb{Z}$, we see that we may take $\alpha = 11$. Then

$$\rho_{K/\mathbb{Q}}(\varphi_3(5)) = \left(\frac{\langle 11\varphi_3(5)\rangle}{K/\mathbb{Q}}\right)$$

$$= \left(\frac{\mathfrak{p}_3^{\mathrm{ord}_3(55)} \prod_{v \neq 3} \mathfrak{p}_v^{\mathrm{ord}_v(11)}}{K/\mathbb{Q}}\right)$$

$$= \left(\frac{11\mathbb{Z}}{K/\mathbb{Q}}\right),$$

which is the automorphism of K sending $\zeta_{15} \mapsto \zeta_{15}^{11}$ (see the example in Chapter 1).

Now suppose the finite place v has $\mathfrak{p}_v$ unramified in K/F. Let $u \in \mathcal{U}_v$, with $\varphi_v(u) = (1, \ldots, 1, u, 1, \ldots, 1)$ as above. Then, since $\mathfrak{p}_v$ is unramified, we have $\mathfrak{p}_v \nmid \mathfrak{m}$, so $\varphi_v(u) \in J_{F,\mathfrak{m}}^+$ (clearly $\varphi_v(u) \equiv 1 \pmod{\mathfrak{q}}$ for all $\mathfrak{q}|\mathfrak{m}$, and $\varphi_v(u) \gg 0$). It follows easily that $\rho_{K/F}(\varphi_v(u)) = \left(\frac{\varphi_v(u)}{K/F}\right) = 1$. Thus, we have shown that for every finite unramified place v, $\left(\frac{\varphi_v(\mathcal{U}_v)}{K/F}\right) = \{1\}$.

Exercise 5.16. We say that an infinite place v of F *ramifies* in K/F if and only if v is real and it extends to some imaginary place of K. With this notion of ramified infinite place, and with the convention $\mathcal{U}_v = F_v^\times$ for infinite places, does the above still hold when v is infinite? $\quad\quad\Diamond$

In the same way, we can view $F_v^\times$ as a subgroup of J_F. Simply extend the map φ_v: for any $x \in F_v^\times$, let $\varphi_v(x) = (1, \ldots, 1, x, 1, \ldots)$, where the x is in the component corresponding to v. Now fix a uniformizer π for $F_v^\times$. Any $x \in F_v^\times$ may be written $x = u\pi^a$ for some $u \in \mathcal{U}_v$, and some integer a. Again suppose v is a finite place of F, with $\mathfrak{p}_v$ unramified in K/F. If $x \in F_v^\times$, then $\varphi_v(x) \in J_{F,\mathfrak{m}}^+$ already, and

$$\langle\varphi_v(x)\rangle = \langle\varphi_v(u\pi^a)\rangle = \mathfrak{p}_v^a.$$

We get

$$\rho_{K/F}(\varphi_v(x)) = \left(\frac{\mathfrak{p}_v}{K/F}\right)^a$$

where $a = \mathrm{ord}_v(x)$.

Since $\left(\frac{\mathfrak{p}_v}{K/F}\right)$ generates the decomposition group $Z(\mathfrak{p}_v)$ when v is finite and $\mathfrak{p}_v$ is unramified, we find in this case

$$\left(\frac{\varphi_v(F_v^\times)}{K/F}\right) = Z(\mathfrak{p}_v).$$

Exercise 5.17. With the conventions discussed in the previous exercise, is the above still true when v is infinite and unramified? To rephrase, for infinite primes, we are taking $\mathcal{U}_v = F_v^\times$, so their images are the same. How would you define the decomposition group of an infinite place? Is it going to be trivial in the case of an unramified infinite place? ◊

What happens when v is ramified? Suppose v corresponds to a finite prime $\mathfrak{p}_v$ that ramifies in K/F. Let K_T denote the fixed field of the inertia subgroup $T(\mathfrak{p}_v)$. Then $\mathfrak{p}_v$ is unramified in K_T/F, so $\left(\frac{\varphi_v(\mathcal{U}_v)}{K_T/F}\right) = 1$. Now $\mathrm{Gal}\,(K_T/F) \cong \mathrm{Gal}\,(K/F)/T(\mathfrak{p}_v)$. The idèlic version of the Consistency Property for Artin symbols gives

$$\left(\frac{\varphi_v(\mathcal{U}_v)}{K/F}\right)\bigg|_{K_T} = \left(\frac{\varphi_v(\mathcal{U}_v)}{K_T/F}\right) = 1.$$

Thus

$$\left(\frac{\varphi_v(\mathcal{U}_v)}{K/F}\right) \subseteq \mathrm{Gal}\,(K/K_T) = T(\mathfrak{p}_v).$$

(They are actually equal, but it will be some time before we can prove it.)

We can also find a result on $\rho_{K/F}(\varphi_v(F_v^\times))$ when v is ramified. For any finite place v of F, let $Z(\mathfrak{p}_v)$, $T(\mathfrak{p}_v)$ be the decomposition and inertia subgroups of $G = \mathrm{Gal}\,(K/F)$ as usual. Let K_Z, K_T be their respective fixed fields.

Since $\mathfrak{p}_v$ splits completely in K_Z, we know that $\varphi_v(F_v^\times) \subseteq N_{K_Z/F}J_{K_Z}$ (see the corollary to Artin Reciprocity). Thus, for $x \in F_v^\times$, we have

$$\rho_{K/F}(\varphi_v(x)) = \left(\frac{\varphi_v(x)}{K/F}\right)$$
$$= \left(\frac{N_{K_Z/F}(\mathbf{b})}{K/F}\right) \text{ for some } \mathbf{b} \in J_{K_Z}$$
$$= \left(\frac{\mathbf{b}}{K/K_Z}\right)$$
$$\in \mathrm{Gal}\,(K/K_Z) = Z(\mathfrak{p}_v).$$

We have shown that for any finite place v of F,

$$\rho_{K/F}(\varphi_v(F_v^\times)) \subseteq Z(\mathfrak{p}_v).$$

As with our result on the image of $\mathcal{U}_v$, more is true. Eventually, we shall prove that the above is in fact an equality.

Example.

2. Suppose we have an abelian extension K/F of number fields that is everywhere unramified (including the infinite places). Let $\mathcal{H} = F^\times N_{K/F}J_K$. By Artin Reciprocity, $\mathcal{H} = \ker \rho_{K/F}$. Since no prime ramifies, we have

$$\left(\frac{\varphi_v(\mathcal{U}_v)}{K/F}\right) = 1$$

and

$$\varphi_v(\mathcal{U}_v) \subseteq \mathcal{H} = F^\times N_{K/F} J_K$$

for every place v. Now $\mathcal{H}$ is open of finite index in J_F, hence $\mathcal{H}$ is also closed (the compliment is open since it is the union of finitely many non-trivial cosets of $\mathcal{H}$). Consider the subgroup of J_F generated by $\{\varphi_v(\mathcal{U}_v) : \text{all } v\}$. It is contained in $\mathcal{H}$, and $\mathcal{H}$ is closed, so its closure is contained in $\mathcal{H}$, i.e., $\prod_v \mathcal{U}_v = \mathcal{E}_F \subseteq \mathcal{H}$. Thus the open subgroup $F^\times \mathcal{E}_F \subseteq \mathcal{H}$.

We shall refer to the above example again; it will help us to begin our study of Hilbert class fields. To do so will require the Existence Theorem, which we prove in the next chapter.

Exercise 5.18. Let K/F be an abelian extension of number fields such that all the principal fractional ideals of F are in the kernel of the Artin map. What can you conclude about the relationship of the class number of F to the degree $[K : F]$? $\Diamond$

Chapter 6
The Existence Theorem, Consequences and Applications

Let K/F be abelian, and let $\mathfrak{m}$ be an ideal of $\mathcal{O}_F$ that is divisible by all the ramifying primes in K/F and is such that $\mathcal{E}_{F,\mathfrak{m}}^+ \subseteq F^\times N_{K/F} J_K$. As we have seen previously, the kernel of the idèlic Artin map $J_F \to \mathrm{Gal}\,(K/F)$ is $F^\times N_{K/F} J_K$, and $F^\times N_{K/F} J_K$ is an open subgroup of J_F containing $F^\times$.

This suggests the following question. Given an open subgroup $\mathcal{H} \subseteq J_F$ with $F^\times \subseteq \mathcal{H}$, when will $\mathcal{H}$ be of the form $F^\times N_{K/F} J_K$ for some finite abelian extension K/F? The answer is provided by the Existence Theorem, which we state here in terms of idèles.

Theorem (Existence). *Every open subgroup $\mathcal{H} \subseteq J_F$ with $\mathcal{H} \supseteq F^\times$ is of the form $\mathcal{H} = F^\times N_{K/F} J_K$ for some (unique) finite abelian extension K/F.*

Once we verify the Existence Theorem, then we shall know that there is a bijective correspondence

$$\{\text{finite abelian extensions } K/F\} \longleftrightarrow \{\text{open subgroups } \mathcal{H} \text{ of } J_F : F^\times \subseteq \mathcal{H}\}.$$

The field K is the class field to $\mathcal{H}$. (This statement combines idèlic versions of the Existence and Completeness Theorems.) We have shown in Chapter 3 that the class fields of Weber (defined in terms of ideals) are unique. Uniqueness of K follows from this, using the relationship between the classical (ideal-based) theory and the idèlic theory that we proved in Chapter 4. But it is also easy to verify directly, which we shall do in the first section.

The correspondence in the Existence/Completeness Theorems is given by

$$\mathcal{H} = F^\times N_{K/F} J_K.$$

Our results from Chapter 5 then imply that we have

$$J_F \big/ \mathcal{H} \cong \mathrm{Gal}\,(K/F)$$

via the Artin map (this is the *Isomorphy Theorem*).

Thus, the Existence Theorem is all that remains in order to complete the proofs of the theorems we discussed in 3.3. One of the main objectives of this chapter is to

N. Childress, *Class Field Theory*, Universitext, DOI 10.1007/978-0-387-72490-4_6, prove the Existence Theorem. The proof we give uses Kummer extensions, so we shall spend a bit of time discussing them. The results on Kummer extensions also will allow us to return to our partially completed study of the Artin map on local fields and prove the Complete Splitting Theorem.

The special case of the Hilbert class field is discussed in the fourth section. In the fifth section, we take a moment to consider extensions of number fields that are perhaps non-abelian. In Section 6, we show how to extend the Artin map to a maximal abelian extension of a number field, and sketch a proof of the Existence Theorem that uses this idea. The remainder of the chapter contains a few applications of class field theory to cyclotomic fields.

1 The Ordering Theorem and the Reduction Lemma

We begin with a consequence of Artin Reciprocity that is related to the Completeness Theorem, and that proves part of the bijective correspondence between finite abelian extensions of a number field F, and open subgroups of J_F that contain $F^\times$.

Proposition 1.1. Let

$$\Phi : \{\text{finite abelian extensions } K \text{ of } F\}$$
$$\longrightarrow \{\text{open subgroups } \mathcal{H} \text{ of } J_F \text{ that contain } F^\times\}$$

be given by $\Phi(K) = F^\times N_{K/F} J_K$. Then:

i. $K \subseteq K'$ if and only if $\Phi(K') \subseteq \Phi(K)$, the ORDERING THEOREM,

ii. $\Phi(KK') = \Phi(K) \cap \Phi(K')$,

iii. $\Phi(K \cap K') = \Phi(K)\Phi(K')$.

iv. If $\mathcal{H} = \Phi(E) = F^\times N_{E/F} J_E$ and $K \supseteq E$, then E is the fixed field of $\rho_{K/F}(\mathcal{H})$.

Proof. Let $\mathcal{H} = \Phi(K)$, $\mathcal{H}' = \Phi(K')$. By Artin Reciprocity, we have $\mathcal{H} = \ker \rho_{K/F}$ and $\mathcal{H}' = \ker \rho_{K'/F}$.

(ii.) Since $\left(\frac{\mathfrak{a}}{KK'/F}\right)\Big|_K = \left(\frac{\mathfrak{a}}{K/F}\right)$ and $\left(\frac{\mathfrak{a}}{KK'/F}\right)\Big|_{K'} = \left(\frac{\mathfrak{a}}{K'/F}\right)$, it follows that $\ker \rho_{KK'/F} \subseteq \mathcal{H} \cap \mathcal{H}'$.

Conversely, if $\mathfrak{a} \in \mathcal{H} \cap \mathcal{H}'$, then there is some element $\sigma = \left(\frac{\mathfrak{a}}{KK'/F}\right)$ in $\text{Gal}(KK'/F)$ that is trivial on both K and K'. Thus $\sigma = 1$ on KK', and we conclude that $\mathfrak{a} \in \ker \rho_{KK'/F}$.

(i.) If $K \subseteq K'$, then $N_{K'/F} J_{K'} = N_{K/F}(N_{K'/K} J_{K'}) \subseteq N_{K/F} J_K$, which gives us $\mathcal{H}' \subseteq \mathcal{H}$.

Conversely, if $\mathcal{H}' \subseteq \mathcal{H}$, then $\mathcal{H}' \cap \mathcal{H} = \mathcal{H}'$, so

$$[K' : F] = \#\left(J_F / \mathcal{H}'\right)$$
$$= \#\left(J_F / \mathcal{H} \cap \mathcal{H}'\right)$$
$$= [KK' : F] \quad \text{by } (ii).$$

But then $K' = KK'$, whence $K \subseteq K'$.

(iv.) We have $K \supseteq E \supseteq F$ and $\mathcal{H} = \ker \rho_{E/F}$ by Artin Reciprocity. If $\mathbf{a} \in \mathcal{H}$, then

$$\left(\frac{\mathbf{a}}{K/F} \right)\Big|_E = \left(\frac{\mathbf{a}}{E/F} \right) = \rho_{E/F}(\mathbf{a}) = 1,$$

so $\rho_{K/F}(\mathcal{H})$ fixes E.

If $x \in K$ is fixed by $\rho_{K/F}(\mathcal{H})$, then $E(x)$ is fixed by $\rho_{K/F}(\mathcal{H})$. Hence, for any $\mathbf{a} \in \mathcal{H}$,

$$\rho_{K/F}(\mathbf{a})\Big|_{E(x)} = \left(\frac{\mathbf{a}}{E(x)/F} \right) = 1,$$

i.e., $\mathbf{a} \in \ker \rho_{E(x)/F}$. We have shown

$$\mathcal{H} = F^\times N_{E/F} J_E \subseteq \ker \rho_{E(x)/F} = F^\times N_{E(x)/F} J_{E(x)}.$$

It follows that $\Phi(E) \subseteq \Phi(E(x))$. Now we may apply (i) to conclude $E(x) \subseteq E$ and hence $x \in E$.

(iii.) Since $K, K' \supseteq K \cap K'$, we have $\mathcal{H}, \mathcal{H}' \subseteq \Phi(K \cap K')$ by (i). But then $\mathcal{H}\mathcal{H}' \subseteq \Phi(K \cap K')$.

Conversely, let E be the fixed field of $\rho_{KK'/F}(\mathcal{H}\mathcal{H}')$. Now $\mathcal{H} \subseteq \mathcal{H}\mathcal{H}'$, so $\rho_{KK'/F}(\mathcal{H}) \subseteq \rho_{KK'/F}(\mathcal{H}\mathcal{H}')$. Hence the fixed field K of $\rho_{KK'/F}(\mathcal{H})$ must contain the fixed field E of $\rho_{KK'/F}(\mathcal{H}\mathcal{H}')$. Similarly, $\mathcal{H}' \subseteq \mathcal{H}\mathcal{H}'$, so $K' \supseteq E$. Thus $K \cap K' \supseteq E$.

On the other hand, if $\mathbf{b} \in \mathcal{H}$ and $\mathbf{b}' \in \mathcal{H}'$, then

$$\left(\frac{\mathbf{bb}'}{KK'/F} \right)\Big|_{K \cap K'} = \left(\frac{\mathbf{b}}{KK'/F} \right)\Big|_{K \cap K'} \left(\frac{\mathbf{b}'}{KK'/F} \right)\Big|_{K \cap K'} = 1.$$

Hence $\rho_{KK'/F}(\mathbf{bb}')$ fixes $K \cap K'$, so that we must have $K \cap K' \subseteq E$. We have $E = K \cap K'$ and $\Phi(E) = F^\times N_{E/F} J_E = \ker \rho_{E/F}$. By (iv), we must have that E is the fixed field of $\rho_{KK'/F}(\Phi(E))$. By the bijectivity of the Galois correspondence, we conclude $\rho_{KK'/F}(\Phi(E)) = \rho_{KK'/F}(\mathcal{H}\mathcal{H}')$.

Now $\mathbf{a} \in \Phi(E)$ implies $\rho_{KK'/F}(\mathbf{a}) = \rho_{KK'/F}(\mathbf{bb}')$, where $\mathbf{b} \in \mathcal{H}$ and $\mathbf{b}' \in \mathcal{H}'$. But then $\mathbf{a}(\mathbf{bb}')^{-1} \in \ker \rho_{KK'/F} = \Phi(KK') = \mathcal{H} \cap \mathcal{H}'$ by (ii), whence $\mathbf{a} \in \mathcal{H}\mathcal{H}'$. □

Note that (i) of the above proposition shows that the map Φ is injective. Once we have shown the Existence Theorem, we shall know that Φ is actually surjective as well, i.e., every open subgroup $\mathcal{H}$ of J_F that contains $F^\times$ is of the form $\mathcal{H} = F^\times N_{K/F} J_K = \ker \rho_{K/F}$ for some finite abelian extension K of F, (this field K is the class field to $\mathcal{H}$ over F). The proof of the bijective correspondence will then be complete.

Corollary 1.2. Suppose K is the class field to the open subgroup $\mathcal{H}$ of J_F, where $\mathcal{H}$ contains $F^\times$, and let $\mathcal{H}' \supseteq \mathcal{H}$ be an open subgroup of J_F. Then $\mathcal{H}'$ has a class field over F.

Proof. Let K' be the fixed field of $\rho_{K/F}(\mathcal{H}')$. We have $F \subseteq K' \subseteq K$. Also,
$\left(\frac{\mathbf{a}}{K'/F}\right) = \left(\frac{\mathbf{a}}{K/F}\right)\Big|_{K'}$, so

$$\mathbf{a} \in \ker \rho_{K'/F} \iff \left(\frac{\mathbf{a}}{K/F}\right)\Big|_{K'} = 1$$

$$\iff \left(\frac{\mathbf{a}}{K/F}\right) \in \rho_{K/F}(\mathcal{H}')$$

$$\iff \text{there is some } \mathbf{b} \in \mathcal{H}' \text{ with } \left(\frac{\mathbf{ab}^{-1}}{K/F}\right) = 1$$

$$\iff \text{there is some } \mathbf{b} \in \mathcal{H}' \text{ with } \mathbf{ab}^{-1} \in \ker \rho_{K/F} = \mathcal{H}$$

$$\iff \mathbf{a} \in \mathcal{H}\mathcal{H}' = \mathcal{H}'. \qquad \square$$

Proposition 1.3 (Reduction Lemma). Let K/F be a cyclic extension of number fields and suppose $\mathcal{H}$ is an open subgroup of J_F that contains $F^\times$. Let $\mathcal{H}_K = \{\mathbf{x} \in J_K : N_{K/F}(\mathbf{x}) \in \mathcal{H}\} = N_{K/F}^{-1}(\mathcal{H})$. If $\mathcal{H}_K$ has a class field over K, then $\mathcal{H}$ has a class field over F.

Proof. Let E be the class field of $\mathcal{H}_K$ over K. Then $\mathrm{Gal}\,(E/K)$ is abelian and $\mathcal{H}_K = K^\times N_{E/K} J_E = \ker \rho_{E/K}$.

We first show that E/F is Galois. Let E' be the Galois closure of E/F, (i.e., the smallest Galois extension of F that contains E). Let $\sigma \in \mathrm{Gal}\,(E'/F)$. Note that for $\mathbf{x} \in J_K$, $N_{K/F}(\sigma(\mathbf{x})) = N_{K/F}(\mathbf{x})$. Also, $\mathbf{x} \in \mathcal{H}_K$ if and only if $N_{K/F}(\mathbf{x}) \in \mathcal{H}$, so if and only if $N_{K/F}(\sigma(\mathbf{x})) \in \mathcal{H}$. Thus $\sigma \mathcal{H}_K = \mathcal{H}_K$. This implies that $\sigma E = E$ (since σE is the class field for $\sigma \mathcal{H}_K$). Thus $E' = E$ and E/F is Galois.

Next we show that E/F is abelian. Let $\sigma \in \mathrm{Gal}\,(E/F)$ with $\mathrm{Gal}\,(K/F) = \langle \sigma|_K \rangle$, (possible since K/F is cyclic). Let $\tau \in \mathrm{Gal}\,(E/K)$. It suffices to show that σ and τ commute.

Since E/K is abelian, it makes sense to talk about the Artin map $\rho_{E/K}$. By Artin Reciprocity, $\rho_{E/K}$ is surjective, so there is some $\mathbf{b} \in J_K$ with $\tau = \rho_{E/K}(\mathbf{b})$. Then

$$\sigma \tau \sigma^{-1} = \left(\frac{\sigma(\mathbf{b})}{\sigma E/\sigma K}\right) = \left(\frac{\sigma(\mathbf{b})}{E/K}\right).$$

Now $N_{K/F}(\sigma(\mathbf{b})/\mathbf{b}) = 1 \in \mathcal{H}$, so $\sigma(\mathbf{b})/\mathbf{b} \in \mathcal{H}_K = \ker \rho_{E/K}$. Thus

$$\left(\frac{\sigma(\mathbf{b})}{E/K}\right) = \left(\frac{\mathbf{b}}{E/K}\right).$$

We get $\sigma \tau \sigma^{-1} = \tau$, whence E/F is abelian as claimed.

Since E is the class field over K to $\mathcal{H}_K$, we have $\mathcal{H}_K = K^\times N_{E/K} J_E$. Also, by definition $N_{K/F} \mathcal{H}_K \subseteq \mathcal{H}$. Now that we know that E/F is abelian, we also find that E is the class field over F to the subgroup $F^\times N_{E/F} J_E$ of J_F. Combining these observations, we obtain

2 Kummer n-extensions and the Proof of the Existence Theorem

$$F^\times N_{E/F} J_E \subseteq F^\times N_{K/F}(N_{E/K} J_E) \subseteq F^\times N_{K/F} \mathcal{H}_K \subseteq F^\times \mathcal{H} = \mathcal{H}.$$

By Corollary 1.2, we conclude that $\mathcal{H}$ has a class field, namely the fixed field of $\rho_{E/F}(\mathcal{H})$. $\square$

2 Kummer n-extensions and the Proof of the Existence Theorem

Let n be a positive integer. An abelian group G is said to have *exponent n* if $g^n = 1$ for every $g \in G$. Similarly, an abelian extension K/F is said to have *exponent n* if the abelian group $\mathrm{Gal}\,(K/F)$ has exponent n.

Let $\mathcal{H}$ be an open subgroup of J_F with $F^\times \subseteq \mathcal{H}$, and suppose that $J_F/\mathcal{H}$ has exponent n. (It certainly has some exponent, since it is a finite group.) We want to show that $\mathcal{H}$ has a class field over F.

Consider the extension $F(\zeta_n)/F$, where as usual ζ_n is a primitive n^th root of unity. We can find a tower of intermediate fields:

$$F = F_0 \subseteq F_1 \subseteq \cdots \subseteq F_t = F(\zeta_n)$$

such that each $\mathrm{Gal}\,(F_{i+1}/F_i)$ is a cyclic group. For each i, let $\mathcal{H}_i = N_{F_i/F}^{-1} \mathcal{H}$. Note that we have $\mathbf{a} \in \mathcal{H}_i$ if and only if $N_{F_{i-1}/F}(N_{F_i/F_{i-1}}(\mathbf{a})) \in \mathcal{H}$, so if and only if $N_{F_i/F_{i-1}}(\mathbf{a}) \in N_{F_{i-1}/F}^{-1} \mathcal{H} = \mathcal{H}_{i-1}$. Thus $\mathcal{H}_i = N_{F_i/F_{i-1}}^{-1} \mathcal{H}_{i-1}$. Our strategy will be first to prove that $\mathcal{H}_t$ has a class field and then to apply the Reduction Lemma to the cyclic extension F_t/F_{t-1} to conclude that $\mathcal{H}_{t-1}$ has a class field. Continuing in (a finite number of) steps, we get eventually that $\mathcal{H} = \mathcal{H}_0$ has a class field.

The above shows that when $J_F/\mathcal{H}$ has exponent n, we are reduced to considering the case where F contains μ_n, the set of all n^th roots of unity. This situation has been the subject of sufficiently much study to have acquired its own set of terminology. A finite abelian extension K/F is called a *Kummer n-extension* if $\mathrm{Gal}\,(K/F)$ is a group with exponent n and F contains all the nth roots of unity. We need some facts from the theory of such extensions ("Kummer theory"). First we show that the Kummer n-extensions correspond to the finite subgroups of $F^\times/(F^\times)^n$.

Theorem 2.1. *Let F be a number field containing all the nth roots of unity. There is a bijective correspondence between the finite Kummer n-extensions K of F and the subgroups W of $F^\times$ with $(F^\times)^n \subseteq W$ and $W/(F^\times)^n$ finite. The correspondence associates W to the field $K = F(W^{1/n})$, for which we have $\mathrm{Gal}\,(K/F) \cong W/(F^\times)^n$.*

Proof. Let K be a Kummer n-extension of F with Galois group G. Let

$$D = \{\alpha \in K^\times : \alpha^n \in F^\times\}.$$

Then $\alpha \in D$ is a root of a polynomial $X^n - a$ for some $a \in F^\times$. Thus for $\sigma \in G$, we have $\sigma(\alpha) = \zeta\alpha$ where ζ is some n^{th} root of unity in F.

Given $\alpha \in D$, let $\psi_\alpha : G \longrightarrow F^\times$ be given by $\psi_\alpha(\sigma) = \frac{\sigma(\alpha)}{\alpha} \in \mu_n \subseteq F^\times$. Note that this makes ψ_α a character of G. Moreover, the map $\psi : \alpha \mapsto \psi_\alpha$ is a homomorphism $D \longrightarrow \widehat{G}$ with $\ker \psi = F^\times$. Thus ψ gives rise to an embedding

$$D\big/_{F^\times} \hookrightarrow \widehat{G}.$$

We claim that this is an isomorphism.

Suppose $\psi(D)$ is a proper subgroup of $\widehat{G}$. Let L be the field associated to $\psi(D)$. Then L is the fixed field of

$$\bigcap_{\chi \in \psi(D)} \ker \chi = \bigcap_{\alpha \in D} \ker \psi_\alpha = \{\sigma \in G : \sigma(\alpha) = \alpha \text{ for all } \alpha \in D\}$$

and $L \subsetneq K$. Note also that $D \subseteq L$. Since $L \neq K$, there is some non-trivial $\sigma \in G$ with $\sigma\big|_L = 1$. But then $\sigma \in \bigcap_{\chi \in \psi(D)} \ker \chi$ and $\psi_\alpha(\sigma) = 1$ for every $\alpha \in D$. It follows that there is some $\sigma \in G$ with $\sigma \neq 1$ and $\psi_\alpha(\sigma) = 1$ for all $\alpha \in D$.

Since G is abelian, we may write $G = \langle \tau \rangle \times G_0$, where we may choose τ and G_0 so that $\sigma \notin G_0$ and $\sigma = \tau^\ell \gamma$, with $\tau^\ell \neq 1$, $\gamma \in G_0$. Let E be the fixed field of G_0; then $\text{Gal}(E/F) \cong \langle \tau \rangle$ is cyclic, say of order t. (Note that $t \mid n$ and $t \nmid \ell$.) Let ξ be a primitive t^{th} root of unity in F. Then $N_{E/F}(\xi) = \xi^t = 1$. Hilbert's Theorem 90 implies that there is some $\alpha \in E^\times$ with $\xi = \frac{\tau(\alpha)}{\alpha}$. Thus $\alpha^t \in F$ and $\alpha \in D$. By our choice of σ, we have $\psi_\alpha(\sigma) = 1$. But $\psi_\alpha(\tau) = \frac{\tau(\alpha)}{\alpha} = \xi$ so that ψ_α is one-to-one on $\langle \tau \rangle$. Also, $\psi_\alpha(G_0) = 1$ since $\alpha \in E$. Thus

$$1 \neq \psi_\alpha(\tau^\ell) = \psi_\alpha(\tau^\ell \gamma) = \psi_\alpha(\sigma) = 1,$$

a contradiction. Thus we must have $\psi(D) = \widehat{G}$, whence

$$D\big/_{F^\times} \cong \widehat{G} \cong G.$$

Since no non-trivial element of G fixes all of D, it follows that $K = F(D)$.

The nth power map gives rise to an epimorphism

$$D\big/_{F^\times} \longrightarrow D^n\big/_{(F^\times)^n}.$$

If two elements of D have equal n^{th} powers, then their quotient is an nth root of unity, so is in $F^\times$. Thus we have an isomorphism

$$D\big/_{F^\times} \cong D^n\big/_{(F^\times)^n}.$$

This shows that all Kummer n-extensions arise as claimed, (take $W = D^n$).

Now let W be any subgroup of $F^\times$ with $(F^\times)^n \subseteq W$ and $W/(F^\times)^n$ finite. Let $\alpha_1, \ldots, \alpha_r$ be representatives for a set of cosets that independently generate $W/(F^\times)^n$. Put

$$K = F(W^{1/n}) = F(\alpha_1^{1/n}, \ldots, \alpha_r^{1/n}).$$

Then K/F is Galois of finite degree. Let $\sigma \in \mathrm{Gal}\,(K/F)$. Then $\sigma(\alpha_i^{1/n}) = \zeta \alpha_i^{1/n}$, where ζ is some n^{th} root of unity in F. We find $\sigma^n = 1$. Since $\mathrm{Gal}\,(K/F)$ is clearly abelian, we have shown that K/F is a finite Kummer n-extension.

It remains only to show that W is uniquely determined by $K = F(W^{1/n})$. Let $D = \{\alpha \in K^\times : \alpha^n \in F^\times\}$. Then $D^n \supseteq W$ and $D^n/(F^\times)^n$ has order $[D : F^\times] = [K : F]$ as before. We need only show $[K : F] \leq [W : (F^\times)^n]$ to get $D^n = W$.

Let $\alpha_i(F^\times)^n$ have order d_i in the group $W/(F^\times)^n$, so that $[W : (F^\times)^n] = d_1 \cdots d_r$. Now $\alpha_i^{d_i} \in (F^\times)^n$, whence

$$[F(\sqrt[n]{\alpha_i}) : F] \leq d_i$$
$$[K : F] = [F(\sqrt[n]{\alpha_1}, \ldots, \sqrt[n]{\alpha_r}) : F] \leq d_1 \cdots d_r. \qquad \square$$

Now that we have some information about Kummer extensions, we need to define certain subgroups of the idèles. They too will play an important role in the proof of the Existence Theorem.

Let F be a number field. Let $\mathcal{S}$ be a finite set of places of F and assume $\mathcal{S} \supseteq \mathcal{S}_\infty = \{\text{infinite places of } F\}$. Define

$$J_{F,\mathcal{S}} = \prod_{v \in \mathcal{S}} F_v^\times \times \prod_{v \notin \mathcal{S}} \mathcal{U}_v \quad \text{an open subgroup of } J_F,$$

$$F_\mathcal{S} = J_{F,\mathcal{S}} \cap F^\times \quad \text{the } \mathcal{S}\text{-}units \text{ of } F, \text{ a discrete subgroup of } J_{F,\mathcal{S}}.$$

Note that $F_\mathcal{S}$ also may be defined without using idèles:

$$F_\mathcal{S} = \{\alpha \in F^\times : \text{the factorization of } \langle \alpha \rangle \text{ involves no prime } \mathfrak{p}_v \text{ with } v \notin \mathcal{S}\}.$$

Exercise 6.1. What are $J_{F,\mathcal{S}_\infty}$ and $F_{\mathcal{S}_\infty}$? $\qquad \Diamond$

Lemma 2.2. There is a finite set of places $\mathcal{S} \supseteq \mathcal{S}_\infty$ such that $J_F = F^\times J_{F,\mathcal{S}}$.

Proof. Since the ideal class group of F is finite, we can choose a set of representatives $\mathfrak{a}_1, \ldots, \mathfrak{a}_h$ for the classes in $\mathcal{C}_F$. There are finitely many prime ideals in $\mathcal{O}_F$ that appear in the factorizations of the $\mathfrak{a}_j$, say $\mathfrak{p}_1, \ldots, \mathfrak{p}_t$ are these primes. Choose $\mathcal{S}$ to be the set of places of F corresponding to $\mathfrak{p}_1, \ldots, \mathfrak{p}_t$ together with the infinite places of F. We claim that this $\mathcal{S}$ satisfies $J_F = F^\times J_{F,\mathcal{S}}$.

That $J_F \supseteq F^\times J_{F,\mathcal{S}}$ is clear. For "$\subseteq$" let $\mathbf{a} = (\ldots, a_v, \ldots) \in J_F$. As in Proposition 3.2 of Chapter 4, we associate a fractional ideal to $\mathbf{a}$ by putting

$$\eta(\mathbf{a}) = \prod_{v \text{ finite}} \mathfrak{p}_v^{\text{ord}_v a_v}.$$

Since the $\mathfrak{a}_j$ comprise a complete set of representatives for $\mathcal{C}_F$, there is $\mathfrak{a}_i$ with $\eta(\mathbf{a}) \in \mathfrak{a}_i \mathcal{P}_F$, say $\eta(\mathbf{a}) = \langle \alpha \rangle \mathfrak{a}_i$, with $\alpha \in F^\times$. Consider the idèle $\alpha^{-1}\mathbf{a}$. We have $\eta(\alpha^{-1}\mathbf{a}) = \mathfrak{a}_i$, so

$$\mathfrak{a}_i = \prod_{v \text{ finite}} \mathfrak{p}_v^{\text{ord}_v(\alpha^{-1}a_v)}.$$

All the primes in the factorization of $\mathfrak{a}_i$ come from $\mathcal{S}$, so $\text{ord}_v(\alpha^{-1}a_v) = 0$ whenever $v \notin \mathcal{S}$. It follows that $\alpha^{-1}\mathbf{a} \in J_{F,\mathcal{S}}$, whence $\mathbf{a} \in F^\times J_{F,\mathcal{S}}$ as desired.   □

Exercise 6.2. The *group of $\mathcal{S}$-idèle classes* of F is

$$C_{F,\mathcal{S}} = {}^{J_{F,\mathcal{S}}}\big/_{F_{\mathcal{S}}}.$$

We have $J_{F,\mathcal{S}} \subseteq J_F$ and $F_{\mathcal{S}} \subseteq F^\times$. so there is a natural monomorphism $C_{F,\mathcal{S}} \hookrightarrow C_F$. Show that there is a topological and algebraic isomorphism

$$ {}^{J_F}\big/_{F^\times J_{F,\mathcal{S}}} \cong {}^{C_F}\big/_{C_{F,\mathcal{S}}}. \qquad \Diamond$$

We are now ready to give a proof of a result on $\mathcal{S}$-units that has its origins in the work of Dirichlet, Chevalley and Hasse.

Theorem 2.3. *Let $\mathcal{S}$ be a finite set of places of F, with $\mathcal{S}_\infty \subseteq \mathcal{S}$. Then $F_{\mathcal{S}}$ is the direct product of the (finite cyclic) group of roots of unity in F, and a free abelian group of rank $\#\mathcal{S} - 1$. That is,*

$$F_{\mathcal{S}} \cong \mathcal{W}_F \times \mathbb{Z}^{\#\mathcal{S}-1}.$$

Proof. Write $\mathcal{S}_0 = \mathcal{S} - \mathcal{S}_\infty$ and let $\mathcal{I}_{\mathcal{S}_0}$ be the group of fractional ideals of F generated by $\{\mathfrak{p}_v : v \in \mathcal{S}_0\}$. We have an exact sequence

$$1 \longrightarrow \mathcal{U}_F \longrightarrow F_{\mathcal{S}} \xrightarrow{\gamma} \mathcal{I}_{\mathcal{S}_0}$$

where $\mathcal{U}_F = \mathcal{O}_F^\times$ and $\gamma : \alpha \mapsto \langle \alpha \rangle$. For each $v \in \mathcal{S}_0$, note that $\mathfrak{p}_v^h$ must be a principal ideal in $\mathcal{I}_{\mathcal{S}_0}$, $(h = \#\mathcal{C}_F)$, so

$$\mathcal{I}_{\mathcal{S}_0}^h \subseteq \gamma(F_{\mathcal{S}}) \subseteq \mathcal{I}_{\mathcal{S}_0}.$$

Since both $\mathcal{I}_{\mathcal{S}_0}$ and $\mathcal{I}_{\mathcal{S}_0}^h$ are free abelian groups of rank $\#\mathcal{S}_0$, we must have that $\gamma(F_{\mathcal{S}})$ is too.

Exercise 6.3. Show that $F_{\mathcal{S}} \cong \mathcal{O}_F^\times \times \gamma(F_{\mathcal{S}})$. $\Diamond$

The theorem now follows from Dirichlet's Unit Theorem.   □

Corollary 2.4. If F contains the nth roots of unity and $\mathcal{S} \supseteq \mathcal{S}_\infty$ is a finite set of places of F, then

$$[F_{\mathcal{S}} : F_{\mathcal{S}}^n] = n^{\#\mathcal{S}}.$$

Proof. $F_{\mathcal{S}} = \langle \zeta \rangle \times A$, where ζ is a root of unity of order divisible by n and A is a free abelian group of rank $\#\mathcal{S} - 1$. Now $\langle \zeta \rangle / \langle \zeta^n \rangle$ has order n and A / A^n has order $n^{\#\mathcal{S}-1}$. $\qquad\qquad\square$

Let us return to the proof of the Existence Theorem. Recall we have reduced the problem to the case where F contains the nth roots of unity (for n equal to the exponent of $J_F / \mathcal{H}$). First we prove a lemma about n^{th} powers in completions of F.

Lemma 2.5. Let v be a finite place of F, and let $n \in \mathbb{Z}_+$. If μ_n denotes the set of all n^{th} roots of unity, then

i. $[\mathcal{U}_v : \mathcal{U}_v^n] = \frac{1}{\|n\|_v} \#(F_v \cap \mu_n)$.
ii. $[F_v^\times : (F_v^\times)^n] = \frac{n}{\|n\|_v} \#(F_v \cap \mu_n)$.

Proof. (ii) follows from (i) since $F_v^\times \cong \mathcal{U}_v \times \mathbb{Z}$ (via the map $x \mapsto (\varepsilon, m)$, where for a fixed uniformizer π, we have $x = \varepsilon \pi^m$).

It remains to prove (i). The proof we give is Artin's.

Let π be a uniformizer for F_v and choose an integer t sufficiently large so that

$$|n\pi^{t+1}| \geq |\pi^{2t}|.$$

We have

$$(1 + x\pi^t)^n \equiv 1 + nx\pi^t \quad (\text{mod } n\pi^{t+1}), \quad \text{for any } x \in \mathcal{O}_v.$$

If $\text{ord}_v n = r$, then

$$(1 + \mathfrak{p}_v^t)^n = 1 + \mathfrak{p}_v^{t+r}.$$

Take t sufficiently large so that no non-trivial n^{th} root of unity lies in $1 + \mathfrak{p}_v^t$. Consider the homomorphism $f : \mathcal{U}_v \longrightarrow \mathcal{U}_v$ given by $f(x) = x^n$. We have

$$[\mathcal{U}_v : 1 + \mathfrak{p}_v^t] = [f(\mathcal{U}_v) : f(1 + \mathfrak{p}_v^t)][\ker f : \ker f\big|_{1+\mathfrak{p}_v^t}]$$

$$= [\mathcal{U}_v^n : (1 + \mathfrak{p}_v^t)^n][F_v \cap \mu_n : 1]$$

$$= [\mathcal{U}_v^n : 1 + \mathfrak{p}_v^{t+r}]\#(F_v \cap \mu_n)$$

$$= \frac{[\mathcal{U}_v : 1 + \mathfrak{p}_v^{t+r}]}{[\mathcal{U}_v : \mathcal{U}_v^n]}\#(F_v \cap \mu_n).$$

Thus

$$[\mathcal{U}_v : \mathcal{U}_v^n] = \frac{[\mathcal{U}_v : 1 + \mathfrak{p}_v^{t+r}]}{[\mathcal{U}_v : 1 + \mathfrak{p}_v^t]} \#(F_v \cap \mu_n)$$

$$= [1 + \mathfrak{p}_v^t : 1 + \mathfrak{p}_v^{t+r}]\#(F_v \cap \mu_n)$$

$$= N\mathfrak{p}_v^r \, \#(F_v \cap \mu_n)$$

$$= \frac{1}{\|n\|_v}\#(F_v \cap \mu_n).$$

$\square$

Note that if F_v contains the nth roots of unity, then $[\mathcal{U}_v : \mathcal{U}_v^n] = \frac{n}{\|n\|_v}$ and $[F_v^\times : (F_v^\times)^n] = \frac{n^2}{\|n\|_v}$.

Theorem 2.6. *Let F be a number field that contains all the nth roots of unity. Let $\mathcal{S}$ be a finite set of places of F containing $\mathcal{S}_\infty$, the places v such that $\mathfrak{p}_v|n$ and sufficiently many finite places so that $J_F = F^\times J_{F,\mathcal{S}}$. Let*

$$B = \prod_{v \in \mathcal{S}}(F_v^\times)^n \times \prod_{v \notin \mathcal{S}}\mathcal{U}_v.$$

Then $F^\times B$ has class field $F(F_\mathcal{S}^{1/n})$ over F.

Proof. Clearly $F_\mathcal{S} \cap (F^\times)^n = F_\mathcal{S}^n$. Also

$$F_\mathcal{S}(F^\times)^n / {}_{(F^\times)^n} \cong F_\mathcal{S} / {}_{F_\mathcal{S} \cap (F^\times)^n} = F_\mathcal{S} / {}_{F_\mathcal{S}^n}$$

is a finite group of order $n^{\#\mathcal{S}}$ by Corollary 2.4. Thus $K = F(F_\mathcal{S}^{1/n})$ is the Kummer n-extension corresponding to $F_\mathcal{S}(F^\times)^n$, and the group of order $n^{\#\mathcal{S}}$ above is isomorphic to $\text{Gal}(K/F)$. Note that K is obtained by adjoining to F the roots of equations $f(X) = X^n - \alpha = 0$ with $\alpha \in F_\mathcal{S}$. If β is such a root, then $f'(\beta) = n\beta^{n-1}$ is divisible only by primes associated to places in $\mathcal{S}$. Hence if $v \notin \mathcal{S}$ then $\mathfrak{p}_v$ cannot ramify in K/F. We must show that $F^\times B = F^\times N_{K/F}J_K$.

For "$\subseteq$", note that since $\text{Gal}(K/F)$ has exponent n, any element $x \in (F_v^\times)^n$ for $v \in \mathcal{S}$ satisfies $\varphi_v(x) \in \ker \rho_{K/F} = F^\times N_{K/F}J_K$. Meanwhile, any $x \in \mathcal{U}_v$ for $v \notin \mathcal{S}$ is a local norm since the norm is surjective on units in unramified local extensions. Thus $\prod_{v \notin \mathcal{S}}\mathcal{U}_v \subseteq N_{K/F}J_K$.

For "$\supseteq$", note that since $[J_F : F^\times N_{K/F}J_K] = \#\text{Gal}(K/F) = [K : F] = n^{\#\mathcal{S}}$ by Artin Reciprocity and Corollary 2.4 above, and since we now have $F^\times B \subseteq F^\times N_{K/F}J_K$, it suffices to show that $[J_F : F^\times B] = n^{\#\mathcal{S}}$. But

$$[J_F : F^\times B] = [F^\times J_{F,\mathcal{S}} : F^\times B]$$

$$= \frac{[J_{F,\mathcal{S}} : B]}{[J_{F,\mathcal{S}} \cap F^\times : B \cap F^\times]}$$

$$= \frac{\prod_{v \in \mathcal{S}} [F_v^\times : (F_v^\times)^n]}{[F_\mathcal{S} : B \cap F^\times]}$$

$$= \frac{\prod_{v \in \mathcal{S}} \frac{n^2}{\|n\|_v}}{[F_\mathcal{S} : B \cap F^\times]} \qquad \text{by Lemma 2.5}$$

$$= \frac{n^{2\#\mathcal{S}}}{[F_\mathcal{S} : B \cap F^\times]} \qquad \text{by the Product Formula.}$$

If we can show that $B \cap F^\times = F_\mathcal{S}^n$, we'll be done. Clearly $F_\mathcal{S}^n \subseteq B \cap F^\times$. Conversely, let $x \in B \cap F^\times$. Then x is a local n^{th} power at all $v \in \mathcal{S}$. Thus $[F_v(x^{1/n}) : F_v] = 1$ for all $v \in \mathcal{S}$, whence $\mathfrak{p}_v$ splits completely in $F(x^{1/n})/F$ for all $v \in \mathcal{S}$. Also, if $v \notin \mathcal{S}$, then $\mathfrak{p}_v$ is unramified in $F(x^{1/n})$. We have shown $J_{F,\mathcal{S}} \subseteq N_{F(x^{1/n})/F} J_{F(x^{1/n})}$ from which we conclude

$$J_F = F^\times J_{F,\mathcal{S}} \subseteq F^\times N_{F(x^{1/n})/F} J_{F(x^{1/n})}.$$

This says that the kernel of the Artin map for $F(x^{1/n})/F$ is all of J_F, whence $\mathrm{Gal}\,(F(x^{1/n})/F)$ is trivial, i.e., $F(x^{1/n}) = F$, and $x^{1/n} \in F$. Now $x \in B$ implies $x \in F_\mathcal{S}^n$. $\qquad \square$

Let F be a number field. Given an open subgroup $\mathcal{H}$ of J_F containing $F^\times$, let $J_F / \mathcal{H}$ have exponent n. As we have observed before, we may assume that F contains the nth roots of unity. Find a set $\mathcal{S}$ of places as in the above theorem, then enlarge $\mathcal{S}$ further to contain all v such that $\mathcal{U}_v \not\subseteq \mathcal{H}$. For this enlarged $\mathcal{S}$, we get $B \subseteq \mathcal{H}$. Since $F^\times B$ has a class field (by the theorem) and $\mathcal{H} = F^\times \mathcal{H} \supseteq F^\times B$, Corollary 1.2 applies and we conclude that $\mathcal{H}$ has a class field too. This completes the proof of the following.

Theorem 2.7 (Existence). *Let F be a number field. Let $\mathcal{H}$ be an open subgroup of J_F with $F^\times \subseteq \mathcal{H}$. Then $\mathcal{H}$ has a class field over F, i.e., there is a finite abelian extension K of F such that $\mathcal{H} = F^\times N_{K/F} J_K$.* $\qquad \square$

With this, we have finished the proofs of all the theorems mentioned at the end of Chapter 3. Also, we have completed the proof of the generalization of Dirichlet's Theorem on Primes in Arithmetic Progressions, as the existence of a class field for $\mathcal{P}_\mathfrak{m}^+$ was all that was needed to show that the Weber L-function satisfies $L_\mathfrak{m}(1, \chi) \neq 0$ for every non-trivial character χ of $\mathcal{R}_\mathfrak{m}^+$. Given Exercise 5.10, the proof of the Chebotarev Density Theorem is also complete.

Using the Existence Theorem, we may also obtain a bit more information about the primes that split completely in abelian extensions K/F. Recall the corollary to Artin Reciprocity:

$$\mathcal{S}_{K/F} = \{\mathfrak{p} \text{ of } \mathcal{O}_F : \mathfrak{p} \text{ splits completely in } K/F\}$$
$$= \{\mathfrak{p} \text{ of } \mathcal{O}_F : \mathfrak{p} \in \mathcal{P}_\mathfrak{m}^+ N_{K/F}(\mathfrak{m})\} = \mathcal{T}_{K/F}.$$

We want to look at $\mathcal{S}_{K/F}$ as it relates to the open subgroups of the idèles. We shall use the embeddings $\varphi_v : F_v^\times \hookrightarrow J_F$ defined in Chapter 5.

Theorem 2.8. *Let $\mathcal{H}$ be an open subgroup of J_F that contains $F^\times$ and let K be the class field of $\mathcal{H}$ over F. Let v be a place of F. If $\mathfrak{p}_v$ splits completely in K/F, then $\varphi_v(F_v^\times) \subseteq \mathcal{H}$.*

Proof. If $\mathfrak{p}_v$ splits completely in K/F, then the local norm at v is surjective on units (because $e = 1$), and there is a uniformizer π for F_v, with π the norm of a uniformizer for K_w, where w is a place of K above v (because $f = 1$). Since every element of $F_v^\times$ has the form $\varepsilon\pi^t$ for some $\varepsilon \in \mathcal{U}_v$ and some $t \in \mathbb{Z}$, we conclude that every element of $F_v^\times$ is a norm. Thus $\varphi_c(F_v^\times) \subseteq F^\times N_{K/F} J_K = \mathcal{H}$. □

The converse of the above proposition is also true (it is called the *Complete Splitting Theorem*), but its proof requires more work; we postpone it until the next section. For now, we shall show a partial converse (for Kummer extensions). Its proof is based on a technique of Herbrand.

Theorem 2.9. *Let $\mathcal{H}$ be an open subgroup of J_F that contains $F^\times$ and let K be the class field over F of $\mathcal{H}$. Suppose $J_F/\mathcal{H}$ has exponent n and that F contains the nth roots of unity. Let v_0 be a place of F with $\varphi_{v_0}(F_{v_0}^\times) \subseteq \mathcal{H}$. Then $\mathfrak{p}_{v_0}$ splits completely in K/F.*

Proof. Let $\mathcal{S}$ be a finite set of places of F chosen so that all of the following are true:

$$\mathcal{S}_\infty \subseteq \mathcal{S},$$
$$v_0 \in \mathcal{S},$$
$$\{v : \mathfrak{p}_v \text{ ramifies in } K/F\} \subseteq \mathcal{S},$$
$$\{v : \mathfrak{p}_v | n\} \subseteq \mathcal{S}, \text{ and}$$
$$J_F = F^\times J_{F,\mathcal{S}}.$$

Let

$$B_1 = F_{v_0}^\times \times \prod_{\substack{v \in \mathcal{S} \\ v \neq v_0}} (F_v^\times)^n \times \prod_{v \notin \mathcal{S}} \mathcal{U}_v$$

$$B_2 = (F_{v_0}^\times)^n \times \prod_{\substack{v \in \mathcal{S} \\ v \neq v_0}} F_v^\times \times \prod_{v \notin \mathcal{S}} \mathcal{U}_v.$$

Both are open subgroups, and we have

$$B_1 \cap B_2 = \prod_{v \in \mathcal{S}} (F_v^\times)^n \times \prod_{v \notin \mathcal{S}} \mathcal{U}_v = B.$$

We claim that $B_1 \subseteq \mathcal{H}$. To prove the claim, note that $\varphi_{v_0}(F_{v_0}^\times) \subseteq \mathcal{H}$ by hypothesis, and since $J_F/\mathcal{H}$ has exponent n, $\varphi_v((F_v^\times)^n) \subseteq \mathcal{H}$. Finally, if $v \notin \mathcal{S}$, then $\mathfrak{p}_v$ is unramified in K/F and the local norm is surjective on $\mathcal{U}_v$, whence $\varphi_v(\mathcal{U}_v) \subseteq \mathcal{H} = F^\times N_{K/F} J_K$.

Since $B_1 \subseteq \mathcal{H}$, $F^\times B_1 \subseteq F^\times \mathcal{H} = \mathcal{H}$. Now $F^\times B_1$ has a class field over F, say K_1. Since $F^\times B_1 \subseteq \mathcal{H}$, we have $K \subseteq K_1$. We'll show that $\mathfrak{p}_{v_0}$ splits completely in K_1/F, whence also in K/F.

Let $W_1 = F^\times \cap B_1$, $W_2 = F^\times \cap B_2$. Then $F_{\mathcal{S}}^n \subseteq W_i \cap (F^\times)^n$ (why?). Conversely, $F_{\mathcal{S}}^n \supseteq W_i \cap (F^\times)^n$ is clear, so that $F_{\mathcal{S}}^n = W_i \cap (F^\times)^n$. Let

$$L_1 = F(W_2^{1/n}), \qquad L_2 = F(W_1^{1/n}).$$

We claim $L_1 = K_1$. Now $L_1 = F((W_2(F^\times)^n)^{1/n})$, and similarly for L_2, so we may apply Kummer theory to conclude

$$[L_1 : F] = [W_2(F^\times)^n : (F^\times)^n]$$
$$= [W_2 : W_2 \cap (F^\times)^n]$$
$$= [W_2 : F_{\mathcal{S}}^n].$$

Similarly, $[L_2 : F] = [W_1 : F_{\mathcal{S}}^n]$.

Let $\mathcal{H}_i = F^\times N_{L_i/F} J_{L_i}$, so that L_i is the class field of $\mathcal{H}_i$ over F. Note that for $v \notin \mathcal{S}$, $\mathfrak{p}_v$ cannot ramify in L_i/F because L_i is obtained by adjoining roots of equations $f(X) = X^n - \alpha = 0$ with $\alpha \in F^\times \cap B_j$. If β is such a root, then $f'(\beta) = n\beta^{n-1}$ is divisible only by primes associated to places in $\mathcal{S}$. Thus the extensions L_i/F are unramified outside $\mathcal{S}$.

If $x \in W_2 = B_2 \cap F^\times$, then x is a local n^{th} power at v_0, so that $x^{1/n} \in F_{v_0}$. It follows that $[F_{v_0}(x^{1/n}) : F_{v_0}] = 1 = ef$ and $\mathfrak{p}_{v_0}$ splits completely in $F(x^{1/n})/F$, hence also in L_1/F. Similarly, if $v \in \mathcal{S}$ and $v \neq v_0$, then $\mathfrak{p}_v$ splits completely in L_2/F.

Now $\varphi_{v_0}(F_{v_0}^\times) \subseteq \mathcal{H}_1$ by Theorem 2.8, and $\varphi_v((F_v^\times)^n) \subseteq \mathcal{H}_1$ for $v \in \mathcal{S}$, $v \neq v_0$, since L_1/F is a Kummer n-extension. Also, $\varphi_v(\mathcal{U}_v) \subseteq \mathcal{H}_1$ for $v \notin \mathcal{S}$ since such v are unramified in L_1/F so all their local units are norms. Thus $B_1 \subseteq \mathcal{H}_1$. Similarly $B_2 \subseteq \mathcal{H}_2$.

We get

$$[L_1 : F] = [J_F : \mathcal{H}_1] \quad \text{by Artin Reciprocity}$$
$$\leq [J_F : F^\times B_1] \quad \text{since } F^\times B_1 \subseteq \mathcal{H}_1$$
$$= [F^\times J_{F,\mathcal{S}} : F^\times B_1]$$
$$= \frac{[J_{F,\mathcal{S}} : B_1]}{[F_{\mathcal{S}} : F^\times \cap B_1]}$$
$$= [F^\times \cap B_1 : F_{\mathcal{S}}^n] \left(\frac{\prod_{\substack{v \in \mathcal{S} \\ v \neq v_0}} [F_v^\times : (F_v^\times)^n]}{[F_{\mathcal{S}} : F_{\mathcal{S}}^n]} \right) \qquad (*)$$
$$= [W_1 : F_{\mathcal{S}}^n] \left(\frac{\prod_{\substack{v \in \mathcal{S} \\ v \neq v_0}} [F_v^\times : (F_v^\times)^n]}{n^{\#\mathcal{S}}} \right)$$
$$= \frac{[L_2 : F]}{n^{\#\mathcal{S}}} \prod_{\substack{v \in \mathcal{S} \\ v \neq v_0}} [F_v^\times : (F_v^\times)^n].$$

Similarly,

$$[L_2 : F] \le [J_F : F^\times B_2] = \frac{[L_1 : F]}{n^{\#S}}[F_{v_0}^\times : (F_{v_0}^\times)^n].$$ (**)

Putting these together, we get

$$[L_1 : F][L_2 : F] \le \frac{\prod_{v \in S}[F_v^\times : (F_v^\times)^n]}{n^{2\#S}}[L_1 : F][L_2 : F]$$

$$= \frac{\prod_{v \in S}\frac{n^2}{\|n\|_v}}{n^{2\#S}}[L_1 : F][L_2 : F]$$

$$= \prod_{v \in S}\frac{1}{\|n\|_v}[L_1 : F][L_2 : F]$$

$$= [L_1 : F][L_2 : F].$$

The last equality follows by the Product Formula (since S contains all v for which $\|n\|_v \ne 1$). We must have equality everywhere in (*) and (**), whence

$$[L_i : F] = [J_F : F^\times B_i].$$

Since $\mathcal{H}_i = F^\times N_{L_i/F} J_{L_i}$, we also have

$$[L_i : F] = [J_F : \mathcal{H}_i].$$

We have shown that $L_1 = K_1$ is class field to $F^\times B_1$ so that $\mathfrak{p}_{v_0}$ splits completely in L_1/F as claimed.   □

3 The Artin Map on Local Fields

We have reached a point where it is possible to revisit our discussion of the Artin map on local fields. In particular, we can prove the following.

Theorem 3.1. *Let K/F be an abelian extension of number fields, and let v be a finite place of F. The Artin map $\rho_{K/F}$ satisfies $\rho_{K/F}(\varphi_v(F_v^\times)) = Z(\mathfrak{p}_v)$, the decomposition group.*

Proof. Previously, we have shown that $\rho_{K/F}(\varphi_v(F_v^\times)) \subseteq Z(\mathfrak{p}_v)$ and that they are equal when $\mathfrak{p}_v$ is unramified in K/F. Since we are concerned only with the completions, and since the completion of K_Z equals the completion of F, we may assume $F = K_Z$, the fixed field of $Z(\mathfrak{p}_v)$, and thus $g(\mathfrak{p}_v) = 1$ for K/F.

Let E be the fixed field of the subgroup $\rho_{K/F}(\varphi_v(F_v^\times))$ of $Z(\mathfrak{p}_v)$. We must show $E = F$.

Suppose $E \neq F$. Then $\mathrm{Gal}(E/F)$ is a non-trivial abelian group, so there is a field L, with $F \subseteq L \subseteq E$ and $\mathrm{Gal}(L/F)$ cyclic of prime order, say of order q. Let ζ be a primitive q^{th} root of unity, and consider the fields below.

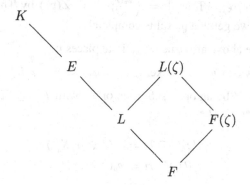

Note that

$$\rho_{L/F}(\varphi_v(F_v^\times)) = \left(\frac{\varphi_v(F_v^\times)}{L/F} \right)$$

$$= \left(\frac{\varphi_v(F_v^\times)}{K/F} \right) \bigg|_L$$

$$= 1$$

since $L \subseteq E$, the fixed field of $\rho_{K/F}(\varphi_v(F_v^\times))$.

Exercise 6.4. Show that this implies $\rho_{L(\zeta)/F(\zeta)}(\varphi_w(F(\zeta)_w^\times)) = 1$, where w is a place of $F(\zeta)$ above v.   ◊

Now we use Exercise 6.4. We have

$$\varphi_w(F(\zeta)_w^\times) \subseteq \ker \rho_{L(\zeta)/F(\zeta)}$$

and

$$J_{F(\zeta)} \big/ \ker \rho_{L(\zeta)/F(\zeta)} \cong \mathrm{Gal}(L(\zeta)/F(\zeta)) \cong \mathrm{Gal}(L/F),$$

(note $[F(\zeta) : F]$ is a divisor of $q - 1$ so is prime to q). Since this has exponent q, we may apply Theorem 2.9 from the previous section to conclude that $\mathfrak{p}_w$ splits completely in $L(\zeta)/F(\zeta)$. Again note that $(q, q - 1) = 1$; since e and f are multiplicative in towers of fields, we must have $e = f = 1$ for L/F also.

We have shown that $\mathfrak{p}_v$ splits completely in L/F. Since $F = K_Z$, this is not possible unless $L = F$, a contradiction. Thus $E = F$ as claimed.   □

Corollary 3.2 (Complete Splitting Theorem). Let $\mathcal{H}$ be an open subgroup of J_F with $F^\times \subseteq \mathcal{H}$, and let K be the class field of $\mathcal{H}$ over F. Let v be a finite place of F. Then $\mathfrak{p}_v$ splits completely in K/F if and only if $\varphi_v(F_v^\times) \subseteq \mathcal{H}$.

Proof. The forward implication follows since every element of $F_v^\times$ is a norm, as we have observed before. We have shown the reverse implication for Kummer n-extensions already; now we prove it in general. Suppose $\varphi_v(F_v^\times) \subseteq \mathcal{H} = F^\times N_{K/F} J_K = \ker \rho_{K/F}$. Then $1 = \left(\frac{\varphi_v(F_v^\times)}{K/F}\right) = Z(\mathfrak{p}_v)$ by Theorem 3.1. Since $\#Z(\mathfrak{p}_v) = ef = 1$, we get that $\mathfrak{p}_v$ splits completely.   □

Exercise 6.5. Is the above also true for infinite places v?   ◇

Exercise 6.6. Show (for $w|v$), that $\ker(\rho_{K/F} \circ \varphi_v) = N_{K_w/F_v} K_w^\times$.   ◇

Corollary 3.3. Let $\mathcal{H}$ be an open subgroup of J_K with $F^\times \subseteq \mathcal{H}$ and let K be the class field of $\mathcal{H}$ over F. Then

$$\varphi_v(F_v^\times) \cap \mathcal{H} = \varphi_v(N_{K_w/F_v} K_w^\times)$$
$$\varphi_v(\mathcal{U}_v) \cap \mathcal{H} = \varphi_v(N_{K_w/F_v} \mathcal{U}_w).$$

Proof. $\mathcal{H} = \ker \rho_{K/F}$.   □

Theorem 3.4. *Let K/F be an abelian extension of number fields, and let $\mathcal{H} = F^\times N_{K/F} J_K$. Let v be a finite place of F. Then $\rho_{K/F}(\varphi_v(\mathcal{U}_v)) = T(\mathfrak{p}_v)$, the inertia subgroup in $\mathrm{Gal}(K/F)$.*

Proof. Recall, we have shown $\rho_{K/F}(\varphi_v(\mathcal{U}_v)) \subseteq T(\mathfrak{p}_v)$ in general, and equality when $\mathfrak{p}_v$ is unramified, i.e., when $T(\mathfrak{p}_v)$ is trivial.

As before, we may assume that $F = K_Z$, so that $\mathrm{Gal}(K/F) = Z(\mathfrak{p}_v)$. Let K_T be the fixed field of $T(\mathfrak{p}_v)$; it is the maximal subfield of K that is unramified over F. We have $\mathrm{Gal}(K/K_T) = T(\mathfrak{p}_v)$.

If w is a place of K_T above v, then

$$\left(\frac{\varphi_w(\mathcal{U}_w)}{K/K_T}\right) = \left(\frac{N_{K_T/F}(\varphi_w(\mathcal{U}_w))}{K/F}\right)$$
$$= \left(\frac{\varphi_v(N_{(K_T)_w/F_v} \mathcal{U}_w)}{K/F}\right)$$
$$= \left(\frac{\varphi_v(\mathcal{U}_v)}{K/F}\right) \quad \text{since } (K_T)_w/F_v \text{ is unramified.}$$

Since $f = 1$ for $\mathfrak{P}_w$ in K/K_T, there is a uniformizer π of $(K_T)_w$ that is a local norm from the completion of K. Thus $\varphi_w(\pi) \in K_T^\times N_{K/K_T} J_K$, whence $\left(\frac{\varphi_w(\pi)}{K/K_T}\right) = 1$. Now $(K_T)_w^\times = \langle \pi \rangle \times \mathcal{U}_w$, and

$$\left(\frac{\varphi_w((K_T)_w^\times)}{K/K_T}\right) = Z(\mathfrak{P}_w) = \mathrm{Gal}(K/K_T) = T(\mathfrak{p}_v).$$

The above implies that we must have $\left(\frac{\varphi_w(\mathcal{U}_w)}{K/K_T}\right) = T(\mathfrak{p}_v)$ since the image of π is trivial. Thus $\left(\frac{\varphi_v(\mathcal{U}_v)}{K/F}\right) = T(\mathfrak{p}_v)$.   □.

Corollary 3.5. If $\mathcal{H}$ is an open subgroup of J_F with $F^\times \subseteq \mathcal{H}$, and K is the class field to $\mathcal{H}$ over F, then for any finite place v of F, the class field to $\mathcal{H}\varphi_v(\mathcal{U}_v)$ is the maximal subfield of K in which $\mathfrak{p}_v$ is unramified, hence it is the field K_T.   □

Recall we have shown (Chapter 4) that $[\mathcal{U}_v : N_{K_w/F_v}\mathcal{U}_w] \leq e(w/v)$, with equality when the extension is cyclic. By Theorem 3.4, we now have equality for any abelian extension.

Corollary 3.6. Let K/F be an abelian extension of number fields, v a finite place of F, and w a place of K above v. Then

$$\mathcal{U}_v \big/ N_{K_w/F_v}\mathcal{U}_w \cong T(\mathfrak{p}_v).$$ □

Now that we have shown that $[\mathcal{U}_v : N_{K_w/F_v}\mathcal{U}_w] = e(w/v)$ for any abelian extension K/F, we can prove a result about the conductor, as was promised in Chapter 5.

Theorem 3.7. *If K/F is an abelian extension of number fields, then $\mathfrak{f}(K/F)$ is divisible by all the ramified primes, and no others.*

Proof. By definition, the conductor is the minimal ideal $\mathfrak{f}$ of $\mathcal{O}_F$ such that $\mathcal{E}_{F,\mathfrak{f}}^+ \subseteq F^\times N_{K/F} J_K = \mathcal{H}$. For any $\mathfrak{m}$, we have

$$\mathcal{E}_{F,\mathfrak{m}}^+ = \prod_{v \text{ imaginary}} \mathbb{C}^\times \times \prod_{v \text{ real}} \mathbb{R}_+^\times \times \prod_{\mathfrak{p}_v | \mathfrak{m}} (1 + \mathfrak{p}_v^{\mathrm{ord}_v \mathfrak{m}}) \times \prod_{\mathfrak{p}_v \nmid \mathfrak{m}} \mathcal{U}_v.$$

Since $N_{K/F}\mathcal{E}_K$ is open in $\mathcal{E}_F$, we can choose $\mathfrak{m}$ so that

$$\mathcal{E}_{F,\mathfrak{m}}^+ \subseteq N_{K/F}\mathcal{E}_K = \prod_v \prod_{w|v} N_{K_w/F_v}\mathcal{U}_w.$$

Since the local norm is surjective on units at unramified places, we have

$$\varphi_v(\mathcal{U}_v) \subseteq N_{K/F}\mathcal{E}_K$$

whenever $\mathfrak{p}_v$ is unramified. Thus the ideal $\mathfrak{m}$ need not be divisible by any unramified prime.

The minimality of $\mathfrak{f}$ implies $\mathfrak{f}|\mathfrak{m}$ for any such $\mathfrak{m}$, so $\mathfrak{f}$ is not divisible by any unramified prime.

On the other hand, suppose $\mathfrak{p}_v$ is ramified in K/F. Recall $\mathcal{H} = F^\times N_{K/F} J_K = \ker \rho_{K/F}$ by Artin Reciprocity. We know (for $\mathfrak{p}_v$ ramified), that $\rho_{K/F}(\varphi_v(\mathcal{U}_v)) = T(\mathfrak{p}_v) \neq 1$, whence $\varphi_v(\mathcal{U}_v) \not\subseteq \ker \rho_{K/F} = \mathcal{H}$. If $\mathfrak{p}_v \nmid \mathfrak{f}$, then the component of $\mathcal{E}_{F,\mathfrak{f}}^+$ at v would be all of $\mathcal{U}_v$, i.e., we would have $\varphi_v(\mathcal{U}_v) \subseteq \mathcal{E}_{F,\mathfrak{f}}^+$. Since $\mathcal{E}_{F,\mathfrak{f}}^+ \subseteq \mathcal{H}$ by the definition of conductor, we get $\varphi_v(\mathcal{U}_v) \subseteq \mathcal{H}$, a contradiction. Thus $\mathfrak{p}_v|\mathfrak{f}$ whenever $\mathfrak{p}_v$ is ramified.   □

Recall, by Corollary 3.3 (to the Complete Splitting Theorem), for $\mathcal{H} = F^\times N_{K/F} J_K$, we have $\mathcal{H} \cap \varphi_v(\mathcal{U}_v) = \varphi_v(N_{K_w/F_v}\mathcal{U}_w)$. We may combine this with Theorem 3.7 to find the conductor of a cyclotomic extension of $\mathbb{Q}$.

Example.

1. Let $F = \mathbb{Q}$, $K = \mathbb{Q}(\zeta_m)$, where $m = p^t$ for some prime p and some positive integer t. Let $\mathfrak{f}$ be the conductor of $K/\mathbb{Q}$. The only ramified prime is p, so that by Theorem 3.7, we have $\mathfrak{f} = p^r\mathbb{Z}$ for some positive integer r. Now p ramifies totally in $K/\mathbb{Q}$, so

$$[\mathbb{Z}_p^\times : N_{K_w/\mathbb{Q}_p}\mathcal{U}_w] = p^{t-1}(p-1).$$

Recall that $\mathbb{Z}_p^\times = \mu_{p-1} \times (1 + p\mathbb{Z}_p)$, so we must have $N_{K_w/\mathbb{Q}_p}\mathcal{U}_w = 1 + p^t\mathbb{Z}_p$, (for $p = 2$ this can be checked directly; for $p > 2$ this is the only subgroup of $\mathbb{Z}_p^\times$ with index $p^{t-1}(p-1)$ — why?).

On the other hand,

$$\mathcal{E}_{\mathbb{Q},\mathfrak{f}}^+ = \mathbb{R}_+^\times \times (1 + p^r\mathbb{Z}_p) \times \prod_{q \neq p} \mathcal{U}_q \subseteq \mathcal{H}.$$

This gives

$$\varphi_p(1 + p^r\mathbb{Z}_p) \subseteq \mathcal{H} \cap \varphi_p(\mathbb{Z}_p^\times) = \varphi_p(N_{K_w.\mathbb{Q}_p}\mathcal{U}_w)$$
$$1 + p^r\mathbb{Z}_p \subseteq 1 + p^t\mathbb{Z}_p$$
$$r \geq t.$$

By the minimality of $\mathfrak{f}$, we must have $r = t$, i.e., $\mathfrak{f} = m\mathbb{Z}$.

Exercise 6.7. Let $F = \mathbb{Q}$, $K = \mathbb{Q}(\zeta_m)$, where m is any positive integer. Show that $\mathfrak{f}(K/\mathbb{Q}) = m\mathbb{Z}$.   ◊

Exercise 6.8. For $K = \mathbb{Q}(\sqrt{d})$, where d is square-free, and $F = \mathbb{Q}$, show that this notion of conductor again agrees with the classical definition of the conductor, i.e., $\mathfrak{f} = f\mathbb{Z}$ where f is the smallest positive integer such that $K \subseteq \mathbb{Q}(\zeta_f)$. (If we had defined the conductor as a divisor, then we would have had $\mathfrak{f} = f\mathbb{Z}$ when K is real and $\mathfrak{f} = \mathfrak{p}_\infty f\mathbb{Z}$ when K is imaginary, where $\mathfrak{p}_\infty$ is the infinite prime of $\mathbb{Q}$.)   ◊

Let us consider cyclotomic fields further. We have shown $\mathcal{E}_{\mathbb{Q},\mathfrak{f}}^+ \subseteq \mathcal{H} = \mathbb{Q}^\times N_{K/\mathbb{Q}} J_K$, where $K = \mathbb{Q}(\zeta_m)$, $\mathfrak{f} = m\mathbb{Z}$, and $m = p^t$ for some prime p and some positive integer t. Exercise 6.7 allows us to extend this result to all positive integers m. Recall $\mathcal{H} = \ker \rho_{K/\mathbb{Q}}$ by Artin Reciprocity. In Proposition 3.4 of Chapter 4 we showed

$$J_{\mathbb{Q}}\big/\mathbb{Q}^\times \mathcal{E}_{\mathbb{Q},\mathfrak{f}}^+ \cong \mathcal{R}_{\mathbb{Q},\mathfrak{f}}^+,$$

the ray class group, and in Chapter 3 we showed

$$\mathcal{R}_{\mathbb{Q},\mathfrak{f}}^+ \cong \left(\mathbb{Z}/m\mathbb{Z}\right)^\times.$$

By Artin Reciprocity,

$$J_\mathbb{Q}/\mathbb{Q}^\times N_{K/\mathbb{Q}} J_K \cong \mathrm{Gal}\,(K/\mathbb{Q}) \cong \left(\mathbb{Z}/m\mathbb{Z}\right)^\times.$$

Putting this together, we conclude

$$\mathbb{Q}^\times \mathcal{E}_{\mathbb{Q},\mathfrak{f}}^+ = \mathbb{Q}^\times N_{K/\mathbb{Q}} J_K.$$

It follows that $K = \mathbb{Q}(\zeta_m)$ is the class field of $\mathbb{Q}^\times \mathcal{E}_{\mathbb{Q},m\mathbb{Z}}^+$, (or, in terms of ideals, of $\mathcal{P}_{\mathbb{Q},m\mathbb{Z}}^+$).

Now suppose $F/\mathbb{Q}$ is any finite abelian extension. Then F is the class field of $\mathbb{Q}^\times N_{F/\mathbb{Q}} J_F$, or, in terms of ideals, it is the class field of some subgroup $\mathcal{H}$ of $\mathcal{I}_\mathbb{Q}(\mathfrak{f})$ with $\mathcal{P}_{\mathbb{Q},\mathfrak{f}}^+ \subseteq \mathcal{H}$. (Here $\mathfrak{f}$ is the conductor of $F/\mathbb{Q}$, an ideal $f\mathbb{Z}$ of $\mathbb{Z}$.) Since $\mathcal{P}_{\mathbb{Q},\mathfrak{f}}^+ \subseteq \mathcal{H}$, we have $\mathbb{Q}^\times \mathcal{E}_{\mathbb{Q},\mathfrak{f}}^+ \subseteq \mathbb{Q}^\times N_{F/\mathbb{Q}} J_F$ in terms of idèles. By the order-reversing property of the correspondence between open subgroups of $J_\mathbb{Q}$ and class fields, the class field of $\mathbb{Q}^\times \mathcal{E}_{\mathbb{Q},\mathfrak{f}}^+$ must contain the class field of $\mathbb{Q}^\times N_{F/\mathbb{Q}} J_F$. It follows that $F \subseteq \mathbb{Q}(\zeta_f)$. We have shown the following.

Theorem 3.8 (Kronecker, Weber). *Every finite abelian extension F of $\mathbb{Q}$ satisfies $F \subseteq \mathbb{Q}(\zeta)$ for some root of unity ζ.* □

4 The Hilbert Class Field

Now that we have proved the Existence Theorem, we know that the group $\mathcal{H} = F^\times J_{F,S_\infty} = F^\times \mathcal{E}_F$ has a class field. Call it F_1. Recall the example at the end of Chapter 5. In it, we showed that if K/F is everywhere unramified, then $F^\times N_{K/F} J_K \supseteq F^\times \mathcal{E}_F$. Thus $K \subseteq F_1$.

The converse is also true. Suppose $\mathcal{H} \supseteq F^\times \mathcal{E}_F$ is a subgroup of J_F. Then $\mathcal{H}$ is open, so it has a class field K over F. Since $\varphi_v(\mathcal{U}_v) \subseteq \mathcal{H} = \ker \rho_{K/F}$ for every v, we have that $\left(\frac{\varphi_v(\mathcal{U}_v)}{K/F}\right) = 1$, whence $T(\mathfrak{p}_v) = 1$. We have shown the following.

Proposition 4.1. *Let F be a number field and let $\mathcal{H}$ be an open subgroup of J_F that contains $F^\times$. Then: $\mathcal{H} \supseteq F^\times \mathcal{E}_F$ if and only if the class field to $\mathcal{H}$ over F is an abelian extension of F that is everywhere unramified.* □

Taking $\mathcal{H} = F^\times \mathcal{E}_F$ and applying Proposition 4.1, we find that the extension F_1/F is abelian and everywhere unramified; it is necessarily the maximal unramified abelian extension of F. F_1 is called the *Hilbert class field* of F.

Note that the Isomorphy Theorem gives $\mathrm{Gal}\,(F_1/F) \cong J_F/F^\times \mathcal{E}_F$. But we showed (much) earlier, that $J_F/F^\times \mathcal{E}_F \cong \mathcal{C}_F$, the ideal class group (see Proposition 4.3.3). Thus

$$\text{Gal}\,(F_1/F) \cong \mathcal{C}_F.$$

In particular, note that $[F_1 : F] = h_F$ and the maximal unramified abelian extension of a number field is a finite extension.

Exercise 6.9. Which primes of $\mathcal{O}_F$ split completely in F_1? $\qquad\qquad\Diamond$

Exercise 6.10. Let F be a number field, and let F_1 be its Hilbert class field. Find the conductor $\mathfrak{f}$ of F_1/F and identify the subgroup $\mathcal{H}$ with $\mathcal{P}_{F,\mathfrak{f}}^+ < \mathcal{H} < \mathcal{I}_F(\mathfrak{f})$ that has F_1 as its class field over F (in the sense of Weber). $\qquad\Diamond$

Exercise 6.11. Let $F \subseteq K \subseteq E \subseteq K_1$ be number fields, where K_1 is the Hilbert class field of K, and suppose E/F is abelian. Show that $\mathfrak{f}(E/F) = \mathfrak{f}(K/F)$ and that an infinite prime of F ramifies in E/F if and only if it ramifies in K/F. $\qquad\Diamond$

Exercise 6.12. Let K be the maximal unramified abelian p-extension of a number field F. Show that $\text{Gal}\,(K/F)$ is isomorphic to the Sylow p-subgroup of the ideal class group $\mathcal{C}_F$. The field K is called the *Hilbert p-class field* of F. Find the conductor $\mathfrak{f}$ of K/F and identify the subgroup $\mathcal{H}$ with $\mathcal{P}_{F,\mathfrak{f}}^+ < \mathcal{H} < \mathcal{I}_F(\mathfrak{f})$ such that the class field over F of $\mathcal{H}$ is the Hilbert p-class field. $\qquad\Diamond$

Example.

2. Let $F = \mathbb{Q}(\sqrt{-5})$. The conductor $\mathfrak{f}(F/\mathbb{Q}) = f\mathbb{Z}$, where f is the smallest positive integer such that $F \subseteq \mathbb{Q}(\zeta_f)$. Thus $\mathfrak{f}(F/\mathbb{Q}) = 20\mathbb{Z}$. We may use Minkowski theory to compute the class number of F: $h_F = 2$. Thus $[F_1 : F] = 2$. We are seeking an extension F_1/F of degree 2 that is everywhere unramified. It will necessarily be the (unique) maximal unramified extension of F. We found such an extension in Chapter 2 (using Dirichlet characters). Hence we conclude $F_1 = \mathbb{Q}(i, \sqrt{-5})$.

Exercise 6.13. Let $F = \mathbb{Q}(\sqrt{-15})$. Find the Hilbert class field F_1. $\qquad\Diamond$

Exercise 6.14. Let $F = \mathbb{Q}(\sqrt{-21})$. Show (without Minkowski theory) that $h_F \geq 4$. Assuming that $h_F = 4$, find the Hilbert class field F_1. $\qquad\Diamond$

The above example and exercises are perhaps misleading in that all share the property that $F_1/\mathbb{Q}$ is abelian. This need not be the case in general. Indeed, $F/\mathbb{Q}$ may not even be Galois. We include the following example from Janusz' book, [J], to illustrate this situation.

Example.

1. Let $F = \mathbb{Q}(\alpha)$, where α is the real cube root of 11. Minkowski theory gives $h_F = 2$. We may factor $2\mathcal{O}_F$ by noting that

$$X^3 - 11 \equiv (X + 1)(X^2 + X + 1) \pmod 2$$

so that $2\mathcal{O}_F = \mathfrak{p}_1\mathfrak{p}_2$ where $N\mathfrak{p}_1 = 2\mathbb{Z}$, $N\mathfrak{p}_2 = 4\mathbb{Z}$. One may show (see Chapter 1 of [J]) that $\mathfrak{p}_1$ is not principal and that $\mathfrak{p}_1^2 = \langle \alpha^2 - 5 \rangle$. The units in $\mathcal{O}_F$ are $\{\pm 1\} \times \langle \varepsilon \rangle$, where $\varepsilon > 0$ is a fundamental unit.

Since $F/\mathbb{Q}$ is not normal, we use Kummer theory to find F_1. Note that $[F_1 : F] = 2$, so F_1/F is a Kummer 2-extension. Thus F_1 is a subfield of the extension of F obtained by adjoining the square roots of all the $\mathcal{S}$-units for a suitable choice of $\mathcal{S}$, as in the proof of the Existence Theorem. Here we may take $\mathcal{S} = \{\mathfrak{p}_1, \mathfrak{p}_2\} \cup \mathcal{S}_\infty$.

If $x \in F^\times$ is divisible only by primes in $\mathcal{S}$, then

$$\langle x \rangle = \mathfrak{p}_1^a \mathfrak{p}_2^b = (\mathfrak{p}_1 \mathfrak{p}_2)^b \mathfrak{p}_1^{a-b} = \langle 2 \rangle^b \mathfrak{p}_1^{a-b}.$$

Only even powers of $\mathfrak{p}_1$ are principal, so $a - b$ is even, and we get

$$\langle x \rangle = \langle 2^b \rangle \langle \alpha^2 - 5 \rangle^k$$

where $k = \frac{a-b}{2}$. Thus

$$x = 2^b(\alpha^2 - 5)^k u$$

for some unit u. We have shown that the $\mathcal{S}$-units are

$$F_\mathcal{S} = \{\pm 1\} \times \langle \varepsilon \rangle \times \langle 2 \rangle \times \langle \alpha^2 - 5 \rangle.$$

To get an unramified extension of F of degree $2 = h_F$, we want to adjoin a root of some (irreducible) polynomial of the form $X^2 - \beta$ with $\beta \in F_\mathcal{S}$ chosen so that no prime ramifies. If 2 or $\alpha^2 - 5$ divides β then $\mathfrak{p}_1$ will ramifiy. Hence we must take $\beta \in \{\pm 1\} \times \langle \varepsilon \rangle$. Since F has a real prime and we do not want it to ramify either, we must take $\beta > 0$. Thus

$$F_1 = F(\sqrt{\varepsilon})$$

where $\varepsilon > 0$ is the fundamental unit in F. Janusz computes this fundamental unit explicitly in his Chapter 1 as $\varepsilon = 89 + 40\alpha + 18\alpha^2$.

As we mentioned in Chapter 3, the work of Weber on class field theory grew out of two main examples. First was the study of the abelian extensions of $\mathbb{Q}$, which are completely described by the Kronecker-Weber Theorem. Second was the study of the abelian extensions of imaginary quadratic fields. For an imaginary quadratic field $F = \mathbb{Q}(\sqrt{-d})$, the theory of complex multiplication gives us a description of the abelian extensions of F. A theorem of Weber ([We3], 1908) and Fueter ([Fue], 1914) tells us that there is a one-to-one correspondence between the ideal classes in $\mathcal{C}_F$ and isomorphism classes of elliptic curves over $\mathbb{C}$ with complex multiplication by $\mathcal{O}_F$; the j-invariant j of such an elliptic curve is an algebraic integer, and the field $F(j)$ is the Hilbert class field F_1. The above imaginary quadratic examples and exercises bear this out. More generally, the maximal abelian extension of F is generated over F by the j-invariant and the values of a certain analytic function (a *Weber function*) at all the torsion points of the elliptic curve, (Takagi, [T], 1920). For details, see Shimura's *Introduction to the Arithmetic Theory of Automorphic Functions*, [Sh], or Lang's *Elliptic Functions*, [L2]. Also see Serre's article

Complex Multiplication in Cassels and Fröhlich, [CF]. Hilbert's Twelfth Problem, [Hi3], is to find for any number field F, functions that play the same role. The theory of elliptic curves with complex multiplication by an imaginary quadratic field has been generalized to higher dimensions: One considers abelian varieties with complex multiplication by a CM-field and automorphic functions. This provides some nice results for CM-fields. However, for other number fields, in general we do not have explicit constructions of their abelian extensions.

A conjecture of Hilbert was that any ideal of a number field F becomes principal in F_1. This was one of the original motivations for class field theory. It was not proved until 1930 by Furtwängler, [Fur2]. Now called the *Principal Ideal Theorem*, its proof was simplified shortly thereafter using an idea of Artin, [A3], which we describe here.

Let F_2 be the Hilbert class field of F_1. We first show that F_2/F is Galois. To this end, suppose λ is an F-isomorphism $F_2 \longrightarrow \lambda(F_2)$. Then $\lambda|_{F_1} \in \text{Gal}(F_1/F)$, so $\lambda(F_1) = F_1$. But $\lambda(F_2)$ is everywhere unramified and abelian over $\lambda(F_1) = F_1$, whence $\lambda(F_2) = F_2$ by the uniqueness of the Hilbert class field.

Let $G = \text{Gal}(F_2/F)$ and let $A = \text{Gal}(F_2/F_1)$, so that $G/A \cong \text{Gal}(F_1/F)$. Clearly, within F_2, the subfield F_1 is maximal abelian over F, so G/A is the largest abelian factor group of G. Thus $A = G'$, the commutator subgroup of G. Note also that since A is abelian, $(G')'$ is trivial.

Consider the map

$$\gamma : \mathcal{I}_F/\mathcal{P}_F \longrightarrow \mathcal{I}_{F_1}/\mathcal{P}_{F_1}$$

given by $\mathfrak{a}\mathcal{P}_F \mapsto (\mathfrak{a}\mathcal{O}_{F_1})\mathcal{P}_{F_1}$. This is readily seen to be a well-defined group homomorphism $\gamma : \mathcal{C}_F \longrightarrow \mathcal{C}_{F_1}$. To prove the Principal Ideal Theorem amounts to showing that this map is trivial.

By Artin Reciprocity, we have

$$\mathcal{C}_F \cong \text{Gal}(F_1/F) \cong G/A = G/G' \text{ and}$$
$$\mathcal{C}_{F_1} \cong \text{Gal}(F_2/F_1) = A = G'.$$

Let $V : G/G' \longrightarrow G'$ be the homomorphism making the following diagram commute.

$$
\begin{array}{ccc}
\mathcal{C}_F & \overset{\cong}{\longrightarrow} & G/G' \\
\gamma \downarrow & & \downarrow V \\
\mathcal{C}_{F_1} & \underset{\cong}{\longrightarrow} & G'
\end{array}
$$

Exercise 6.15. Given any group G, put $G^{\text{ab}} = G/G'$. Suppose $H < G$ with $[G : H]$ finite. We can define $V : G^{\text{ab}} \longrightarrow H^{\text{ab}}$ as follows. Choose a set $\mathcal{R}$ of representatives for the left cosets of H in G. Let $g \in G$. For each $r \in \mathcal{R}$ there is a unique pair $\bar{r} \in \mathcal{R}$ and h_r such that $gr = \bar{r}h_r$. Put $V(gG') = \prod_{r \in \mathcal{R}} h_r H'$. Show that this map V is well-defined and that it is the desired map to make the above diagram commute. $\Diamond$

Artin's idea was to show that the map V, the group theoretical map called the *transfer* (or *Verlagerung*), must be trivial. This reduced the problem to a purely group theoretical result (also called the "principal ideal theorem," even though it does not appear to be about ideals at all!):

If G is a finite group and G' is its commutator subgroup, then the transfer $V : {}^{G}/_{G'} \longrightarrow$

${}^{G'}/_{(G')'}$ is the trivial map.

We omit the proof, but the interested reader may consult the 1934 paper of Iyanaga, [Iy], the 1954 paper of Witt, [Wi], or Neukirch's book, [N]. We record the result below. (For a result that has the Principal Ideal Theorem as a corollary, see Suzuki's 1991 article, [Su].)

Theorem 4.2 (Principal Ideal Theorem). *Every fractional ideal $\mathfrak{a}$ of a number field F becomes principal in F_1, i.e., $\mathfrak{a}\mathcal{O}_{F_1}$ is principal.* $\square$

Consider the tower

$$F \subseteq F_1 \subseteq F_2 \subseteq \cdots$$

called the *Hilbert class field tower* of F. If this tower is finite, then every ideal of some F_i is principal, so that the number field F is contained in a number field F_i of class number 1. It was an open question for a long time as to whether the Hilbert class field tower could ever be infinite. In the early 1960s, Golod and Shafarevich, [GS], were able to give examples of number fields F having infinite class field towers. For example, the field $\mathbb{Q}(\sqrt{d})$, where $d = 2^2 \cdot 3 \cdot 7 \cdot 11 \cdot 13 \cdot 19 \cdot 23$ is such a field. (See Roquette's article in Cassels and Fröhlich, [CF], for details.)

Suppose F is a number field that can be embedded in a number field L of class number 1.

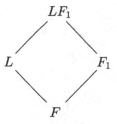

Since $h_L = 1$, we must have $L_1 = L$. Since LF_1/L is abelian and everywhere unramified (because F_1/F is), we must have $LF_1 \subseteq L_1$. Thus $F_1 \subseteq L$.

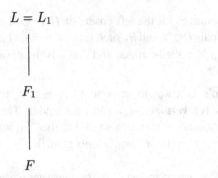

$$L = L_1$$

$$F_1$$

$$F$$

Now consider F_2. By the same reasoning, LF_2/L is everywhere unramified and abelian, so $F_2 \subseteq L$. Continuing by induction, we find that the entire class field tower for F is contained in the number field L. This forces the class field tower of F to be finite. The result of Golod and Shafarevich shows that there are number fields F that cannot be embedded into a number field of class number 1. In such a tower, for any i the ideals of F_i will all become principal in F_{i+1}, but there will always be other ideals of F_{i+1} that are non-principal.

The above argument is based on the fact that if K/F is an extension of number fields, then KF_1/K is abelian and everywhere unramified so that $KF_1 \subseteq K_1$. We can exploit this idea further to obtain information about class numbers in certain situations.

Theorem 4.3. *Let K/F be an extension of number fields and suppose $K \cap F_1 = F$. Then*

i. $h_F | h_K$.
ii. *the map $N_{K/F} : \mathcal{C}_K \longrightarrow \mathcal{C}_F$ is surjective.*

Proof. Note that $\operatorname{Gal}(KF_1/K) \cong \operatorname{Gal}(F_1/F) \cong \mathcal{C}_F$ is abelian. Also KF_1/K is everywhere unramified (since F_1/F is). Thus $K \subseteq KF_1 \subseteq K_1$. We have the following picture.

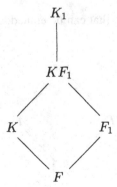

$$K_1$$

$$KF_1$$

$$K \qquad\qquad F_1$$

$$F$$

But then

$$\mathcal{C}_K \cong \operatorname{Gal}(K_1/K) \overset{\text{restr.}}{\twoheadrightarrow} \operatorname{Gal}(KF_1/K) \cong \operatorname{Gal}(F_1/F) \cong \mathcal{C}_F.$$

This implies that $h_F | h_K$.

Apply the Consistency Property of the Artin symbol, and it follows that for any $\mathfrak{a} \in \mathcal{I}_K$,

$$\left(\frac{\mathfrak{a}}{K_1/K} \right) \Big|_{F_1} = \left(\frac{N_{K/F}\mathfrak{a}}{F_1/F} \right).$$

Note that the Artin symbol is defined for *any* ideal of K, since K_1/K is unramified everywhere. This gives us a commutative diagram:

$$
\begin{array}{ccc}
\mathcal{C}_K & \xrightarrow{\cong} & \mathrm{Gal}\,(K_1/K) \\
N_{K/F} \downarrow & & \downarrow \text{restriction to } F_1 \\
\mathcal{C}_F & \xrightarrow[\cong]{} & \mathrm{Gal}\,(F_1/F)
\end{array}
$$

The horizontal arrows are the isomorphisms arising from Artin Reciprocity.

Since $\mathrm{Gal}\,(KF_1/K) \cong \mathrm{Gal}\,(F_1/F)$ via restriction, the vertical arrow on the right is an epimorphism. This implies that the map $N_{K/F} : \mathcal{C}_K \longrightarrow \mathcal{C}_F$ is surjective. $\square$

Proposition 4.4. *If K/F is an extension of number fields, and there is some prime $\mathfrak{p}$ of $\mathcal{O}_F$ that is totally ramified in K/F, then $h_F | h_K$.*

Proof. Let K/F be an extension of number fields and suppose $\mathfrak{p}$ is a prime of $\mathcal{O}_F$ that is totally ramified in K/F. Now $\mathfrak{p}$ is totally ramified in $K \cap F_1/F$ because it is totally ramified in K/F. But also $\mathfrak{p}$ is unramified in F_1/F, so it is unramified $K \cap F_1/F$. The only way that $\mathfrak{p}$ can be simultaneously totally ramified *and* unramified in $K \cap F_1/F$ is if $K \cap F_1 = F$. Now apply (i) of the Theorem. $\square$

5 Arbitrary Finite Extensions of Number Fields

The ideas from the previous section can be used to obtain some information on non-abelian extensions of number fields. For example, if E/F is a (not necessarily abelian) Galois extension of number fields, let $G = \mathrm{Gal}\,(E/F)$ and let G' be the commutator subgroup of G. Put $G^{\mathrm{ab}} = G/G'$ as before. If K denotes the fixed field of G', then K/F is abelian with Galois group isomorphic to G^{ab}. By Artin Reciprocity, we have an isomorphism

$$J_F / F^\times N_{K/F} J_K \cong G^{\mathrm{ab}}.$$

It is easy to show that $F^\times N_{E/F} J_E \subseteq F^\times N_{K/F} J_K$ since $K \subseteq E$, but in fact we can say more. The following theorem gives $F^\times N_{E/F} J_E = F^\times N_{K/F} J_K$, since it shows that they have the same class field.

Theorem 5.1. *Let E/F be an extension of number fields and let $\mathcal{H} = F^\times N_{E/F} J_E$, an open subgroup of J_F that contains $F^\times$. Let K be the class field of $\mathcal{H}$ over F. Then K/F is the maximal abelian subextension of E/F.*

Proof. We know that $\mathcal{H}$ is an open subgroup of J_F by Exercise 4.24. Thus it has a class field K over F. Now K/F is abelian, and $F^\times N_{K/F} J_K = \mathcal{H}$. If K/F is a subextension of E/F, then it is clearly the maximal abelian subextension of E/F. (By the Ordering Theorem, any properly larger abelian subextension of E/F would correspond to a smaller subgroup $\tilde{\mathcal{H}}$ with $\mathcal{H} = F^\times N_{E/F} J_E \subseteq \tilde{\mathcal{H}} \subsetneqq \mathcal{H}$.) Hence it suffices to show that $K \subseteq E$. We do this by considering the extension KE/E. By Artin Reciprocity $\mathcal{H}$ is the kernel of the Artin map for K/F. If $\mathbf{a} \in J_E$, then $N_{E/F}(\mathbf{a}) \in \mathcal{H}$, whence

$$ 1 = \left(\frac{N_{E/F}(\mathbf{a})}{K/F} \right) = \left(\frac{\mathbf{a}}{KE/E} \right) \bigg|_K $$

by the Consistency Property. Since an automorphism of KE/E is completely determined by its action on K, we must have that $\mathbf{a}$ is in the kernel of the Artin map for KE/E. It follows that the kernel of the Artin map for KE/E is all of J_E. Thus (by Artin Reciprocity again), $\mathrm{Gal}\,(KE/E)$ is trivial, i.e., $KE = E$ and $K \subseteq E$. $\square$

Exercise 6.16. Let F be a number field and let $\mathcal{H}$ be an open subgroup of J_F that contains $F^\times$. Show that if E/F is a finite extension, then KE is the class field over E to $N_{E/F}^{-1} \mathcal{H}$. $\Diamond$

Exercise 6.17. Let E/F be a finite extension of number fields. Show that $[E : F] = [J_F : F^\times N_{E/F} J_E]$ if and only if E/F is abelian. $\Diamond$

At the present time, efforts to extend class field theory to arbitrary Galios extensions of number fields remain largely incomplete. However, there is at least one idea that generalizes quite nicely, via the Chebotarev Density Theorem. Let K/F be a Galois extension of number fields and as usual let $\mathcal{S}_{K/F}$ be the set of prime ideals of $\mathcal{O}_F$ that split completely in K/F. If K/F is abelian, then we know that $\mathcal{S}_{K/F}$ uniquely determines K. Using the Chebotarev Density Theorem, (see Exercise 5.10), we may extend this result to arbitrary finite Galois extensions of the number field F.

Let $G = \mathrm{Gal}\,(K/F)$, a possibly non-abelian group, where K/F is a Galois extension of number fields. For $\sigma \in G$, put

$$ \mathcal{S}_\sigma = \{ \text{unramified primes } \mathfrak{p} \text{ of } \mathcal{O}_F : \left(\frac{\mathfrak{P}}{K/F} \right) \in [\sigma]_G \text{ for } \mathfrak{P} | \mathfrak{p} \mathcal{O}_K \} $$

(as before $[\sigma]_G$ denotes the conjugacy class of σ in G). Note that if σ and τ are conjugate in G, then $\mathcal{S}_\sigma = \mathcal{S}_\tau$, while if they are not conjugate, then $\mathcal{S}_\sigma \cap \mathcal{S}_\tau = \emptyset$. The Chebotarev Density Theorem says that $\delta_F(\mathcal{S}_\sigma) = \frac{c}{n}$, where $c = \#[\sigma]_G$ and $n = \#G$.

Now let E/F be an arbitrary (not necessarily Galois) extension of number fields and put

$$ \mathcal{S}_{E/F}^1 = \{ \text{unramified primes } \mathfrak{p} \text{ of } \mathcal{O}_F : f(\mathfrak{P}/\mathfrak{p}) = 1 \text{ for some prime } \mathfrak{P} | \mathfrak{p} \mathcal{O}_E \}. $$

Note we have $\mathcal{S}^1_{E/F} = \mathcal{S}_{E/F}$ when E/F is Galois. In general, suppose K/F is the normal closure of E/F, and put $G = \mathrm{Gal}\,(K/F)$, $H = \mathrm{Gal}\,(K/E)$. Then, up to a set of Dirichlet density zero, we may write $\mathcal{S}^1_{E/F}$ as a disjoint union over the conjugacy classes $[\sigma]_G$ in G:

$$\mathcal{S}^1_{E/F} \approx \bigcup_{[\sigma]_G \cap H \neq \emptyset} \mathcal{S}_\sigma .$$

Observe that we have $H \subseteq \bigcup_{[\sigma]_G \cap H \neq \emptyset} [\sigma]_G$ with equality if and only if E/F is Galois. Thus

$$\frac{1}{[E:F]} = \frac{\#H}{\#G}$$

$$\leq \frac{1}{\#G} \sum_{[\sigma]_G \cap H \neq \emptyset} \#[\sigma]_G$$

$$= \sum_{[\sigma]_G \cap H \neq \emptyset} \delta_F(\mathcal{S}_\sigma) \qquad \text{by Chebotarev}$$

$$= \delta_F \left(\bigcup_{[\sigma]_G \cap H \neq \emptyset} \mathcal{S}_\sigma \right)$$

$$= \delta_F(\mathcal{S}^1_{E/F}),$$

with equality if and only if E/F is Galois.

Exercise 6.18. Let E/F be an extension of number fields. Show that E/F is Galois if and only if every prime in $\mathcal{S}^1_{E/F}$ splits completely in E/F. ◊

Theorem 5.2. *Suppose K/F is a Galois extension of number fields and L/F is any finite extension. Then $\mathcal{S}^1_{L/F} \prec \mathcal{S}_{K/F}$ if and only if $K \subseteq L$.*

Proof. We show the forward implication (the converse is clear). Suppose $\mathcal{S}^1_{L/F} \prec \mathcal{S}_{K/F}$ and let N/F be the Galois closure of KL/F. Put $G = \mathrm{Gal}\,(N/F)$ and $H = \mathrm{Gal}\,(N/K)$, $H' = \mathrm{Gal}\,(N/L)$.

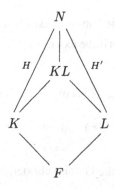

Then

$$\mathcal{S}_{L/F}^1 \approx \bigcup_{[\sigma]_G \cap H' \neq \emptyset} \mathcal{S}_{\sigma,N/F} \prec \mathcal{S}_{K/F} = \mathcal{S}_{K/F}^1 \approx \bigcup_{[\tau]_G \cap H \neq \emptyset} \mathcal{S}_{\tau,N/F}.$$

Let $\sigma \in H'$. By Chebotarev, the set $\mathcal{S}_{\sigma,N/F}$ has positive Dirichlet density. So there is some some prime $\mathfrak{p} \in \mathcal{S}_{\sigma,N/F}$ that occurs in one of the $\mathcal{S}_{\tau,N/F}$, where $\tau \in G$ satisfies $[\tau]_G \cap H \neq \emptyset$. We have shown that $\mathcal{S}_{\sigma,N/F} \cap \mathcal{S}_{\tau,N/F}$ contains the prime $\mathfrak{p}$. But the only way to have $\mathcal{S}_{\sigma,N/F} \cap \mathcal{S}_{\tau,N/F} \neq \emptyset$ is if $[\sigma]_G = [\tau]_G$, and since H is normal in G, we also have $[\tau]_G \subseteq H$. It follows that $\sigma \in H$. We have shown that $H' \subseteq H$, and thus their fixed fields satisfy the reverse containment: $K \subseteq L$.   □

Corollary 5.3. A Galois extension of number fields K/F is uniquely determined by the set $\mathcal{S}_{K/F}$ of primes that split completely.   □

Note that when E/F is Galois and non-abelian, and K/F is its maximal abelian subextension, then on the one hand we have $F^\times N_{E/F} J_E = F^\times N_{K/F} J_K$ by Theorem 5.1, but on the other we have $\mathcal{S}_{E/F} \subsetneq \mathcal{S}_{K/F}$ by Corollary 5.3. This behavior is illustrated for example in the case when $F = \mathbb{Q}$ and E is the splitting field of $X^3 - 11$ over $\mathbb{Q}$. The reader may verify that the prime $7\mathbb{Z}$ splits completely in $K = \mathbb{Q}(\zeta_3)/\mathbb{Q}$ but not in $E/\mathbb{Q}$, while the prime $19\mathbb{Z}$ splits completely in $E/\mathbb{Q}$ (so also in $K/\mathbb{Q}$).

6 Infinite Extensions and an Alternate Proof of the Existence Theorem

We begin with a brief review of infinite Galois theory. Let M/F be a (possibly infinite) Galois extension with Galois group G. In order to describe the Galois correspondence it is necessary to topologize G. For each $\sigma \in G$, we take the cosets

$$\{\sigma \, \mathrm{Gal}\,(M/K) : \ K/F \text{ is a finite subextension of } M/F\}$$

as a basis of open neighborhoods of σ. The resulting topology is called the *Krull topology* on G.

Exercise 6.19. Let $G = \mathrm{Gal}\,(M/F)$ as above.

a. Show that the topology defined above makes G a Hausdorff topological group.
b. What are the open subgroups if the extension M/F is finite? What are the closed subgroups?   ◊

Exercise 6.20. We can show that G is compact as follows. Consider the map

$$\vartheta : G \prod_K \mathrm{Gal}\,(K/F) \quad \text{given by} \quad \vartheta : \sigma \mapsto \prod_K \sigma\big|_K$$

where K/F runs through all finite Galois subextensions of M/F.

a. Show that $\prod_K \mathrm{Gal}\,(K/F)$ is compact.

b. Show that ϑ is injective and continuous.

c. Show that $\vartheta(G)$ is closed in $\prod_K \mathrm{Gal}\,(K/F)$.   ◊

Theorem 6.1 (Main Theorem of Galois Theory – General Case). *Let M/F be a Galois extension with Galois group G. The map $L \mapsto \mathrm{Gal}\,(M/L)$ is a bijective correspondence between the subextensions L/F of M/F and the closed subgroups of $\mathrm{Gal}\,(M/F)$. Moreover, in this correspondence the open subgroups of $\mathrm{Gal}\,(M/F)$ are paired with the finite subextensions of M/F, (this is because any open subgroup is also closed, so has finite index in $\mathrm{Gal}\,(M/F)$ by Exercise 6.20).*

Proof. We must show that if L/F is a subextension of M/F, then $\mathrm{Gal}\,(M/L)$ is a closed subgroup of G. Certainly it is a subgroup. Also, we may write

$$\mathrm{Gal}\,(M/L) = \bigcap_{\substack{K \subseteq L \\ [K:F]<\infty}} \mathrm{Gal}\,(M/K).$$

Thus it suffices to show that $\mathrm{Gal}\,(M/K)$ is closed for all finite subextensions K/F. But if K/F is a finite subextension and N is its normal closure in M, then N/F is finite, and for any $\sigma \in \mathrm{Gal}\,(M/K)$, we may see that σ lies inside an open neighborhood within $\mathrm{Gal}\,(M/K)$ by noting we have $\sigma \in \sigma\,\mathrm{Gal}\,(M/N) \subseteq \mathrm{Gal}\,(M/K)$. It follows that $\mathrm{Gal}\,(M/K)$ is open, and hence it is also closed.

It is clear that the map $L \mapsto \mathrm{Gal}\,(M/L)$ is injective, since the fixed field of a subgroup of G is uniquely determined. We must show that it is also surjective, i.e., if H is a closed subgroup of G, then there is some subextension L/F such that $H = \mathrm{Gal}\,(M/L)$. If this is to succeed, it must be that L is the fixed field of H. Thus we must show that if H is a closed subgroup of G and L is the fixed field of H, then $\mathrm{Gal}\,(M/L) = H$.

That $\mathrm{Gal}\,(M/L) \supseteq H$ follows from the definitions, so we consider an arbitrary element $\tau \in \mathrm{Gal}\,(M/L)$. Let E/L be any finite subextension of M/L and apply the natural map $H \to \mathrm{Gal}\,(E/L)$, (given by restriction to E). The image of H in $\mathrm{Gal}\,(E/L)$ has fixed field L, and since E/L is a finite extension, we may apply finite Galois theory to conclude that the image of H is all of $\mathrm{Gal}\,(E/L)$. Thus we may find $\sigma \in H$ such that $\sigma\big|_E = \tau\big|_E$. But then $\sigma \in \tau\,\mathrm{Gal}\,(M/E) \cap H$, and hence τ lies inside the topological closure of H. Since H is closed, we have $\tau \in H$ as needed.   □

Much in the same way as for finite extensions one may also show that closed normal subgroups of $\mathrm{Gal}\,(M/F)$ correspond to intermediate fields L such that L/F is Galois, and in this case we have ${\mathrm{Gal}\,(M/F)}\big/{\mathrm{Gal}\,(M/L)} \cong \mathrm{Gal}\,(L/F)$.

Exercise 6.21. Let $G = \mathrm{Gal}\,(M/F)$ and let $\sigma_0 \in G$ denote the identity. Show that σ_0 has a basis of neighborhoods consisting of *normal* subgroups of G. A compact Hausdorff topological group with this property is called a *profinite group*. (HINT: Consider the groups $\mathrm{Gal}\,(M/K)$, where K varies through the finite Galois subextensions of M/F.)   ◊

Exercise 6.22. Show that a compact Hausdorff topological group is profinite if and only if it is totally disconnected. (Hence every finite group is a profinite group.) ◊

Exercise 6.23. Show that the group of units of the ring of integers in a finite extension of $\mathbb{Q}_p$ is a profinite group.   ◊

Next, we discuss how profinite groups can be constructed from finite groups. Recall, a *directed set* is an ordered set $\mathcal{I}$ that has the property that every pair of elements $i_1, i_2 \in \mathcal{I}$ is dominated by some element of $\mathcal{I}$, i.e., there is some element $i_3 \in \mathcal{I}$ such that $i_1 \leq i_3$ and $i_2 \leq i_3$. A *projective system* over a directed set $\mathcal{I}$ is a family of sets (groups for us) G_i and morphisms (group homomorphisms for us) $\varphi_{ij} : G_j \to G_i$ defined whenever $i \leq j$ in $\mathcal{I}$, such that $\varphi_{ii} = \mathrm{id}_{G_i}$ for any i, and whenever $i \leq j \leq k$ we have $\varphi_{ij} \circ \varphi_{jk} = \varphi_{ik}$. Given a projective system, we define its *projective limit* to be

$$G = \varprojlim_{i \in \mathcal{I}} G_i = \Big\{ \prod_{i \in \mathcal{I}} \sigma_i \in \prod_{i \in \mathcal{I}} G_i : \varphi_{ij}(\sigma_j) = \sigma_i \text{ whenever } i \leq j \Big\}.$$

When the G_i are groups and the φ_{ij} are homomorphisms, G is a group. When the G_i are topological spaces and the φ_{ij} are continuous, we get that G is a closed subspace of the product of the G_i, (see Exercise 6.24).

Exercise 6.24. Let $\{G_i, \varphi_{ij}\}$ be a projective system of finite groups over the directed set $\mathcal{I}$. Show that $\varprojlim_{i \in \mathcal{I}} G_i$ is a closed subgroup of $\prod_{i \in \mathcal{I}} G_i$. (Consider the G_i as discrete topological spaces.)   ◊

Exercise 6.24 tells us that the projective limit of finite groups $\varprojlim_{i \in \mathcal{I}} G_i$ is a compact Hausdorff toplogical group (here we are using the subspace topology induced from the product topology on $\prod_{i \in \mathcal{I}} G_i$). Hence $\varprojlim_{i \in \mathcal{I}} G_i$ will be a profinite group if it has a basis of open neighborhoods of 1 consisting of normal subgroups.

For a finite subset $\mathcal{J}$ of $\mathcal{I}$, consider the subgroups of $\prod_{i \in \mathcal{I}} G_i$ of the form

$$U_{\mathcal{J}} = \prod_{i \notin \mathcal{J}} G_i \prod_{i \in \mathcal{J}} H_i,$$

where each H_i is a normal subgroup of G_i. The intersections $U_{\mathcal{J}} \cap \varprojlim_{i \in \mathcal{I}} G_i$ are normal subgroups of $\varprojlim_{i \in \mathcal{I}} G_i$. It is straightforward to verify that they form a basis of open neighborhoods of 1 in $\varprojlim_{i \in \mathcal{I}} G_i$. Hence the projective limit of finite groups is a profinite group (thus the terminology "profinite")!

Exercise 6.25. Let p be a prime. Order $\mathbb{N}$ in the usual way, and (for $n \leq m$) use the natural maps $\mathbb{Z}/p^m\mathbb{Z} \longrightarrow \mathbb{Z}/p^n\mathbb{Z}$ to form a projective system. Show that

$$\varprojlim_{n \in \mathbb{N}} \mathbb{Z}/p^n\mathbb{Z} \cong \mathbb{Z}_p.$$

◊

Example.

4. Fix a prime p and consider the extension $\mathbb{F}_p^{\text{alg}}/\mathbb{F}_p$. For each positive integer n, we have a finite subextension $\mathbb{F}_{p^n}/\mathbb{F}_p$ of degree n inside $\mathbb{F}_p^{\text{alg}}/\mathbb{F}_p$. Let φ_n denote the Frobenius automorphism in $\text{Gal}\,(\mathbb{F}_{p^n}/\mathbb{F}_p)$; then there is an isomorphism

$$\text{Gal}\,(\mathbb{F}_{p^n}/\mathbb{F}_p) \cong \mathbb{Z}/n\mathbb{Z}$$

given by $\varphi_n \mapsto 1 + n\mathbb{Z}$. The Galois groups $\text{Gal}\,(\mathbb{F}_{p^n}/\mathbb{F}_p)$ comprise a projective system, indexed by the positive integers (ordered by "divides"). The maps $\text{Gal}\,(\mathbb{F}_{p^m}/\mathbb{F}_p) \longrightarrow \text{Gal}\,(\mathbb{F}_{p^n}/\mathbb{F}_p)$ in this projective system are the canonical ones. We may form the group

$$\varprojlim_{n \in \mathbb{Z}_+} \text{Gal}\,(\mathbb{F}_{p^n}/\mathbb{F}_p)$$

$$= \{ \prod_{n \in \mathbb{Z}_+} \sigma_n \in \prod_{n \in \mathbb{Z}_+} \text{Gal}\,(\mathbb{F}_{p^n}/\mathbb{F}_p) : \sigma_m \big|_{\mathbb{F}_{p^n}} = \sigma_n \text{ whenever } n|m \}.$$

It is clear that we have an isomorphism

$$\varprojlim_{n \in \mathbb{Z}_+} \text{Gal}\,(\mathbb{F}_{p^n}/\mathbb{F}_p) \cong \varprojlim_{n \in \mathbb{Z}_+} \mathbb{Z}/n\mathbb{Z}$$

where in the limit on the right, the ordering on $\mathbb{Z}_+$ is again "divides," and where the maps $\mathbb{Z}/m\mathbb{Z} \longrightarrow \mathbb{Z}/n\mathbb{Z}$ for the limit on the right (when $n|m$) are the natural ones.

Exercise 6.26. Show that

$$\text{Gal}\,(\mathbb{F}_p^{\text{alg}}/\mathbb{F}_p) \cong \varprojlim_{n \in \mathbb{Z}_+} \text{Gal}\,(\mathbb{F}_{p^n}/\mathbb{F}_p).$$

Then use the Chinese Remainder Theorem to show that $\varprojlim_{n \in \mathbb{N}} \mathbb{Z}/n\mathbb{Z}$, which is usually denoted $\hat{\mathbb{Z}}$, is isomorphic to the product $\prod_p \mathbb{Z}_p$. $\diamond$

Continue to let φ_n denote the Frobenius automorphism in $\text{Gal}\,(\mathbb{F}_{p^n}/\mathbb{F}_p)$. Under the isomorphism $\text{Gal}\,(\mathbb{F}_p^{\text{alg}}/\mathbb{F}_p) \cong \varprojlim_{n \in \mathbb{N}} \mathbb{Z}/n\mathbb{Z} = \hat{\mathbb{Z}}$, the element $\varphi = (\ldots, \varphi_n, \ldots) \in \text{Gal}\,(\mathbb{F}_p^{\text{alg}}/\mathbb{F}_p)$ is mapped to $(\ldots, 1+n\mathbb{Z}, \ldots) = 1 \in \hat{\mathbb{Z}}$. Under the isomorphism with $\prod_p \mathbb{Z}_p$, the element φ corresponds to $(\ldots, 1, \ldots)$.

Exercise 6.27. Let p be a prime and let $G = \text{Gal}\,(\mathbb{F}_p^{\text{alg}}/\mathbb{F}_p)$. Show that the element $\varphi \in G$ as discussed above, satisfies $\varphi : x \mapsto x^p$. Hence φ is the Frobenius automorphism in G. $\diamond$

By Exercise 6.27, the Frobenius automorphism $\varphi \in G = \mathrm{Gal}(\mathbb{F}_p^{\mathrm{alg}}/\mathbb{F}_p)$ is mapped to $1 \in \hat{\mathbb{Z}}$ under the isomorphism of Exercise 6.26. It follows that the image of $\langle \varphi \rangle$ in $\hat{\mathbb{Z}}$ is $\langle 1 \rangle \cong \mathbb{Z}$. Hence, (in contrast to the situation for finite extensions of $\mathbb{F}_p$), $\langle \varphi \rangle$ is a *proper* subgroup of G. Note however, that $\langle 1 \rangle \cong \mathbb{Z}$ is dense in $\hat{\mathbb{Z}}$ (why?), so that $\langle \varphi \rangle$ is dense in G.

Exercise 6.28. Let G be a profinite group and let H vary through the open normal subgroups of G. Show that $G \cong \varprojlim_{H} G/H$ topologically and algebraically. ◊

Exercise 6.29. Show that if M/F is a Galois extension, then its Galois group (with the Krull topology) is a profinite group. ◊

Example.

5. For a natural number n, we let $\mathbb{Q}^{(n)}$ denote the unique subfield of $\mathbb{Q}(\zeta_{p^{n+1}})$ that is cyclic of degree p^n over $\mathbb{Q}$. Then

$$\mathbb{Q} = \mathbb{Q}^{(0)} \subseteq \mathbb{Q}^{(1)} \subseteq \mathbb{Q}^{(2)} \subseteq \cdots$$

and if $\mathbb{Q}^{(\infty)} = \cup \mathbb{Q}^{(n)}$ then $\mathbb{Q}^{(\infty)}/\mathbb{Q}$ is infinite Galois, with Galois group isomorphic to $\mathbb{Z}_p$.

More generally, if F is a number field, we put $F^{(n)} = F\mathbb{Q}^{(n)}$ and $F^{(\infty)} = F\mathbb{Q}^{(\infty)} = \cup F^{(n)}$. Then $F^{(\infty)}/F$ is infinite Galois with Galois group isomorphic to $\mathbb{Z}_p$. (This extension $F^{(\infty)}/F$ is called the *cyclotomic* $\mathbb{Z}_p$-*extension* of F; any infinite Galois extension of F with Galois group isomorphic to $\mathbb{Z}_p$ is called a $\mathbb{Z}_p$-*extension* of F.)

Exercise 6.30. Suppose that K/F is a $\mathbb{Z}_p$-extension. Show that for every $n \in \mathbb{N}$ there is a unique intermediate field $K^{(n)}$ with $[K^{(n)} : F] = p^n$. Moreover, show that these intermediate fields $K^{(n)}$ are the only proper intermediate fields in the extension K/F. ◊

If G is any group, then the *profinite completion* of G is

$$\varprojlim_{H} G/H$$

where H varies through all the normal subgroups of G of finite index in G. For example, the profinite completion of $\mathbb{Z}$ is $\hat{\mathbb{Z}}$.

Exercise 6.31. What is the profinite completion of a finite group? ◊

Exercise 6.32. Suppose A is a G-module, where G is a profinite group. Show that the following are equivalent.

i. $\cup_H A^H = A$, where H runs through the open subgroups of G.

ii. The map $G \times A \longrightarrow A$ defined by $(g, x) \mapsto gx$ is continuous, where A is given the discrete topology. ◊

At this point, we may return to our study of class field theory. We want to sketch an alternate proof of the Existence Theorem. For this proof, we need to have an analogue of the Artin map for an infinite abelian extension M of a number field F. Using the above discussion of infinite Galois theory, we are led to consider projective limits as we study $\mathrm{Gal}\,(M/F)$. It is the Consistency Property of the idèlic Artin symbol that allows us to construct an Artin map for an infinite extension using the Artin maps for its finite subextensions.

From the Consistency Property, we see that if K/F and L/E are abelian extensions of number fields, with $F \subseteq E$ and $K \subseteq L$, then the diagram below commutes.

$$
\begin{array}{ccc}
J_E & \xrightarrow{\ \rho_{L/E}\ } & \mathrm{Gal}\,(L/E) \\
{\scriptstyle N_{E/F}}\downarrow & & \downarrow \\
J_F & \xrightarrow{\ \rho_{K/F}\ } & \mathrm{Gal}\,(K/F)
\end{array}
$$

(The vertical map on the right is the natural one.)

Suppose M/F is an infinite abelian extension, where F is a number field. We may define a homomorphism

$$
\rho_{M/F} : J_F \longrightarrow \mathrm{Gal}\,(M/F)
$$

using a projective limit. If K/F runs through the finite subextensions of M/F then the elements $\rho_{K/F}(\mathbf{a})$ can be pasted together to form an element of $\varprojlim_{K} \mathrm{Gal}\,(K/F)$. Since we can identify $\varprojlim_{K} \mathrm{Gal}\,(K/F)$ with $\mathrm{Gal}\,(M/F)$, we may set the corresponding element of $\mathrm{Gal}\,(M/F)$ equal to $\rho_{M/F}(\mathbf{a})$. Note that if $\mathbf{a} \in F^{\times}$, then for any of the finite extensions K/F, the image $\rho_{K/F}(\mathbf{a})$ is the identity in $\mathrm{Gal}\,(K/F)$. Hence $\rho_{M/F}(\mathbf{a})$ must be the identity in $\mathrm{Gal}\,(M/F)$, i.e., $F^{\times} \subseteq \ker \rho_{M/F}$. We continue to use the term "Artin map" for $\rho_{M/F}$, even when M/F is infinite Galois.

Continuing to let F be a number field, we let $M = F_{\mathrm{ab}}$ be the maximal abelian extension of F, an infinite Galois extension. The group $\mathrm{Gal}\,(F_{\mathrm{ab}}/F)$ is profinite, as we have seen in Exercise 6.29. Recall that when K/F is finite abelian, the Artin map gives rise to a surjective homomorphism

$$
C_F \twoheadrightarrow \mathrm{Gal}\,(K/F)
$$

with kernel $N_{K/F} C_K$. Taking the projective limit of the set of all finite abelian groups $\mathrm{Gal}\,(K/F)$ as discussed above, we construct the Artin map $\rho_{M/F}$ in the case $M = F_{\mathrm{ab}}$, (typically, $\rho_{F_{\mathrm{ab}}/F}$ is denoted simply ρ_F). Now let ω_F denote the map on C_F that corresponds to the map ρ_F on J_F (possible since $F^{\times} \subseteq \ker \rho_F$). The map

$$
\omega_F : C_F \to \mathrm{Gal}\,(F_{\mathrm{ab}}/F)
$$

is called the *norm residue symbol* for F.

Exercise 6.33. Show that the norm residue symbol is a continuous surjective homomorphism, and its kernel is contained in every open subgroup of finite index in C_F.   ◊

Now we sketch an alternate proof of the Existence Theorem, as it appears in the lecture notes of Artin and Tate, [AT]. Define an absolute value on J_F by

$$|\mathbf{a}| = \prod_v \|a_v\|_v$$

and put $J_F^0 = \{\mathbf{a} \in J_F \text{ of absolute value } 1\}$ and $C_F^0 = \{[\mathbf{a}] \in C_F \text{ where } \mathbf{a} \in J_F^0\}$.

Exercise 6.34. Show that the above satisfies the axioms for an absolute value on J_F. Then show that $F^\times \subseteq J_F^0$, from which we see that C_F^0 is well-defined.   ◊

Exercise 6.35. Show that C_F^0 is compact and that $\omega_F(C_F^0) = \omega_F(C_F)$.   ◊

Let B be an open subgroup of finite index in C_F. Then B is also closed in C_F. By Exercise 6.35, C_F^0 is compact. Since $B^0 = B \cap C_F^0$ is closed in C_F^0, it is compact as well. Note that $[C_F^0 : B^0] = [C_F : B]$.

Put $H = \omega_F(B^0)$; this is a closed subgroup of $G = \mathrm{Gal}(F_{\mathrm{ab}}/F)$. Since $B \supseteq \ker \omega_F$, we may conclude that $B^0 = \omega_F^{-1}(H) \cap C_F^0$. Thus $[G : H] = [C_F^0 : B^0] = [C_F : B]$.

Let K be the fixed field of H. Then (from the definition of ω_F) $N_{K/F}C_K \subseteq \omega_F^{-1}(H) \subseteq B$. But then $[C_F : N_{K/F}C_K] = [K : F] = [G : H] = [C_F : B]$ gives $B = N_{K/F}C_K$.

The argument sketched above shows that every open subgroup B of finite index in C_F is of the form $N_{K/F}C_K$ for some finite abelian extension K/F. For uniqueness of the extension K/F, see the Ordering Theorem or Theorem 3.2.2.

7 An Example: Cyclotomic Fields

Let $p > 2$ be prime, let $E = \mathbb{Q}(\zeta_p)$, and let $E^+ = \mathbb{Q}(\zeta_p + \zeta_p^{-1})$, the totally real subfield of index 2 in E.

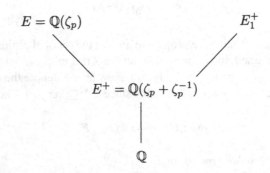

We have a Hilbert class field E_1^+ of E^+ that is everywhere unramified over E^+ and also we know that $p\mathbb{Z}$ is totally ramified in $E/\mathbb{Q}$, so also in E/E^+. Thus $E \cap E_1^+ = E^+$ and we may apply Proposition 4.4 from the section on the Hilbert class field to get that h_{E^+} divides h_E. (Note that the infinite primes of E^+ are all real and the infinite primes of E are all imaginary, so that the infinite primes ramify in E/E^+ too.)

Let $h = h_E$ and $h^+ = h_{E^+}$. Define

$$h^- = {}^h/_{h^+}, \quad \text{the } relative\ class\ number \text{ of } E.$$

How can we interpret h^- algebraically? Since $E_1^+ \cap E = E^+$, we have the following picture.

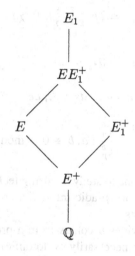

Thus, we may apply (ii) of Theorem 4.3 to conclude that $N_{E/E^+} : \mathcal{C}_E \longrightarrow \mathcal{C}_{E^+}$ is surjective. Let

$$\mathcal{C}_E^* = \{[\mathfrak{a}] \in \mathcal{C}_E : N_{E/E^+}\mathfrak{a} \in \mathcal{P}_{E^+}\} = \ker_{\mathcal{C}_E} N_{E/E^+}.$$

We have shown that

$$1 \longrightarrow \mathcal{C}_E^* \xrightarrow{\subseteq} \mathcal{C}_E \xrightarrow{N_{E/E^+}} \mathcal{C}_{E^+} \longrightarrow 1$$

is exact, whence

$$\mathcal{C}_E / \mathcal{C}_E^* \cong \mathcal{C}_{E^+}.$$

From this we conclude

$$\text{Gal}(E_1/E) \cong \mathcal{C}_E$$
$$\text{Gal}(EE_1^+/E) \cong \text{Gal}(E_1^+/E^+) \cong \mathcal{C}_{E^+}$$
$$\text{Gal}(E_1/EE_1^+) \cong \mathcal{C}_E^*.$$

Exercise 6.36. How much of the above can be generalized to the cyclotomic field $E = \mathbb{Q}(\zeta_m)$ where m is a prime power? What if m is an arbitrary positive integer?   ◊

For $E = \mathbb{Q}(\zeta_p)$, we record the following result of Kummer.

Theorem 7.1. *Let $p > 2$ be a prime, let $E = \mathbb{Q}(\zeta_p)$ have class number h, and let $E^+ = \mathbb{Q}(\zeta_p + \zeta_p^{-1})$ have class number h^+. Let $h^- = {}^h/_{h^+}$. Then*

$$h^- = 2p \prod_{\substack{\chi \text{ odd} \\ f_\chi = p}} \frac{1}{2} L(0, \chi)$$

$$h^+ = [\mathcal{U}_E : \mathcal{Y}_E]$$

where $\mathcal{Y}_E$ is the group of cyclotomic units of E, i.e.,

$$\mathcal{Y}_E = \left\{ \frac{1 - \zeta_p^a}{1 - \zeta_p^b} : a, b \not\equiv 0 \pmod{p} \right\}.$$
                                                                                        □

The above theorem allows one to study h^- using techniques from complex analysis, and also, (because there are p-adic interpolations of Dirichlet L-functions), techniques from p-adic analysis.

The result that says h^+ divides h comes from a proposition that applies more generally (to fields that are not necessarily cyclotomic). In particular, we could use it to show that h_F divides h_E whenever E is a totally imaginary extension of a totally real field F, with $[E : F] = 2$, (a *CM*-field). In the case when E is cyclotomic, we actually have a stronger result; its proof makes use of the roots of unity in E.

Theorem 7.2. *Let $p > 2$ be a prime, let $E = \mathbb{Q}(\zeta_p)$, and let $E^+ = \mathbb{Q}(\zeta_p + \zeta_p^{-1})$ as before. The map $\mathcal{C}_{E^+} \longrightarrow \mathcal{C}_E$ given by $[\mathfrak{a}]_{E^+} \mapsto [\mathfrak{a}\mathcal{O}_E]_E$ is injective.*

Proof. We must show that if $\mathfrak{a}\mathcal{O}_E$ is principal in E then $\mathfrak{a}$ was principal in E^+ already. Say

$$\mathfrak{a}\mathcal{O}_E = \langle \alpha \rangle$$

for $\alpha \in E$ and note that $\bar{\alpha}/\alpha$ is real and

$$\langle \bar{\alpha}/\alpha \rangle = \bar{\mathfrak{a}}\mathfrak{a}^{-1}\mathcal{O}_E = \mathcal{O}_E = \langle 1 \rangle,$$

(because $\mathfrak{a}$ is real so that $\bar{\mathfrak{a}} = \mathfrak{a}$). Thus $\bar{\alpha}/\alpha$ is a unit in $\mathcal{O}_E$. Also $\bar{\alpha}/\alpha$ has absolute value 1 in $\mathbb{C}$, as do all of its conjugates (over $\mathbb{Q}$).

Exercise 6.37. Show that if ε is an algebraic integer all of whose conjugates have absolute value 1 in $\mathbb{C}$, then ε is a root of unity, (a theorem of Kronecker). ◇

By Exercise 6.37, $\bar{\alpha}/\alpha$ is a root of unity. Let $\pi = \zeta_p - 1$. Then $\mathfrak{P} = \langle \pi \rangle$ is the prime ideal of E above $p\mathbb{Z}$ and

$$\pi/\bar{\pi} = \zeta_p - 1 / \zeta_p^{-1} - 1 = -\zeta_p$$

so $\pi/\bar{\pi}$ generates the group $\mathcal{W}_E$. Thus

$$\bar{\alpha}/\alpha = (\pi/\bar{\pi})^t$$

for some t, whence $\alpha\pi^t = \bar{\alpha}\bar{\pi}^t \in \mathbb{R}$. Also $\mathfrak{a} \subseteq \mathbb{R}$. Now if $x \in \mathbb{R} \cap E$, then $\mathrm{ord}_\pi(x) \in 2\mathbb{Z}$. Similarly, since $\mathfrak{a} \subseteq \mathbb{R}$, we have $\mathrm{ord}_\mathfrak{P}(\mathfrak{a}) \in 2\mathbb{Z}$. Thus

$$t = \mathrm{ord}_\pi(\alpha\pi^t) - \mathrm{ord}_\pi(\alpha)$$
$$= \mathrm{ord}_\pi(\alpha\pi^t) - \mathrm{ord}_\mathfrak{P}(\mathfrak{a}) \in 2\mathbb{Z}.$$

We get

$$\bar{\alpha}/\alpha = (-\zeta_p)^t \in \mathcal{W}_E^2.$$

Thus

$$\bar{\alpha}/\alpha = (-\zeta_p)^{2d} = \zeta_p^d/\bar{\zeta}_p^d$$

whence $\alpha\zeta_p^d \in \mathbb{R}$. This gives

$$\mathfrak{a}\mathcal{O}_E = \alpha\mathcal{O}_E = \alpha\zeta_p^d\mathcal{O}_E = \langle\alpha\zeta_p^d\rangle\mathcal{O}_E$$

where $\langle\alpha\zeta_p^d\rangle$ is a principal ideal of E^+. By Exercise 6.38 below, we must have $\mathfrak{a} = \langle\alpha\zeta_p^d\rangle$. □

Exercise 6.38. Let E/F be an extension of number fields and let $\mathfrak{a}$, $\mathfrak{b}$ be fractional ideals of F. Show that if $\mathfrak{a}\mathcal{O}_E = \mathfrak{b}\mathcal{O}_E$ then $\mathfrak{a} = \mathfrak{b}$. ◇

Exercise 6.39. Can the previous theorem be generalized to the field $\mathbb{Q}(\zeta_m)$, where $m = p^t$? ◇

We record the following two theorems of Kummer, which are related to the one we already mentioned. The first comes from results on special values of L-functions. The second represents one of the classical approaches to the search for a proof of Fermat's Last Theorem; however it was an incomplete approach, as we discuss below.

Theorem 7.3. *Let* $E = \mathbb{Q}(\zeta_p)$ *and let* B_n *denote the* n^{th} *Bernoulli number, i.e.,*

$$\frac{t}{e^t - 1} = \sum_{n=0}^{\infty} B_n \frac{t^n}{n!}.$$

Then $p | h_E$ *if and only if* p *divides the numerator of some* B_{2k}, *where* $1 \leq k < \frac{p-1}{2}$. $\square$

Theorem 7.4. *If* $p > 2$ *is prime and* $p \nmid h_E$, *where* $E = \mathbb{Q}(\zeta_p)$, *then*

$$x^p + y^p = z^p, \qquad (xyz, p) = 1$$

has no non-trivial solution in integers. $\square$

If $p \nmid h_E$, we say p is *regular*. Otherwise p is *irregular*. There are infinitely many irregular primes (the proof is by contradiction, using congruences amongst Bernoulli numbers). The first few examples of irregular primes are 37, 59, 67, 101, 103, ... Also, h_E grows rapidly as p increases; $\mathcal{O}_E$ is a p.i.d. in only a small number of cases, (a result obtained independently by Montgomery and Uchida in 1971 says that $h_E = 1$ if and only if $p \leq 19$).

A conjecture known as *Vandiver's conjecture*, (although its origin seems to date to Kummer's work), says that if $E = \mathbb{Q}(\zeta_p)$, then $p \nmid h^+$. It has been verified computationally for primes up to several digits. There are probabilistic heuristics that seem to indicate that it should be true for a large majority of primes. See Chapter 8 of [Wa] for a discussion of this.

We know it is very possible to have $p | h$. And certainly, if $p | h^-$ then $p | h$. If we want to prove the converse, then showing that $p | h^+$ implies $p | h^-$ will suffice. The result that $p | h$ if and only if $p | h^-$ allows one to study regular versus irregular primes using only h^-, (which is generally more accessible, since one may apply analytic techniques using L-functions). This is exactly the approach taken in the proof of the theorem of Kummer mentioned above, where the Bernoulli numbers arise. The reader is encouraged to consult a text on analytic number theory, (or on cyclotomic fields), for a detailed account of the relationship between L-functions and Bernoulli numbers and a proof of Kummer's result. See Washington's *Introduction to Cyclotomic Fields*, [Wa], and Lang's *Cyclotomic Fields I and II*, [L3], for much more on these and other related ideas. We shall be content here to prove that $p | h$ if and only if $p | h^-$. First we need a lemma.

Lemma 7.5. *If* K/E *is everywhere unramified and Galois with* $\text{Gal}(K/E) = G$, *then* $\mathcal{I}_K^G = \mathcal{I}_E$.

Proof. That $\mathcal{I}_K^G \supseteq \mathcal{I}_E$ is clear. For "$\subseteq$" suppose $\mathfrak{A} \in \mathcal{I}_K^G$ and factor $\mathfrak{A}$ as a product of prime ideals in $\mathcal{O}_K$. For $\mathfrak{P} | \mathfrak{A}$, let $\mathfrak{p} = \mathfrak{P} \cap E$. Since G transitively permutes the primes above $\mathfrak{p}$, but fixes $\mathfrak{A}$, we see that every prime of $\mathcal{O}_K$ that divides $\mathfrak{p}$ must occur in the factorization of $\mathfrak{A}$ and that all occur with equal multiplicity. Thus $\mathfrak{A}$ has the form

$$\mathfrak{A} = \mathfrak{p}_1^{r_1} \cdots \mathfrak{p}_t^{r_t} \mathcal{O}_K,$$

which we identify with an ideal of E in the obvious way. $\square$

Theorem 7.6 (Kummer). *Let $E = \mathbb{Q}(\zeta_p)$ where $p > 2$ is prime. If $p|h^+$, then $p|h^-$.*

Proof. Assume $p|h^+$. Since $h^+ = [E_1^+ : E^+]$, we see that $G = \text{Gal}(E_1^+/E^+)$ has a quotient of order p, whence there is an intermediate field L with $E^+ \subseteq L \subseteq E_1^+$, and $[L : E^+] = p$. Note that L/E^+ is everywhere unramified and abelian (in fact cyclic). Let $K = EL$.

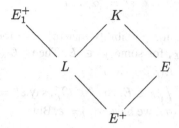

Now K/E is also unramified everywhere and cyclic of degree p (because $E \cap L = E^+$ since $(2, p) = 1$). Say $\text{Gal}(K/E) = \langle \sigma \rangle$.

Note that $N_{K/E}(\zeta_p) = \zeta_p^p = 1$, so we may apply Hilbert's Theorem 90 to conclude that $\zeta_p = \frac{\beta}{\sigma(\beta)}$ for some $\beta \in K^\times$. Now $\beta \notin E$, as $\zeta_p \neq 1$.

We get $\sigma(\beta) = \beta\zeta_p^{-1}$ and also that σ fixes $\beta^p = \alpha$, say. We have $\alpha \in E$ and

$$K \supseteq E(\sqrt[p]{\alpha}) = E(\beta) \supsetneq E.$$

Since $[K : E]$ is prime, we must have $K = E(\beta)$.

Now $\text{Gal}(E/E^+) = \langle j \rangle$, say, where j is essentially complex conjugation. Thus $\text{Gal}(K/L)$ is also generated by complex conjugation; we use j to denote the generator of $\text{Gal}(K/L)$ as well. Write $\bar{\beta} = j(\beta)$ as usual. Then in $\text{Gal}(K/E^+)$, we have $\sigma j = j\sigma$ (why?) and $\sigma(\beta\bar{\beta}) = \beta\zeta_p^{-1}\overline{\beta\zeta_p^{-1}} = \beta\bar{\beta}$. Thus $\beta\bar{\beta} \in E$. Of course, $\beta\bar{\beta} \in \mathbb{R}$ too, so $\beta\bar{\beta} \in E^+$.

Consider the ideal $\beta\mathcal{O}_K$. Since $\sigma(\beta)$ and β are associates in $\mathcal{O}_K$, the ideal $\beta\mathcal{O}_K$ is fixed by σ, hence by $\text{Gal}(K/E)$.

Since K/E is everywhere unramified, $\mathcal{I}_K^{\text{Gal}(K/E)} = \mathcal{I}_E$ by Lemma 7.5, whence

$$\beta\mathcal{O}_K = \mathfrak{a}\mathcal{O}_K$$

for some $\mathfrak{a} \in \mathcal{I}_E$. We claim that $[\mathfrak{a}]$ has order p in $\mathcal{C}_E^*$.

To show that $[\mathfrak{a}] \in \mathcal{C}_E^*$, note that since $\beta\mathcal{O}_K = \mathfrak{a}\mathcal{O}_K$, we have

$$(N_{K/L}\mathfrak{a}\mathcal{O}_K) = (N_{K/L}\beta\mathcal{O}_K),$$
$$\mathfrak{a}j(\mathfrak{a})\mathcal{O}_K = \beta\bar{\beta}\mathcal{O}_K.$$

But $\beta\bar{\beta} \in E^+$. Also $(N_{E/E^+}\mathfrak{a})\mathcal{O}_E = \mathfrak{a}j(\mathfrak{a})$, an ideal of E. By Exercise 6.38,

$$\mathfrak{a}j(\mathfrak{a}) = \beta\bar{\beta}\mathcal{O}_E$$
$$N_{E/E^+}\mathfrak{a} = \langle\beta\bar{\beta}\rangle \text{ in } E^+.$$

To show that $[\mathfrak{a}]$ has order p, note that

$$\mathfrak{a}^p \mathcal{O}_K = \beta^p \mathcal{O}_K = \alpha \mathcal{O}_K$$

and $\alpha \in E$. By Exercise 6.38 again,

$$\mathfrak{a}^p = \langle \alpha \rangle \text{ in } E.$$

It remains to eliminate the possibility that $[\mathfrak{a}]$ has order 1, i.e., $\mathfrak{a}$ is principal in E. Suppose $\mathfrak{a} = \gamma \mathcal{O}_E$, for some $\gamma \in E$. Then $\gamma \mathcal{O}_K = \mathfrak{a}\mathcal{O}_K = \beta \mathcal{O}_K$, so $\beta/\gamma = \varepsilon \in \mathcal{O}_K^\times$ and $K = E(\beta) = E(\varepsilon)$.

Now $\varepsilon^p = \beta^p/\gamma^p = \alpha/\gamma^p \in E$, so $\varepsilon^p \in \mathcal{O}_E^\times$, say $\varepsilon^p = \delta$. Since ε generates K over E and $\mathrm{Gal}\,(K/E) = \langle \sigma \rangle$, we have $\sigma(\varepsilon) \neq \varepsilon$. But

$$\left(\sigma(\varepsilon)/\varepsilon\right)^p = \sigma(\delta)/\delta = \delta/\delta = 1$$

so $\sigma(\varepsilon)/\varepsilon$ is a primitive p^{th} root of unity, say ζ_p^a, with $(a, p) = 1$. We have $\sigma(\varepsilon) = \zeta_p^a \varepsilon$.

Let $\eta = \varepsilon/\bar{\varepsilon}$, an element of $\mathcal{O}_K^\times$. Since $\mathrm{Gal}\,(K/L) \cong \mathrm{Gal}\,(E/E^+) = \langle j \rangle$ we see that L is totally real and K is totally imaginary. If $\lambda : K \hookrightarrow \mathbb{C}$ is any embedding, then $\lambda(L) \subseteq \mathbb{R}$ and

$$\lambda(\eta) = \lambda(\varepsilon)/\lambda(\bar{\varepsilon}) = \lambda(\varepsilon)/\overline{\lambda(\varepsilon)}$$

whence $|\lambda(\eta)|_c = 1$ for every embedding λ of K. Thus η is a root of unity.

Also, since $\eta = \varepsilon/\bar{\varepsilon}$, we may compute

$$\sigma(\eta) = \sigma(\varepsilon)/\sigma(\bar{\varepsilon}) = \zeta_p^a \varepsilon / \overline{\zeta_p^a \varepsilon} = \zeta_p^{2a} \varepsilon / \bar{\varepsilon} = \zeta_p^{2a} \eta.$$

Since $(a, p) = 1$ and $p > 2$, we have $(2a, p) = 1$ and $\zeta_p^{2a} \neq 1$. This gives $\sigma(\eta) \neq \eta$, so $\eta \notin E$.

Now $E \subsetneq E(\eta) \subseteq K$ and $[K : E] = p$, so $K = E(\eta) = \mathbb{Q}(\zeta_p, \eta)$, a cyclotomic extension of E. All cyclotomic extensions are ramified (**Exercise 6.40**). Thus K/E is ramified, a contradiction. We have shown that $[\mathfrak{a}]$ cannot have order 1 in $\mathcal{C}_E^*$, so must have order p and the claim is proved.

Now $p | \# \mathcal{C}_E^*$ and

$$\# \mathcal{C}_E^* = \# \mathcal{C}_E / \# \mathcal{C}_{E^+} = h/h^+ = h^-.$$

Thus $p | h^-$. $\square$

We can also apply what we know about infinite Galois extensions to cyclotomic fields, yielding some ideas about $\mathbb{Z}_p$-extensions that are important in Iwasawa theory. To do this, we must first understand ramification in infinite extensions. The notion of ramification index is troublesome, since the upper field in our extension is not a number field. Without factorization of ideals, we cannot determine a ramification index using the methods employed for number fields. Instead, we shall approach the issue by considering inertia groups. Recall in the case of a finite Galois extension, the ramification index of a prime ideal is simply the order of its inertia group. In the infinite Galois case, we want to use an analog of this to define the ramification index. Of course, first we must have a suitable definition of the inertia subgroup associated to a prime ideal in the case of an infinite Galois group.

If K is an algebraic extension of $\mathbb{Q}$ (finite or infinite), then we let $\mathcal{O}_K$ denote the ring of algebraic integers in K. If $\mathfrak{P}$ is a prime ideal of $\mathcal{O}_K$, then $\mathfrak{P} \cap \mathbb{Z}$ is a prime ideal of $\mathbb{Z}$. Moreover, if $\mathfrak{P}$ is non-zero, then so is $\mathfrak{P} \cap \mathbb{Z}$ (why?). Thus $\mathfrak{P} \cap \mathbb{Z} = p\mathbb{Z}$ for some prime p.

Exercise 6.41. Show that $\mathcal{O}_K / \mathfrak{P}$ is a field, and is an algebraic extension of $\mathbb{F}_p$; then show that it is Galois over $\mathbb{F}_p$ with abelian Galois group. $\Diamond$

Now suppose that K/F is a (finite or infinite) Galois extension. Let $\mathfrak{P}$, $\mathcal{O}_K$ be as before and put $\mathfrak{p} = \mathfrak{P} \cap \mathcal{O}_F$, a prime ideal of $\mathcal{O}_F$.

Exercise 6.42. Show that $\mathrm{Gal}\,(K/F)$ acts transitively on the set of prime ideals of $\mathcal{O}_K$ above $\mathfrak{p}$. (HINT: You can assume this result for finite extensions of F, then use the profiniteness of the Galois group for K/F to relate the infinite extension K/F to an appropriately chosen tower of finite subextensions.) $\Diamond$

We define the decomposition group for $\mathfrak{P}/\mathfrak{p}$ as we did for finite extensions of number fields:

$$Z(\mathfrak{P}/\mathfrak{p}) = \{\sigma \in \mathrm{Gal}\,(K/F) : \sigma(\mathfrak{P}) = \mathfrak{P}\}.$$

We also define the inertia subgroup:

$$T(\mathfrak{P}/\mathfrak{p}) = \{\sigma \in Z(\mathfrak{P}/\mathfrak{p}) : \sigma(x) \equiv x \pmod{\mathfrak{P}} \text{ for all } x \in \mathcal{O}_K\}.$$

The decomposition and inertia subgroups are closed subgroups of $\mathrm{Gal}\,(K/F)$ so correspond to intermediate fields. And just as in the case of a finite extension of number fields we have an exact sequence

$$1 \longrightarrow T(\mathfrak{P}/\mathfrak{p}) \longrightarrow Z(\mathfrak{P}/\mathfrak{p}) \longrightarrow \mathrm{Gal}\left(\mathcal{O}_K/\mathfrak{P} \big/ \mathcal{O}_F/\mathfrak{p}\right) \longrightarrow 1.$$

When K/F is Galois we can define the ramification index for $\mathfrak{P}/\mathfrak{p}$ by

$$e(\mathfrak{P}/\mathfrak{p}) = \#T(\mathfrak{P}/\mathfrak{p}).$$

In the case when K/F is an algebraic but not necessarily Galois extension, we must find a way to use inertia subgroups if we want a definition of the ramification index that does not rely on factorization of ideals. But to have an inertia subgroup we must first have a Galois extension. We accomplish this by considering an algebraic closure $\mathbb{Q}^{alg}$. Let Ω be a prime ideal in the ring of algebraic integers of $\mathbb{Q}^{alg}$ that lies above $\mathfrak{P}$. Since the extensions $\mathbb{Q}^{alg}/K$ and $\mathbb{Q}^{alg}/F$ are Galois, we may consider the inertia groups $T(\Omega/\mathfrak{P})$ and $T(\Omega/\mathfrak{p})$. Note that $T(\Omega/\mathfrak{P}) = T(\Omega/\mathfrak{p}) \cap \text{Gal}(\mathbb{Q}^{alg}/K)$. We define the ramification index for $\mathfrak{P}/\mathfrak{p}$ as

$$e(\mathfrak{P}/\mathfrak{p}) = [T(\Omega/\mathfrak{p}) : T(\Omega/\mathfrak{P})].$$

Note that if K/F is Galois, this definition gives $e(\mathfrak{P}/\mathfrak{p}) = \#T(\mathfrak{P}/\mathfrak{p})$ so is consistent with the conventional definition of ramification index for extensions of number fields. Also note that the ramification index need not be finite when K/F is an infinite extension.

Exercise 6.43. How would you define the ramification index for an Archimedean place in the (possibly infinite) algebraic extension K/F? (HINT: Consider extensions to K of the embeddings of F, but take care to show that in a Galois extension, the Galois group acts transitively on the extensions of a given infinite place of F; then define decomposition and inertia groups for these infinite places and follow the ideas we used above.)   ◊

Proposition 7.7. Suppose K/F is a $\mathbb{Z}_p$-extension, where F is a number field. Let $\mathfrak{q}$ be a prime of F that does not divide $p\mathcal{O}_F$. Then K/F is unramified at $\mathfrak{q}$.

Proof. Let T be the inertia subgroup for $\mathfrak{q}$ in $\text{Gal}(K/F)$. We know T is a closed subgroup of $\text{Gal}(K/F) \cong \mathbb{Z}_p$, so we must have $T = 0$ or $T \cong p^m\mathbb{Z}_p$ for some m. Suppose it is the latter. Then we may choose primes $\mathfrak{q}_n$ of $K^{(n)}$ above $\mathfrak{q}$ recursively as follows. Put $\mathfrak{q}_0 = \mathfrak{q}$, and let $\mathfrak{q}_{n+1}$ be a prime of $K^{(n+1)}$ above $\mathfrak{q}_n$.

Let $K_{\mathfrak{q}}^{(n)}$ denote the completion of $K^{(n)}$ at $\mathfrak{q}_n$ and let $K_{\mathfrak{q}}^{(\infty)} = \cup K_{\mathfrak{q}}^{(n)}$. Let $\mathcal{U}_{\mathfrak{q}}^{(n)}$, $\mathcal{U}_{\mathfrak{q}}^{(\infty)}$ be the units in the integer rings of $K_{\mathfrak{q}}^{(n)}$, $K_{\mathfrak{q}}^{(\infty)}$ respectively. By Theorem 3.4, there is a surjective homomorphism $\mathcal{U}_{\mathfrak{q}}^{(n)} \longrightarrow T(\mathfrak{q}_{n+1}/\mathfrak{q}_n)$, where $T(\mathfrak{q}_{n+1}/\mathfrak{q}_n)$ is the inertia subgroup for $\mathfrak{q}_n$ in $\text{Gal}(K^{(n+1)}/K^{(n)})$. We leave it as **Exercise 6.44** to show that there is a surjective homomorphism $\mathcal{U}_{\mathfrak{q}}^{(\infty)} \longrightarrow T$. Since $T \cong p^m\mathbb{Z}_p$, we have a surjective homomorphism $\mathcal{U}_{\mathfrak{q}}^{(\infty)} \longrightarrow p^m\mathbb{Z}_p$.

On the other hand, if q is the prime number such that $\mathfrak{q} \cap \mathbb{Z} = q\mathbb{Z}$, then one can use the q-adic logarithm to show that $\mathcal{U}_{\mathfrak{q}}^{(\infty)} \cong$ (a finite group) $\times \mathbb{Z}_q^a$ for some positive integer a.

Note that the torsion part of $p^m\mathbb{Z}_p$ is trivial. Hence the above implies that there is a continuous surjective homomorphism $\mathbb{Z}_q^a \longrightarrow p^m\mathbb{Z}_p$. Composing with the natural map, (reduction modulo p^{m+1}), we find there is a continuous surjective homomorphism $\mathbb{Z}_q^a \longrightarrow p^m\mathbb{Z}_p / {}_{p^{m+1}\mathbb{Z}_p}$ so there is a closed subgroup of index p in $\mathbb{Z}_q^a$. Since no such subgroup exists, we have a contradiction. Thus $T = 0$ and $\mathfrak{q}$ is unramified in K/F.   □

Proposition 7.8. Let K/F be a $\mathbb{Z}_p$-extension, where F is a number field. Some prime of $\mathcal{O}_F$ ramifies in K/F and moreover there is some level m such that every prime that ramifies in $K/K^{(m)}$ is totally ramified.

Proof. The maximal abelian unramified extension of F (i.e., the Hilbert class field of F) is a finite extension, so some prime must ramify in K/F. By Proposition 7.7, only primes above p can ramify in K/F. Call them $\mathfrak{p}_1, \ldots, \mathfrak{p}_s$ and let $T_1, \ldots, T_s$ be their inertia groups. Then $\cap\, T_j = p^m \mathbb{Z}_p$ for some m. The fixed field of $p^m \mathbb{Z}_p$ is $K^{(m)}$ and $\mathrm{Gal}\,(K/K^{(m)}) \subseteq T_j$ for each j. Thus the primes above $\mathfrak{p}_j$ in $K^{(m)}$ are all totally ramified in $K/K^{(m)}$. $\qquad\square$

Next we show how class field theory is used to define a certain Galois action that is central to Iwasawa theory. Those wishing to read more about Iwasawa theory should consult Washington's book, [Wa], for a nice introduction to the subject.

Let K/F be a $\mathbb{Z}_p$-extension, where $\Gamma = \mathrm{Gal}\,(K/F) \cong \mathbb{Z}_p$. Since the closed subgroups of $\mathbb{Z}_p$ are precisely 0 and $p^n \mathbb{Z}_p$ for natural numbers n, it follows that for each n there is a unique intermediate field $K^{(n)}$ such that $[K^{(n)} : F] = p^n$, and also that these $K^{(n)}$, together with K, are the only intermediate fields in K/F. We have

$$F = K^{(0)} \subseteq K^{(1)} \subseteq K^{(2)} \subseteq \cdots \subseteq K.$$

For each n, let $M^{(n)}$ be the maximal unramified abelian p-extension of $K^{(n)}$, and put

$$X_n = \mathrm{Gal}\,(M^{(n)}/K^{(n)}).$$

Recall from Exercise 6.12 on the Hilbert p-class field that X_n is isomorphic to the p-Sylow subgroup of the ideal class group of $K^{(n)}$. Let $M = \cup\, M^{(n)}$ and let $X = \mathrm{Gal}\,(M/K)$.

Exercise 6.45. Show that each (finite) extension $M^{(n)}/F$ is Galois, then show that the (infinite) extension M/F is also Galois. $\qquad\Diamond$

Let $G = \mathrm{Gal}\,(M/F)$. We have the following diagrams.

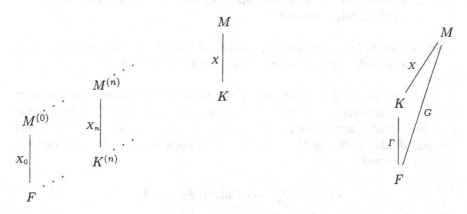

One of the central ideas in Iwasawa theory is that X can be viewed as a Γ- module in a natural way. (In fact, although we don't show it here, it turns out to be finitely generated and $\mathbb{Z}_p[[\Gamma]]$-torsion.) Structure theorems about such modules then lead to a rich theory, which includes a surprising relationship to p-adic L-functions. See Washington [Wa] or Lang [L3] for (many) more details. We shall be content here simply to describe the action of Γ on X.

Replacing F by some $K^{(m)}$ if necessary, we shall assume for simplicity that all primes that ramify in K/F are totally ramified. With this assumption, we have that

$$K^{(n+1)} \cap M^{(n)} = K^{(n)}$$

and

$$\mathrm{Gal}\,(M^{(n)}/K^{(n)}) \cong \mathrm{Gal}\,(K^{(n+1)}M^{(n)}/K^{(n+1)})$$

is a quotient of X_{n+1}. We have a map $X_{n+1} \longrightarrow X_n$, which, via class field theory, is seen to arise from the norm map on the ideal class groups. But also $X_n \cong \mathrm{Gal}\,(KM^{(n)}/K)$. Thus

$$\varprojlim X_n \cong \mathrm{Gal}\,\left((\underset{n \geq 0}{\cup} KM^{(n)})/K \right) = \mathrm{Gal}\,(M/K) = X.$$

For $\gamma \in \Gamma_n = {}^{\Gamma}/_{\Gamma^{p^n}} = \mathrm{Gal}\,(K^{(n)}/F)$, we may extend γ to $\tilde{\gamma} \in \mathrm{Gal}\,(M^{(n)}/F)$. For $x \in X_n$, we have an action of γ via

$$x^\gamma = \tilde{\gamma} x \tilde{\gamma}^{-1}.$$

This makes X_n a $\mathbb{Z}_p[\Gamma_n]$-module. Suppose we are given an element $\gamma \in \Gamma$ and an element $x = (x_0, x_1, \ldots) \in X = \varprojlim X_n$. For each n we let the coset of γ in $\mathbb{Z}_p[\Gamma_n]$ act on the nth component of x by conjugation as described above. We leave it as **Exercise 6.46** to show that the result, which we denote x^γ, is in X, and hence that X is a $\mathbb{Z}_p[[\Gamma]]$-module as desired.

Exercise 6.47. In fact, show that the above action of Γ on X is given by $x^\gamma = \tilde{\gamma} x \tilde{\gamma}^{-1}$, where in this case $\tilde{\gamma}$ is an extension of γ to $G = \mathrm{Gal}\,(M/F)$.   ◊

Finally we include a brief discussion of another infinite extension of a number field F that is important in Iwasawa theory. Fix a prime p and let M be the maximal p-extension that is unramified except at primes above p. Let N/F be the maximal unramified subextension of M/F, i.e., N is the Hilbert p-class field of F. We have an exact sequence

$$1 \longrightarrow \mathrm{Gal}\,(M/N) \longrightarrow \mathrm{Gal}\,(M/F) \longrightarrow A \longrightarrow 1$$

where A is the Sylow p-subgroup of the ideal class group $\mathcal{C}_F$. Thus we shall have described $\mathrm{Gal}\,(M/F)$ once we have described $\mathrm{Gal}\,(M/N)$. It is possible to do this in terms of certain groups of units.

Let $\mathcal{U}_{\mathfrak{p}} = \mathcal{O}_{\mathfrak{p}}^{\times}$ and let $\mathcal{U}_{\mathfrak{p}}^1 = \{x \in \mathcal{U}_{\mathfrak{p}} : x \equiv 1 \pmod{\mathfrak{p}}\}$. Let $\bar{\mathcal{U}}$ denote the topological closure of the image of $\mathcal{O}_F^{\times}$ in $\mathcal{U}_p = \prod_{\mathfrak{p}|p} \mathcal{U}_{\mathfrak{p}}$ under the diagonal embedding. Let $\mathcal{U}_p^1 = \prod_{\mathfrak{p}|p} \mathcal{U}_{\mathfrak{p}}^1$.

Exercise 6.48. Show that $\mathrm{Gal}\,(M/N) \cong \mathcal{U}_p^1 \big/ \mathcal{U}_p^1 \cap \bar{\mathcal{U}}.$ $\qquad\qquad \Diamond$

The above is related to an important conjecture, due to Leopoldt. Note that $\mathcal{U}_p^1$ is a $\mathbb{Z}_p$-module of rank $[F : \mathbb{Q}]$ so $\mathcal{U}_p^1 \cap \bar{\mathcal{U}}$ is also a $\mathbb{Z}_p$-module. By Exercise 6.48, we have

$$\mathrm{rank}_{\mathbb{Z}_p}(\mathrm{Gal}\,(M/F)) = [F : \mathbb{Q}] - \mathrm{rank}_{\mathbb{Z}_p}(\mathcal{U}_p^1 \cap \bar{\mathcal{U}}).$$

Leopoldt's Conjecture says that $\mathrm{rank}_{\mathbb{Z}_p}(\mathcal{U}_p^1 \cap \bar{\mathcal{U}}) = r_1 + r_2 - 1$ where r_1 and r_2 are the number of real embeddings of F and the number of conjugate pairs of imaginary embeddings of F, respectively. There is a proof of this conjecture, given by A. Brumer, in the case when F is an abelian extension of $\mathbb{Q}$. In the general case it remains an open problem. If Leopoldt's Conjecture is true, then we have $\mathrm{rank}_{\mathbb{Z}_p}(\mathrm{Gal}\,(M/F)) = r_2 + 1$. This would give that the $\mathbb{Z}_p$-rank of the Galois group of the compositum of all the $\mathbb{Z}_p$-extensions of F is $r_2 + 1$. See Washington's *Introduction to Cyclotomic Fields*, [Wa], for more on the implications of Leopoldt's Conjecture.

Chapter 7
Local Class Field Theory

Many of the main theorems on the class field theory of local fields were first proved by Hasse ([Has], 1930), but his proofs relied on connections to global class field theory. Schmidt ([Sc], 1930) and Chevalley ([Ch1], 1933) were able to give an approach that did not rely on the global theory. Indeed, once this had been accomplished, it became apparent that the proofs of many of the results for the global case could be reinterpreted using the analogous local results. A cohomological approach to local class field theory was crystallized in the work of Hochschild and Nakayama ([HN], 1952). See also the book by Artin and Tate, [AT], Serre's book, [Se2], and Serre's article in Cassels and Fröhlich, [CF].

One may rephrase the Kronecker-Weber Theorem to say that a maximal abelian extension of $\mathbb{Q}$ is generated by the torsion points of the action of $\mathbb{Z}$ on $\mathbb{C}^\times$, (where $n \in \mathbb{Z}$ sends $x \in \mathbb{C}^\times$ to x^n). Similarly, a maximal abelian extension of a local field K is generated by the torsion points of an action of the ring of integers of K on a module that arises via the formal group laws of Lubin and Tate. We discuss here the approach of Lubin and Tate ([LT], 1965), especially as treated in a paper of Hazewinkel ([Haz2], 1975), who was able to adapt the ideas from [LT] so that the theory of formal groups was not explicitly present in the exposition. While we take Hazewinkel's point of view, we also include a bit about the underlying formal groups, so that the module mentioned above can be described.

This chapter is intended as an introduction to the subject; for a more complete exposition of local class field theory describing in detail the relationship to formal groups, see Iwasawa's *Local Class Field Theory* ([I], 1986), wherein work of Coleman ([Col], 1979) plays an important role. An extensive treatment of local class field theory from a different point of view may be found in Serre's *Local Fields*, [Se2].

In the first section, we discuss some preliminary results on local fields and their extensions, including some important infinite extensions. The second section is devoted to the study of extensions of a complete discretely valued field with algebraically closed residue field. In the third section, we return to local fields (where the residue field is finite) and prove some key results on units and their norms. Formal group laws and the Lubin-Tate formal group laws are introduced in sections four and five, respectively. In section six we see how the ideas of Lubin and Tate

N. Childress, *Class Field Theory*, Universitext, DOI 10.1007/978-0-387-72490-4_7,
© Springer Science+Business Media, LLC 2009

lead to certain totally ramified extensions of a local field, from which its maximal abelian extension is constructed. We conclude with an explicit construction of the local Artin map and a local version of Artin Reciprocity in section seven.

The reader may notice that for completions of number fields, some of the results in this chapter follow from the class field theory for number fields, (in particular, see sections 5.4 and 6.4). However, we prove them here without relying on global class field theory.

1 Some Preliminary Facts About Local Fields

In this section, we let K denote a field that is complete with respect to a normalized discrete valuation $v_K : K^\times \longrightarrow \mathbb{Z}$. If desired, the reader may assume that K is an extension field of $\mathbb{Q}_p$ (of possibly infinite degree). Let

$$\mathcal{O}_K = \{x \in K : v_K(x) \geq 0\}$$
$$\mathcal{U}_K = \{x \in K : v_K(x) = 0\}$$
$$\pi_K, \text{ a uniformizer in } K, \text{ so } v_K(\pi_K) = 1$$
$$\mathcal{P}_K = \{x \in K : v_K(x) > 0\} = \pi_K \mathcal{O}_K$$
$$\mathcal{U}_K^m = \{x \in \mathcal{U}_K : x \equiv 1 \pmod{\mathcal{P}_K^m}\}$$
$$\mathbb{F}_K = \mathcal{O}_K / \mathcal{P}_K, \text{ the residue field of } K.$$

We do not necessarily assume K is a local field (where the residue field $\mathbb{F}_K$ is finite), however we do assume that $\mathbb{F}_K$ is perfect. If $\mathbb{F}_K$ is either finite or algebraically closed, and L/K is a finite Galois extension, then $\mathrm{Gal}\,(L/K)$ is solvable (see Serre's *Local Fields*, [Se2], for a proof).

Exercise 7.1. Suppose L/K is a finite Galois extension. Let K_L denote the maximal unramified subextension in L/K. Put $G = \mathrm{Gal}\,(L/K)$, and $G_{\mathrm{ram}} = \mathrm{Gal}\,(L/K_L)$. (Most authors use G_0 instead of G_{ram}, but we shall follow Hazewinkel here.)

a. Show: G_{ram} is a normal subgroup of G and G/G_{ram} is a cyclic group. Does G_{ram} have an analogue in the global theory (of number fields)?

b. Suppose E is an intermediate field in L/K, and E/K is Galois. Show that the image of G_{ram} under the natural map $\mathrm{Gal}\,(L/K) \longrightarrow \mathrm{Gal}\,(E/K)$ is contained in $\mathrm{Gal}\,(E/K)_{\mathrm{ram}}$.   ◊

We shall be interested in several extensions of K. Denote by

K_{ur} a maximal unramified extension of K,

$\hat{K}_{ur}$ the completion of K_{ur},

$\mathbb{F}_{ur}$ the residue field of $\hat{K}_{ur}$,

Ω a complete, algebraically closed extension of $\hat{K}_{ur}$.

Consider the case when K is local. Let $K_{(t)}$ denote the unramified extension of K of degree t, i.e., the splitting field over K of the polynomial $X^{q^t} - X$, where $q = \#\mathbb{F}_K$. Clearly $K_{(t)} \subseteq K_{(n)}$ if and only if $t|n$. The field K_{ur} is simply the union $\bigcup_t K_{(t)}$. (It is straightforward to show that any unramified extension of K, finite or not, must be contained in this union.)

Proposition 1.1. Let K be a local field. The field $\mathbb{F}_{ur}$ is an algebraic closure of $\mathbb{F}_K$. Moreover, there is a natural isomorphism $\mathrm{Gal}\,(K_{ur}/K) \cong \mathrm{Gal}\,(\mathbb{F}_{K_{ur}}/\mathbb{F}_K)$.

Proof. First note that $\mathcal{O}_{K_{ur}} = \bigcup_t \mathcal{O}_{K_{(t)}}$ and $\mathcal{P}_{K_{ur}} = \bigcup_t \mathcal{P}_{K_{(t)}}$ so that $\mathbb{F}_{K_{ur}} = \bigcup_t \mathbb{F}_{K_{(t)}}$. But $[\mathbb{F}_{K_{(t)}} : \mathbb{F}_K] = [K_{(t)} : K] = t$, so $\mathbb{F}_{K_{(t)}}$ is just the unique degree t extension of the finite field $\mathbb{F}_K$. Thus, $\mathbb{F}_{ur}$ is an algebraic closure of $\mathbb{F}_K$.

For the assertion about the Galois groups, note that

$$\mathrm{Gal}\,(K_{ur}/K) = \varprojlim \mathrm{Gal}\,(K_{(t)}/K)$$

and

$$\mathrm{Gal}\,(\mathbb{F}_{K_{ur}}/\mathbb{F}_K) = \varprojlim \mathrm{Gal}\,(\mathbb{F}_{K_{(t)}}/\mathbb{F}_K),$$

where the maps for the limits are the canonical ones. Since $\mathrm{Gal}\,(K_{(t)}/K) \cong \mathrm{Gal}\,(\mathbb{F}_{K_{(t)}}/\mathbb{F}_K)$, the result follows. $\qquad\square$

We have a notion of Frobenius automorphism for finite unramified extensions L/K of local fields; namely the Frobenius automorphism is just the lift to $\mathrm{Gal}\,(L/K)$ of the map $x \mapsto x^q$ from $\mathrm{Gal}\,(\mathbb{F}_L/\mathbb{F}_K)$. (Here q is the cardinality of the residue field of K.) But the map $x \mapsto x^q$ also can be viewed as belonging to $\mathrm{Gal}\,(\mathbb{F}_{K_{ur}}/\mathbb{F}_K)$; its lift to $\mathrm{Gal}\,(K_{ur}/K)$ will be called the *Frobenius automorphism of K*, denoted φ.

Exercise 7.2. Suppose $\mathbb{F}_K$ is a finite field.

a. Show that K_{ur}/K is abelian, so that we may write $K_{ab} \supset K_{ur}$, where K_{ab} denotes a maximal abelian extension of K.

b. Show that $\mathrm{Gal}\,(K_{ur}/K) \cong \hat{\mathbb{Z}}$, (see Exercise 6.26).

c. Show that the subgroup $\langle \varphi \rangle$ is dense in $\mathrm{Gal}\,(K_{ur}/K)$ and that K is the fixed field of $\langle \varphi \rangle$ in K_{ur}.

d. Let L/K be a finite extension with residue field degree f. Let φ_L denote the Frobenius automorphism of L. Show that $L_{ur} = LK_{ur}$ and $\varphi_L\big|_{K_{ur}} = \varphi^f$.

e. What elements must be adjoined to K to obtain K_{ur} when $K = \mathbb{Q}_p$? $\diamondsuit$

Proposition 1.2. Let K be a local field and suppose L/K is Galois with $K_{\mathrm{ur}} \subseteq L$. Let $\sigma \in \mathrm{Gal}\,(L/K)$ be such that $\sigma\big|_{K_{\mathrm{ur}}} = \varphi$. Let F be the fixed field of σ in L. Then $FK_{\mathrm{ur}} = L$ and $F \cap K_{\mathrm{ur}} = K$, so $\mathrm{Gal}\,(L/F) \cong \mathrm{Gal}\,(K_{\mathrm{ur}}/K)$. Note this implies F/K is totally ramified.

Proof. The fixed field of $\sigma\big|_{K_{\mathrm{ur}}}$ must be $F \cap K_{\mathrm{ur}}$. Since K is the fixed field of φ, we get $F \cap K_{\mathrm{ur}} = K$. Recall, for a positive integer t, we let $K_{(t)}$ denote the (unique) unramified extension of K of degree t. To show that $FK_{\mathrm{ur}} = L$, it suffices to show for any t that $FK_{(t)} = E$ whenever $F \subseteq E \subseteq L$ and $[E : F] = t$. Since $F \cap K_{\mathrm{ur}} = K$, we have $[FK_{(t)} : F] = t$, from which it follows that $EK_{(t)}/F$ is a finite subextension of L/F.

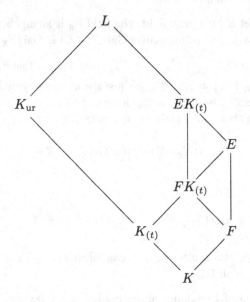

Since $\langle \sigma \rangle$ is dense in $\mathrm{Gal}\,(L/F)$, (F is the fixed field of σ), we find that $\langle \sigma\big|_{EK_{(t)}} \rangle = \mathrm{Gal}\,(EK_{(t)}/F)$. Because this is cyclic, there is only one intermediate field of degree t over F. Hence $E = FK_{(t)}$.   □

Exercise 7.3. Suppose L/K is a finite Galois extension. Show that $\hat{L}_{\mathrm{ur}}/\hat{K}_{\mathrm{ur}}$ is Galois, with $\mathrm{Gal}\,(\hat{L}_{\mathrm{ur}}/\hat{K}_{\mathrm{ur}}) \cong \mathrm{Gal}\,(L/K)_{\mathrm{ram}}$.   ◇

   One may show that when $\mathbb{F}_K$ is a finite field and L/K is an unramified Galois extension, the norm map $N_{L/K} : \mathcal{U}_L \longrightarrow \mathcal{U}_K$ is surjective. In Chapter 4 (see Lemma 4.5.3) we proved this when L and K are finite extensions of $\mathbb{Q}_p$ for some prime p. In the case where $\mathbb{F}_K$ is algebraically closed and L/K is a finite extension, the norm map $N_{L/K} : \mathcal{U}_L \longrightarrow \mathcal{U}_K$ is again surjective, as is $N_{L/K} : L^\times \longrightarrow K^\times$, (see Serre's *Local Fields*, [Se2]).

Theorem 1.3 (Decomposition Theorem). *Let L/K be a finite Galois extension, and suppose $\mathbb{F}_K$ is a finite field. There is a totally ramified extension L'/K such that*

$L'_{ur} = L'K_{ur} = LK_{ur} = L_{ur}$. *Moreover, if* $\text{Gal}(L/K)_{ram}$ *is central in* $\text{Gal}(L/K)$, *then we can take* L'/K *to be abelian.*

Proof. Letting K_L be as in Exercise 7.1, we have that $\text{Gal}(K_L/K)$ is cyclic, generated by Frobenius φ. Lift φ to $\varphi' \in \text{Gal}(L/K)$. The order of φ is $[K_L : K]$, which must divide the order of φ'. Thus for t equal to the order of φ' in $\text{Gal}(L/K)$, we get $K_L \subseteq K_{(t)}$.

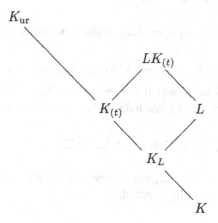

Let $\varphi'' \in \text{Gal}(LK_{(t)}/K)$ be determined by the following conditions:

$$\varphi''\Big|_{K_{(t)}} = \text{Frobenius in } \text{Gal}(K_{(t)}/K),$$

$$\varphi''\Big|_{L} = \varphi'.$$

This uniquely determines φ''. Denote the fixed field of φ'' by L'. Then L'/K is totally ramified (why?) and $L'K_{(t)} = LK_{(t)}$. But $K_{(t)} \subset K_{ur}$, and this implies $L'K_{ur} = LK_{ur}$.

Now suppose $\text{Gal}(L/K)_{ram}$ is central in $\text{Gal}(L/K)$. Then $\text{Gal}(LK_{(t)}/K)_{ram}$ is central in $\text{Gal}(LK_{(t)}/K)$, so $\langle\varphi''\rangle$ is a normal subgroup. Thus L'/K is Galois (and abelian). □

Example.

1. Let $K = \mathbb{Q}_3$ and let L be the splitting field of the polynomial $X^4 - 3X^2 + 18$ over $\mathbb{Q}_3$. Then $L/\mathbb{Q}_3$ is a cyclic Galois extension of degree 4 and ramification index 2; the intermediate field K_L is the only quadratic extension of $\mathbb{Q}_3$ contained in L. Note that $K_L/\mathbb{Q}_3$ is unramified with $\text{Gal}(K_L/\mathbb{Q}_3) = \langle\varphi\rangle$, where $\varphi : x \mapsto x^3$ is the Frobenius automorphism. The lift φ' of φ to $\text{Gal}(L/\mathbb{Q}_3)$ has order 4, so $t = 4$. The field $LK_{(4)}$ is a degree-8 extension of $\mathbb{Q}_3$, with $\text{Gal}(LK_{(4)}/\mathbb{Q}_3) \cong \mathbb{Z}/2\mathbb{Z} \times \mathbb{Z}/4\mathbb{Z}$. The automorphism φ'' has order 4 in $\text{Gal}(LK_{(4)}/\mathbb{Q}_3)$; its fixed field L' is a totally ramified quadratic extension of $\mathbb{Q}_3$.

Exercise 7.4. Suppose $\mathrm{Gal}(L/K) \cong \mathcal{Q}_8$ (the quaternions) and $\mathrm{Gal}(L/K)_{\mathrm{ram}} = \mathrm{Gal}(L/K_L)$ is cyclic of order 4. (For example, this occurs when $K = \mathbb{Q}_3$ and L is a splitting field of $X^8 + 9X^4 + 36$. See the database of Jones and Roberts, [JR], for other such examples.) Find the order of φ' in $\mathrm{Gal}(L/K)$, the degree $[LK_{(t)} : K]$ and the degree $[L' : K]$. What can you say about $L' \cap L$? Is L'/K a Galois extension? Prove that your answers are correct. $\Diamond$

Exercise 7.5. Suppose $\mathbb{F}_K$ is a finite field.

a. Let M be a maximal totally ramified abelian extension of K. Show that $K_{\mathrm{ab}} = K_{\mathrm{ur}}M$.

b. Show that $\mathrm{Gal}(K_{\mathrm{ab}}/K)_{\mathrm{ram}} \cong \varprojlim \mathrm{Gal}(L/K)_{\mathrm{ram}}$, where L varies over the finite abelian extensions of K and the maps $\mathrm{Gal}(L/K)_{\mathrm{ram}} \longrightarrow \mathrm{Gal}(E/K)_{\mathrm{ram}}$ for the limit (where $E \subseteq L$) are induced by the natural maps $\mathrm{Gal}(L/K) \longrightarrow \mathrm{Gal}(E/K)$.

c. Show that $\mathrm{Gal}(K_{\mathrm{ab}}/K) \cong \mathrm{Gal}(K_{\mathrm{ab}}/K)_{\mathrm{ram}} \times \hat{\mathbb{Z}}$. $\Diamond$

Exercise 7.6. Suppose $\mathbb{F}_K$ is algebraically closed, and let L/K be an abelian extension. Show that L/K is totally ramified. $\Diamond$

2 A Fundamental Exact Sequence

In this section, we prove a result for abelian extensions L/K, where K is complete with respect to a normalized discrete valuation and has an algebraically closed residue field $\mathbb{F}_K$. This result will play a crucial role as we progress toward the local version of Artin Reciprocity. Before continuing, we must take a moment to prove a lemma on finite abelian groups.

Lemma 2.1. Let G be a finite abelian group and let $g \in G$. Then G contains a subgroup H such that G/H is cyclic and the order of gH in G/H is the same as the order of g in G.

Proof. Decompose the abelian group G as a direct sum $G \cong \bigoplus_p G_p$, where G_p is the Sylow p-subgroup of G. Identify $g \in G$ with the element $(\dots, g_p, \dots) \in \bigoplus_p G_p$.

Now write each G_p as a direct sum of cyclic groups, say $G_p \cong \mathbb{Z}/p^{i_1}\mathbb{Z} \oplus \cdots \oplus \mathbb{Z}/p^{i_r}\mathbb{Z}$ and identify $g_p \in G_p$ with $(g_{p1}, \dots, g_{pr})$ in this direct sum. Let

$$d_p(g_p) = \max_{1 \le n \le r}\{i_n - \mathrm{ord}_p(g_{pn})\},$$

so the order of g_p in G_p is $p^{d_p(g_p)}$. Let j be an index where the maximum is attained, i.e., $d_p(g_p) = i_j - \mathrm{ord}_p(g_{pj})$. Let $H_p = \bigoplus_{n \ne j} \mathbb{Z}/p^{i_n}\mathbb{Z}$ (a subgroup of G_p) and let

$H = \bigoplus_p H_p$ (a subgroup of G). For any p, the order of $g_p H_p$ in G_p/H_p is the same as the order of g_p in G_p. Thus H is the desired subgroup of G. $\qquad\square$

Throughout this section, we suppose K is complete with respect to a discrete valuation and $\mathbb{F}_K$ is algebraically closed. For the extension L/K, we put

$$\mathcal{V}(L/K) = \left\langle \frac{\sigma(u)}{u} : u \in \mathcal{U}_L, \, \sigma \in \mathrm{Gal}(L/K) \right\rangle,$$

a subgroup of $\mathcal{U}_L$.

Exercise 7.7. Let L/K be as above, and suppose $\sigma, \tau \in \mathrm{Gal}(L/K)$.

a. Show that $\frac{\sigma\tau(\pi_L)}{\pi_L}$ and $\frac{\sigma(\pi_L)}{\pi_L} \frac{\tau(\pi_L)}{\pi_L}$ represent the same coset in $\mathcal{U}_L/\mathcal{V}(L/K)$.

b. Let $r \in \mathbb{Z}$. Show that $\frac{\sigma(\pi_L^r)}{\pi_L^r}$ and $\frac{\sigma^r(\pi_L)}{\pi_L}$ represent the same coset in $\mathcal{U}_L/\mathcal{V}(L/K)$. $\Diamond$

Exercise 7.8. Show that the map $i : \mathrm{Gal}(L/K) \longrightarrow \mathcal{U}_L/\mathcal{V}(L/K)$ given by $i(\sigma) = \frac{\sigma(\pi_L)}{\pi_L}\mathcal{V}(L/K)$ is a well-defined homomorphism of groups. $\Diamond$

Proposition 2.2. Let L/K be a finite abelian extension. The group homomorphism i of Exercise 7.8 is injective.

Proof. Let $\tau \in G = \mathrm{Gal}(L/K)$ with $\tau \neq 1$ and let H be a subgroup of G such that G/H is cyclic and the order of τH in G/H is the same as the order of τ in G. Let $\sigma \in G$ be such that σH generates G/H and let $r \in \mathbb{Z}_+$ with $\tau H = \sigma^r H$. Since $\tau H \neq H$, we may take $r < |\sigma|$. We have $\tau = \sigma^r h_0$ for some $h_0 \in H$. Suppose $i(\tau) = \mathcal{V}(L/K)$. Then, using Exercise 7.7 and considering the form of a typical element of $\mathcal{V}(L/K)$, we have

$$\frac{\sigma(\pi_L^r)}{\pi_L^r} \frac{h_0(\pi_L)}{\pi_L} = \prod_{s=1}^{|\sigma|} \prod_{h \in H} \frac{\sigma^s h(u_{sh})}{u_{sh}}$$

where each $u_{sh} \in \mathcal{U}_L$. Now replace each factor on the right-hand side with the expression derived from the following equation

$$
\begin{aligned}
\frac{\sigma^s h(u_{sh})}{u_{sh}} &= \left[\frac{\sigma^s h(u_{sh})}{\sigma^{s-1} h(u_{sh})} \frac{\sigma^{s-1} h(u_{sh})}{\sigma^{s-2} h(u_{sh})} \cdots \frac{\sigma^2 h(u_{sh})}{\sigma h(u_{sh})} \frac{\sigma h(u_{sh})}{h(u_{sh})} \right] \frac{h(u_{sh})}{u_{sh}} \\
&= \frac{\sigma\left[\sigma^{s-1} h(u_{sh}) \, \sigma^{s-2} h(u_{sh}) \cdots \sigma h(u_{sh}) \, h(u_{sh}) \right]}{\sigma^{s-1} h(u_{sh}) \, \sigma^{s-2} h(u_{sh}) \cdots \sigma h(u_{sh}) \, h(u_{sh})} \frac{h(u_{sh})}{u_{sh}} \\
&= \frac{\sigma(w_{sh})}{w_{sh}} \frac{h(u_{sh})}{u_{sh}}, \qquad \text{say,}
\end{aligned}
$$

yielding

$$\frac{\sigma(\pi_L^r)}{\pi_L^r} \frac{h_0(\pi_L)}{\pi_L} = \prod_{s=1}^{|\sigma|} \prod_{h \in H} \frac{\sigma(w_{sh})}{w_{sh}} \frac{h(u_{sh})}{u_{sh}}$$

where each $w_{sh} \in \mathcal{U}_L$. Collecting, we obtain

$$\frac{\sigma(\pi_L^r)}{\pi_L^r} \frac{h_0(\pi_L)}{\pi_L} = \frac{\sigma(w)}{w} \prod_{h \in H} \frac{h(u_h)}{u_h}$$

for $w \in \mathcal{U}_L, u_h \in \mathcal{U}_L$. Let E be the fixed field of H and apply $N_{L/E}$ to both sides of this equation to obtain

$$\frac{\tilde{\sigma}(\pi_E^r)}{\pi_E^r} = \frac{\tilde{\sigma}(\tilde{w})}{\tilde{w}}$$

where $\pi_E = N_{L/E}(\pi_L)$, $\tilde{w} = N_{L/E}(w)$, and $\tilde{\sigma} \in \operatorname{Gal}(E/K)$ corresponds to $\sigma H \in G/H$. Since G/H is cyclic generated by σH, $\operatorname{Gal}(E/K)$ is cyclic generated by $\tilde{\sigma}$, so the above implies that $\pi_E^r \tilde{w}^{-1} \in K$. But, since E/K is totally ramified (see Exercise 7.6) and $r < |\tilde{\sigma}| = [E : K]$, this cannot happen. Thus, for $\tau \neq 1$, we have $i(\tau) \neq \mathcal{V}(L/K)$.   □

We have reached the first step in our effort to define a local analogue of the norm residue symbol (see section 6). Namely, we are able to show, for a finite abelian extension L/K where $\mathbb{F}_K$ is algebraically closed, a certain sequence is exact. This will be accomplished by first proving the result for cyclic extensions, then using induction and the cyclic result to deduce exactness of the sequence for abelian extensions.

Theorem 2.3. *Suppose $\mathbb{F}_K$ is algebraically closed, and L/K is a finite cyclic extension. Let $N : \mathcal{U}_L / \mathcal{V}(L/K) \longrightarrow \mathcal{U}_K$ be the map that sends the coset $u\mathcal{V}(L/K)$ to $N_{L/K}(u)$. Since elements of $\mathcal{V}(L/K)$ have norm 1, this is a well-defined homomorphism. The sequence*

$$1 \longrightarrow \operatorname{Gal}(L/K) \overset{i}{\longrightarrow} \mathcal{U}_L / \mathcal{V}(L/K) \overset{N}{\longrightarrow} \mathcal{U}_K \longrightarrow 1$$

is exact.

Proof. We have shown that i is injective and we also know that N is surjective since $\mathbb{F}_K$ is algebraically closed. It remains to show that the image of i and the kernel of N are equal.

It is clear that the composition Ni is the trivial map, as $N_{L/K}(\sigma(\pi_L)) = N_{L/K}(\pi_L)$ for any $\sigma \in \operatorname{Gal}(L/K)$. Thus im $i \subseteq \ker N$. For the reverse containment, suppose $u\mathcal{V}(L/K) \in \ker N$, i.e., $N_{L/K}(u) = 1$. By Hilbert's Theorem 90, there is some $y \in L^{\times}$ such that $u = \frac{\sigma(y)}{y}$, where σ is a generator of $\mathrm{Gal}\,(L/K)$. We can write $y = \pi_L^r v$ for $v \in \mathcal{U}_L$, $r \in \mathbb{Z}$. Using Exercise 7.7, we conclude

$$u \mathcal{V}(L/K) = \frac{\sigma(\pi_L^r)}{\pi_L^r} \mathcal{V}(L/K) = \frac{\sigma^r(\pi_L)}{\pi_L} \mathcal{V}(L/K) = i(\sigma^r) \in \mathrm{im}\, i. \qquad \square$$

Now we begin the process of generalizing Theorem 2.3 to finite abelian extensions L/K. We continue to assume that $\mathbb{F}_K$ is algebraically closed.

Lemma 2.4. Let L/K be a finite Galois extension, and let $K \subseteq E \subseteq L$, where E/K is Galois. Then $N_{L/E} \mathcal{V}(L/K) = \mathcal{V}(E/K)$.

Proof. Let $H = \mathrm{Gal}\,(L/E)$, a subgroup of $G = \mathrm{Gal}\,(L/K)$. For "$\supseteq$" we must show that if $\tilde{g} \in \mathrm{Gal}\,(E/K) \cong {}^G\!/_H$ and if $u \in \mathcal{U}_E$, then $\frac{\tilde{g}(u)}{u} \in N_{L/E} \mathcal{V}(L/K)$. Now we know that $N_{L/E} : \mathcal{U}_L \longrightarrow \mathcal{U}_E$ is surjective, so there is some $v \in \mathcal{U}_L$ such that $N_{L/E}(v) = u$. Let $\tilde{g} \in \mathrm{Gal}\,(E/K)$. Then $\tilde{g}$ corresponds to some $gH \in {}^G\!/_H$, where $g \in G$. We have

$$\begin{aligned}
N_{L/E}\Big(\frac{g(v)}{v}\Big) &= \prod_{h \in H} \frac{hg(v)}{v} \\
&= \frac{\prod_h g(g^{-1}hg)(v)}{\prod_h h(v)} \\
&= \frac{\tilde{g}(\prod_h (g^{-1}hg)(v))}{\prod_h h(v)} \\
&= \frac{\tilde{g}(u)}{u}
\end{aligned}$$

as needed. Note by starting with $v \in \mathcal{U}_L$ and $g \in G$, the above equation also gives us "$\subseteq$". $\qquad \square$

Lemma 2.5. Let L/K be a finite abelian extension, let E be an intermediate field such that L/E is cyclic. Then

$$1 \longrightarrow \mathrm{Gal}\,(L/E) \xrightarrow{\;i\;} {}^{\mathcal{U}_L}\!/_{\mathcal{V}(L/K)} \xrightarrow{\;\tilde{N}\;} {}^{\mathcal{U}_E}\!/_{\mathcal{V}(E/K)} \longrightarrow 1$$

is exact, where the map $\tilde{N}$ is induced by $N_{L/E}$.

Proof. First, i is injective because it is the restriction of an injective map on $\mathrm{Gal}\,(L/K)$. Also, $\tilde{N}$ is surjective because $N_{L/E} : \mathcal{U}_L \longrightarrow \mathcal{U}_E$ is surjective. We have a commutative diagram

$$
\begin{array}{ccccccccc}
1 & \longrightarrow & \mathrm{Gal}\,(L/E) & \xrightarrow{\;i\;} & {}^{\mathcal{U}_L}\!/_{\mathcal{V}(L/E)} & \xrightarrow{\;N\;} & \mathcal{U}_E & \longrightarrow & 1 \\
& & {\scriptstyle=}\big\downarrow & & \big\downarrow & & \big\downarrow & & \\
1 & \longrightarrow & \mathrm{Gal}\,(L/E) & \xrightarrow{\;i\;} & {}^{\mathcal{U}_L}\!/_{\mathcal{V}(L/K)} & \xrightarrow{\;\tilde{N}\;} & {}^{\mathcal{U}_E}\!/_{\mathcal{V}(E/K)} & \longrightarrow & 1
\end{array}
$$

where the vertical maps are the natural ones (note $\mathcal{V}(L/E) \subseteq V(L/K)$), and the top row is exact. We must show ker $\tilde{N}$ = image i in the bottom row. Let $u\mathcal{V}(L/K) \in$ ker $\tilde{N}$. Then $u \in \mathcal{U}_L$ satisfies $N_{L/E}(u) \in \mathcal{V}(E/K)$. The previous lemma implies that there is $v \in \mathcal{V}(L/K)$ such that $N_{L/E}(v) = N_{L/E}(u)$, whence $N_{L/E}(uv^{-1}) =$ 1. Exactness of the top row gives $uv^{-1}\mathcal{V}(L/E) = \frac{\sigma(\pi_L)}{\pi_L}\mathcal{V}(L/E)$ for some $\sigma \in$ Gal (L/E). But then $u\mathcal{V}(L/K) = \frac{\sigma(\pi_L)}{\pi_L}\mathcal{V}(L/K) \in$ image i. We have shown ker $\tilde{N} \subseteq$ image i. The reverse containment is clear.   □

Theorem 2.6. *If $\mathbb{F}_K$ is algebraically closed and L/K is a finite abelian extension, then*

$$1 \longrightarrow \text{Gal}(L/K) \overset{i}{\longrightarrow} \mathcal{U}_L\big/\mathcal{V}(L/K) \overset{N}{\longrightarrow} \mathcal{U}_K \longrightarrow 1$$

is an exact sequence.

Proof. We induct on the degree $[L : K]$. The base step is trivial. For the induction step, let E/K be a proper subextension of L/K (so the induction hypothesis applies to E/K), chosen so that L/E is cyclic. Consider the following diagram.

$$
\begin{array}{ccc}
1 & & 1 \\
\downarrow & & \downarrow \\
\text{Gal}(L/E) & \overset{=}{\longrightarrow} & \text{Gal}(L/E) \\
\text{incl.}\downarrow & & i\downarrow \\
1 \longrightarrow \text{Gal}(L/K) \overset{i}{\longrightarrow} & \mathcal{U}_L\big/\mathcal{V}(L/K) & \overset{N}{\longrightarrow} \mathcal{U}_K \longrightarrow 1 \\
\text{nat.}\downarrow & \tilde{N}\downarrow & =\downarrow \\
1 \longrightarrow \text{Gal}(E/K) \overset{i}{\longrightarrow} & \mathcal{U}_E\big/\mathcal{V}(E/K) & \overset{N}{\longrightarrow} \mathcal{U}_K \longrightarrow 1 \\
\downarrow & \downarrow & \\
1 & & 1
\end{array}
$$

It is straightforward to check that this commutes. The previous lemma makes the second column exact. The induction hypothesis makes the third row exact. Clearly the first column is also exact. Hence the second row must be exact too.   □

Exercise 7.9. Extend the previous theorem to the case where L/K is totally ramified non-abelian to obtain an exact sequence

$$1 \longrightarrow \text{Gal}(L/K)^{\text{ab}} \longrightarrow \mathcal{U}_L\big/\mathcal{V}(L/K) \longrightarrow \mathcal{U}_K \longrightarrow 1.$$ ◇

3 Local Units Modulo Norms

Throughout this section we suppose K is a local field (so $\mathbb{F}_K$ is finite). Let L/K be a finite abelian extension that is totally ramified. Then the extension $\hat{L}_{\mathrm{ur}}/\hat{K}_{\mathrm{ur}}$ is abelian and totally ramified with $\mathrm{Gal}\,(\hat{L}_{\mathrm{ur}}/\hat{K}_{\mathrm{ur}}) \cong \mathrm{Gal}\,(L/K)$. Recall we let $\mathbb{F}_{\mathrm{ur}}$ denote the residue field of $\hat{K}_{\mathrm{ur}}$ (and of K_{ur}), so that $\mathbb{F}_{\mathrm{ur}}$ is an algebraic closure of $\mathbb{F}_K$. We use φ to denote the Frobenius automorphism in $\mathrm{Gal}\,(\mathbb{F}_{\mathrm{ur}}/\mathbb{F}_K)$; also we continue to use φ to denote its lifts in $\mathrm{Gal}\,(K_{\mathrm{ur}}/K)$ and $\mathrm{Gal}\,(L_{\mathrm{ur}}/L)$, and their extensions to $\hat{K}_{\mathrm{ur}}$ and $\hat{L}_{\mathrm{ur}}$. We have a homomorphism

$$\varphi - 1 : \mathcal{U}_{\hat{K}_{\mathrm{ur}}} \longrightarrow \mathcal{U}_{\hat{K}_{\mathrm{ur}}}$$

given by $u \mapsto \varphi(u)u^{-1}$. With it, we obtain the following commutative diagram, where A, B, C, D are the appropriate kernels and cokernels, so that the rows and columns are exact.

$$
\begin{array}{ccc}
1 & & 1 \\
\downarrow & & \downarrow \\
A & \xrightarrow{\;\beta\;} & B \\
\downarrow & & \downarrow \\
\end{array}
$$

$$1 \longrightarrow \mathrm{Gal}\,(L/K) \longrightarrow \mathcal{U}_{\hat{L}_{\mathrm{ur}}}\big/\mathcal{V}(\hat{L}_{\mathrm{ur}}/\hat{K}_{\mathrm{ur}}) \xrightarrow{\;N\;} \mathcal{U}_{\hat{K}_{\mathrm{ur}}} \longrightarrow 1$$

$$\varphi-1 \Big\downarrow \qquad \varphi-1 \Big\downarrow \qquad \varphi-1 \Big\downarrow \qquad\qquad (*)$$

$$1 \longrightarrow \mathrm{Gal}\,(L/K) \longrightarrow \mathcal{U}_{\hat{L}_{\mathrm{ur}}}\big/\mathcal{V}(\hat{L}_{\mathrm{ur}}/\hat{K}_{\mathrm{ur}}) \longrightarrow \mathcal{U}_{\hat{K}_{\mathrm{ur}}} \longrightarrow 1$$

$$
\begin{array}{ccc}
\downarrow & & \downarrow \\
C & \xrightarrow{\;\gamma\;} & D \\
\downarrow & & \downarrow \\
1 & & 1
\end{array}
$$

Lemma 3.1. In the situation described above, we have:

i. $\varphi - 1 : \mathcal{U}_{\hat{K}_{\mathrm{ur}}} \longrightarrow \mathcal{U}_{\hat{K}_{\mathrm{ur}}}$ is surjective, as is $\varphi - 1 : \mathcal{O}_{\hat{K}_{\mathrm{ur}}} \longrightarrow \mathcal{O}_{\hat{K}_{\mathrm{ur}}}$,
ii. $\varphi - 1 : \mathcal{V}(\hat{L}_{\mathrm{ur}}/\hat{K}_{\mathrm{ur}}) \longrightarrow \mathcal{V}(\hat{L}_{\mathrm{ur}}/\hat{K}_{\mathrm{ur}})$ is surjective,
iii. $\ker\left(\varphi - 1 : \mathcal{U}_{\hat{K}_{\mathrm{ur}}} \longrightarrow \mathcal{U}_{\hat{K}_{\mathrm{ur}}}\right) = \mathcal{U}_K$.

Proof. For (i), let $\mathcal{U}_{\hat{K}_{\mathrm{ur}}}^n = \{u \in \mathcal{U}_{\hat{K}_{\mathrm{ur}}} : u \equiv 1 \mod \pi_K^n\}$, where n is any positive integer. We have induced homomorphisms

$$\varphi - 1 : \mathcal{U}_{\hat{K}_{\mathrm{ur}}} \Big/ \mathcal{U}^1_{\hat{K}_{\mathrm{ur}}} \longrightarrow \mathcal{U}_{\hat{K}_{\mathrm{ur}}} \Big/ \mathcal{U}^1_{\hat{K}_{\mathrm{ur}}}$$

and, for $n \geq 1$,

$$\varphi - 1 : \mathcal{U}^n_{\hat{K}_{\mathrm{ur}}} \Big/ \mathcal{U}^{n+1}_{\hat{K}_{\mathrm{ur}}} \longrightarrow \mathcal{U}^n_{\hat{K}_{\mathrm{ur}}} \Big/ \mathcal{U}^{n+1}_{\hat{K}_{\mathrm{ur}}}.$$

But

$$\mathcal{U}_{\hat{K}_{\mathrm{ur}}} \Big/ \mathcal{U}^1_{\hat{K}_{\mathrm{ur}}} \cong \mathbb{F}^\times_{\mathrm{ur}} \qquad \text{(the multiplicative group)}$$

and, for $n \geq 1$,

$$\mathcal{U}^n_{\hat{K}_{\mathrm{ur}}} \Big/ \mathcal{U}^{n+1}_{\hat{K}_{\mathrm{ur}}} \cong \mathbb{F}_{\mathrm{ur}} \qquad \text{(the additive group)}.$$

Thus we have homomorphisms

$$\varphi - 1 : \mathbb{F}^\times_{\mathrm{ur}} \to \mathbb{F}^\times_{\mathrm{ur}} \qquad \text{given by} \qquad x \mapsto x^{q-1},$$
$$\varphi - 1 : \mathbb{F}_{\mathrm{ur}} \to \mathbb{F}_{\mathrm{ur}} \qquad \text{given by} \qquad x \mapsto x^q - x.$$

Since $\mathbb{F}_{\mathrm{ur}}$ is algebraically closed, these are surjective. We leave it as **Exercise 7.10** to show that this implies $\varphi - 1 : \mathcal{U}_{\hat{K}_{\mathrm{ur}}} \longrightarrow \mathcal{U}_{\hat{K}_{\mathrm{ur}}}$ is surjective. A similar argument, (using $\pi^n_K \mathcal{O}_{\hat{K}_{\mathrm{ur}}}$ instead of $\mathcal{U}^n_{\hat{K}_{\mathrm{ur}}}$), gives that $\varphi - 1 : \mathcal{O}_{\hat{K}_{\mathrm{ur}}} \longrightarrow \mathcal{O}_{\hat{K}_{\mathrm{ur}}}$ is also surjective.

For (ii), let $\tau(x)x^{-1} \in \mathcal{V}(\hat{L}_{\mathrm{ur}}/\hat{K}_{\mathrm{ur}})$ and choose $y \in \mathcal{U}_{\hat{L}_{\mathrm{ur}}}$ such that $(\varphi - 1)(y) = x$. Then

$$(\varphi - 1)\left(\frac{\tau(y)}{y}\right) = \frac{\varphi\tau(y)}{\varphi(y)}\left(\frac{\tau(y)}{y}\right)^{-1} = \frac{\tau\varphi(y)}{\tau(y)}\left(\frac{\varphi(y)}{y}\right)^{-1} = \frac{\tau(x)}{x}$$

(we have used that L/K is totally ramified to deduce that φ and τ commute).

For (iii), suppose $u \in \mathcal{U}_{\hat{L}_{\mathrm{ur}}}$ satisfies $\varphi(u) = u$. We may write $u = \varepsilon_0 + \pi_K v$, where $\varepsilon_0 \in K_{\mathrm{ur}}$. Since $\varphi(u) = u$, we have $\varphi(\varepsilon_0) \equiv \varepsilon_0 \mod \pi_K$, so we may find $u_0 \in K$ with $u = u_0 + \pi_K w_1$. But now $\varphi(u) = u$ gives $\varphi(w_1) = w_1$. Write $w_1 = \pi_K^{n_1} \varepsilon_1$, where $\varepsilon_1 \in \mathcal{U}_{\hat{K}_{\mathrm{ur}}}$; then $\varphi(\varepsilon_1) = \varepsilon_1$. Repeat the above with ε_1 in place of u to get

$$u = u_0 + \pi_K^{n_1+1} u_1 + \pi_K^{n_1+2} w_2, \qquad \text{where } u_0, u_1 \in K.$$

By induction we find that u is congruent to an element of K modulo π_K^n for any n. Since K is complete, this implies $u \in \mathcal{U}_K$ (we already knew $u \in \mathcal{U}_{\hat{K}_{\mathrm{ur}}}$). Thus $\ker(\varphi - 1) \subseteq \mathcal{U}_K$. The reverse containment is clear. $\qquad \square$

Now we want to define a map

$$\theta_{L/K} : \mathcal{U}_K \longrightarrow \mathrm{Gal}\,(L/K)$$

for abelian extensions L/K that are totally ramified. In the diagram $(*)$, the rows are exact by Theorem 2.6, so we can apply the Snake Lemma to obtain a homomorphism $\delta : B \longrightarrow C$. By (iii) of the previous lemma, $B = \mathcal{U}_K$. Also, since L/K is totally ramified, φ commutes with any $\tau \in \mathrm{Gal}\,(L/K)$. It follows that $\varphi - 1 : \mathrm{Gal}\,(L/K) \longrightarrow \mathrm{Gal}\,(L/K)$ is the "zero" map, so we have identified $C = \mathrm{Gal}\,(L/K)$. We let $\theta_{L/K}(u)$ be the element of $\mathrm{Gal}\,(L/K)$ that corresponds to $\delta(u) \in C$.

Proposition 3.2. Let $\theta_{L/K}$ be as described above. Then

i. $\theta_{L/K}$ is surjective,
ii. $\ker \theta_{L/K} = N_{L/K}\mathcal{U}_L$.

Proof. For (i) it suffices to show $D = 0$ in $(*)$. But we know by (i) of Lemma 3.1 that $\varphi - 1 : \mathcal{U}_{\hat{L}_{\mathrm{ur}}} \longrightarrow \mathcal{U}_{\hat{L}_{\mathrm{ur}}}$ is surjective.

For (ii), the diagram $(*)$ implies that $\beta(A) \supseteq N_{L/K}\mathcal{U}_L$. Let $\tilde{x} = x\mathcal{V}(\hat{L}_{\mathrm{ur}}/\hat{K}_{\mathrm{ur}}) \in A$, where $x \in \mathcal{U}_{\hat{L}_{\mathrm{ur}}}$. Since $A = \ker(\varphi - 1)$ we have $\varphi(x)x^{-1} \in \mathcal{V}(\hat{L}_{\mathrm{ur}}/\hat{K}_{\mathrm{ur}})$. By (ii) of Lemma 3.1, there is $y \in \mathcal{V}(\hat{L}_{\mathrm{ur}}/\hat{K}_{\mathrm{ur}})$ such that $\varphi(y)y^{-1} = \varphi(x)x^{-1}$. But then $\varphi(xy^{-1}) = xy^{-1}$ so that $xy^{-1} \in \mathcal{U}_L$ by (iii) of Lemma 3.1. Since $x\mathcal{V}(\hat{L}_{\mathrm{ur}}/\hat{K}_{\mathrm{ur}}) = xy^{-1}\mathcal{V}(\hat{L}_{\mathrm{ur}}/\hat{K}_{\mathrm{ur}})$, we also have

$$
\begin{aligned}
N_{\hat{L}_{\mathrm{ur}}/\hat{K}_{\mathrm{ur}}}(x) &= N_{\hat{L}_{\mathrm{ur}}/\hat{K}_{\mathrm{ur}}}(xy^{-1}) \\
&= N_{L/K}(xy^{-1}) \qquad \text{(why?)} \\
&\in N_{L/K}\mathcal{U}_L.
\end{aligned}
$$

It follows that $\beta(\tilde{x}) \in N_{L/K}\mathcal{U}_L$. $\square$

Theorem 3.3. *For any finite totally ramified abelian extension L/K, there is an isomorphism*

$$\tilde{\theta}_{L/K} : \mathcal{U}_K \Big/ N_{L/K}\mathcal{U}_L \longrightarrow \mathrm{Gal}\,(L/K).$$

Moreover, if E/K is a subextension of the totally ramified abelian extension L/K, then the following diagram commutes.

$$
\begin{array}{ccc}
\mathcal{U}_K \big/ N_{L/K}\mathcal{U}_L & \xrightarrow{\tilde{\theta}_{L/K}} & \mathrm{Gal}\,(L/K) \\
{\scriptstyle \mathrm{nat.}}\Big\downarrow & & \Big\downarrow{\scriptstyle \mathrm{nat.}} \\
\mathcal{U}_K \big/ N_{E/K}\mathcal{U}_E & \xrightarrow{\tilde{\theta}_{E/K}} & \mathrm{Gal}\,(E/K)
\end{array}
$$

Proof. The isomorphism follows from the proposition. The commutativity of the diagram follows from (∗).   □

We want to consider finite abelian extensions L/K that are not necessarily totally ramified. If φ is the Frobenius automorphism in $\mathrm{Gal}\,(\mathbb{F}_{\mathrm{ur}}/\mathbb{F}_K)$, let φ' be any lift of φ to $\mathrm{Gal}\,(L_{\mathrm{ur}}/K)$. Let L' be the fixed field of φ'. We know that L'/K is abelian and totally ramified. Also $L'_{\mathrm{ur}} = L_{\mathrm{ur}}$. We may identify $\mathrm{Gal}\,(L/K)_{\mathrm{ram}}$ and $\mathrm{Gal}\,(L'/K)$ and obtain the following diagram.

$$
\begin{array}{ccc}
& 1 & \quad\quad 1 \\
& \downarrow & \quad\quad \downarrow \\
& A \xrightarrow{\ \beta\ } B & \\
& \downarrow & \quad\quad \downarrow \\
1 \longrightarrow \mathrm{Gal}\,(L/K)_{\mathrm{ram}} \longrightarrow \mathcal{U}_{\hat{L}_{\mathrm{ur}}}\big/\mathcal{V}(\hat{L}_{\mathrm{ur}}/\hat{K}_{\mathrm{ur}}) \longrightarrow \mathcal{U}_{\hat{K}_{\mathrm{ur}}} \longrightarrow 1 \\
\varphi'-1\downarrow \quad\quad \varphi'-1\downarrow \quad\quad \varphi-1\downarrow \\
1 \longrightarrow \mathrm{Gal}\,(L/K)_{\mathrm{ram}} \longrightarrow \mathcal{U}_{\hat{L}_{\mathrm{ur}}}\big/\mathcal{V}(\hat{L}_{\mathrm{ur}}/\hat{K}_{\mathrm{ur}}) \longrightarrow \mathcal{U}_{\hat{K}_{\mathrm{ur}}} \longrightarrow 1 \\
\downarrow \quad\quad \downarrow \\
C \xrightarrow{\ \gamma\ } D \\
\downarrow \quad\quad \downarrow \\
1 \quad\quad 1
\end{array}
$$

As before, it follows that we have an isomorphism

$$
\mathcal{U}_K\big/N_{L'/K}\mathcal{U}_{L'} \longrightarrow \mathrm{Gal}\,(L'/K) = \mathrm{Gal}\,(L/K)_{\mathrm{ram}}.
$$

There is some finite unramified extension E/K such that $L'E = LE$ and LE/L is unramified. Moreover, since the norm is surjective on units in an unramified extension, we have $N_{LE/K}\mathcal{U}_{LE} = N_{L/K}N_{LE/L}\mathcal{U}_{LE} = N_{L/K}\mathcal{U}_L$, and similarly $N_{L'E/K}\mathcal{U}_{L'E} = N_{L'/K}\mathcal{U}_{L'}$. Thus $N_{L'/K}\mathcal{U}_{L'} = N_{L/K}\mathcal{U}_L$, and we have an isomorphism

$$
\tilde{\theta}_{L/K} : \mathcal{U}_K\big/N_{L/K}\mathcal{U}_L \longrightarrow \mathrm{Gal}\,(L/K)_{\mathrm{ram}}.
$$

The following theorem gives us a starting point for the definition of (the totally ramified part of) the local Artin map. (See the proof of Theorem 6.8 for the extension of this theorem to $\mathrm{Gal}\,(K_{\mathrm{ab}}/K)$.) If the local field K is a completion of the global field F, then the Artin map for K and the Artin map for F are connected as in

Section 3 of Chapter 6. However, soon we shall be able to give a definition of the local Artin map that does not rely on the global Artin map. Compare the theorem below with Corollary 3.6, whose proof uses the global Artin map.

Theorem 3.4. *If L/K is a finite abelian extension, then there is a canonical isomorphism $\tilde{\theta}_{L/K} : \mathcal{U}_K / N_{L/K}\mathcal{U}_L \longrightarrow \mathrm{Gal}\,(L/K)_{ram}$ and moreover, if $K \subseteq M \subseteq L$, then the following diagram commutes.*

$$
\begin{array}{ccc}
\mathcal{U}_K\big/N_{L/K}\mathcal{U}_L & \xrightarrow{\;\cong\;} & \mathrm{Gal}\,(L/K)_{\mathrm{ram}} \\[2mm]
\Big\downarrow{\scriptstyle\text{nat.}} & & \Big\downarrow{\scriptstyle\text{nat.}} \\[2mm]
\mathcal{U}_K\big/N_{M/K}\mathcal{U}_M & \xrightarrow{\;\cong\;} & \mathrm{Gal}\,(M/K)_{\mathrm{ram}}
\end{array}
$$

Proof. We have already shown the isomorphism; commutativity of the diagram follows from $(*)$. $\qquad\qquad\square$

4 One-Dimensional Formal Group Laws

Formal groups were first defined by Bochner ([Bo], 1946) in characteristic zero. The theory in characteristic p was developed by Chevalley and Dieudonné in the 1950s. An interpretation of formal groups using power series was developed by Lazard ([Laz1], [Laz2], 1955). Lazard's is the approach that we use here. We begin with a few general facts about one-dimensional formal group laws. There are higher dimensional formal group laws, but we shall not need them. For a complete treatment of formal group laws, including higher dimensions, see Hazewinkel's *Formal Groups and Applications*, [Haz1].

Let R be a commutative ring with identity, and denote by $R[[X]]$, $R[[X, Y]]$, etc., the rings of formal power series with coefficients from R. For two such formal power series F, G we write $F \equiv G$ (mod deg d) to mean that F and G coincide in terms of degree less than d.

A *one-dimensional formal group law* over R is a power series, $F \in R[[X, Y]]$ such that

i. $F(X, 0) = X$, $F(0, Y) = Y$, and
ii. $F(X, F(Y, Z)) = F(F(X, Y), Z)$.

If we also have $F(X, Y) = F(Y, X)$, then F is said to be a *commutative formal group law*.

Exercise 7.11. Show that (i) in the above definition may be replaced by the requirement $F(X, Y) \equiv X + Y$ (mod deg 2). $\qquad\qquad\diamond$

Example.

2. Over an arbitrary commutative ring R with identity, we may define

$$G_a(X, Y) = X + Y \text{ and } G_m(X, Y) = X + Y + XY.$$

Each is easily seen to be a commutative formal group law. G_a is called the *additive formal group law* and G_m is called the *multiplicative formal group law*.

All one-dimensional formal group laws over a "nice" ring R are commutative. To have a non-commutative one-dimensional formal group law over R, there must be some non-zero torsion nilpotent element in R (a result of Connell and Lazard).

Exercise 7.12. Let $R = \mathbb{F}_p[t]\big/\langle t^2 \rangle$ and let $F(X, Y) = X + Y + tXY^p$. Show that F is a non-commutative one-dimensional formal group law over R.   ◊

Exercise 7.13. Let F be a one-dimensional formal group law over R. Show that there is a power series $i(X) \in R[[X]]$ that functions as an "inverse" for F, i.e., so that $F(X, i(X)) = 0$.   ◊

In certain circumstances, a formal group law over R may be used to define an actual group operation. For example, if R is a complete local ring with maximal ideal $\mathfrak{p}$, then it is easy to see that for $x, y \in \mathfrak{p}$, the formal group law $F(X, Y)$ over R converges $\mathfrak{p}$-adically at $X = x$, $Y = y$ to an element of $\mathfrak{p}$. Thus we may define a group operation $+_F$ on $\mathfrak{p}$ by setting

$$x +_F y = F(x, y).$$

Exercise 7.14. Let F be a one-dimensional commutative formal group law over R. For $f, g \in XR[[X]]$, define

$$f +_F g = F(f(X), g(X)).$$

Show that with this operation, $XR[[X]]$ is an abelian group.   ◊

We may define a notion of "homomorphism" between formal group laws as follows. Suppose F and G are one-dimensional formal group laws over R. A power series $\theta \in R[[X]]$ that satisfies

i. $\theta(X) \equiv 0 \pmod{\deg 1}$, and
ii. $\theta(F(X, Y)) = G(\theta(X), \theta(Y))$

is called an *R-homomorphism* from F to G. Denote the set of all R-homomorphisms from F to G by $\mathrm{Hom}_R(F, G)$. An R-homomorphism from F to F is called an *R-endomorphism* of F. Denote the set of all R-endomorphisms of F by $\mathrm{End}_R(F)$.

Exercise 7.15. Show that $\mathrm{Hom}_R(F, G)$ is a subgroup of $XR[[X]]$ under the operation $+_G$ defined earlier.   ◊

Exercise 7.16. Show that $\text{End}_R(F)$ is a ring under $+_F$ and "formal composition" (of power series): $(\theta \circ \theta')(X) = \theta(\theta'(X))$. ◇

There is, of course also a notion of isomorphism for formal group laws. Suppose $\theta \in \text{Hom}_R(F, G)$. If there is some $\theta' \in \text{Hom}_R(G, F)$ so that θ and θ' satisfy $\theta'(\theta(X)) = \theta(\theta'(X)) = X$, then we say θ is an *R-isomorphism* and θ' is the inverse isomorphism to θ.

Example.

3. The formal power series $\exp(X)$ and $\log(1+X)$ are mutually inverse $\mathbb{Q}$-isomorphisms between the formal group laws G_a and G_m of Example 2.

Given $\theta \in \text{Hom}_R(F, G)$, we denote by $J(\theta)$ the element $a \in R$ such that

$$\theta(X) \equiv aX \quad (\text{mod } \deg 2).$$

$J(\theta)$ is called the *Jacobian* of θ.

Exercise 7.17. Show that θ is an isomorphism if and only if the Jacobian $J(\theta)$ is a unit in R. ◇

Example.

4. Take $R = \mathbb{F}_3$. Then

$$\begin{aligned}
G_a(X, G_a(X, X)) &= G_a(X, X + X) \\
&= X + X + X \\
&= 0, \\
G_m(X, G_m(X, X)) &= G_m(X, X + X + X^2) \\
&= X + X + X + X^2 + X^2 + X^2 + X^3 \\
&= X^3.
\end{aligned}$$

If $\theta(X) = a_1 X + a_2 X^2 + \cdots$ were an $\mathbb{F}_3$-isomorphism from G_m to G_a, then its Jacobian would have to be a unit, whence $a_1 \neq 0$ in $\mathbb{F}_3$. But also, it would have to satisfy $\theta(G_m(X, Y)) = G_a(\theta(X), \theta(Y))$, which implies

$$\begin{aligned}
\theta(X^3) &= \theta(G_m(X, G_m(X, X))) \\
&= G_a(\theta(X), \theta(G_m(X, X))) \\
a_1 X^3 + a_2 X^6 + \cdots &= G_a(\theta(X), G_a(\theta(X), \theta(X))) \\
&= 0,
\end{aligned}$$

a contradiction. Hence G_a and G_m are *not* $\mathbb{F}_3$-isomorphic.

Exercise 7.18. Show that G_a and G_m are not $\mathbb{F}_p$-isomorphic for any prime p. ◇

5 The Formal Group Laws of Lubin and Tate

The formal group laws we study in this section were introduced in 1965 by Lubin and Tate, [LT]. Following the approach in Hazewinkel's paper, [Haz2], we shall use them to prove the main results of local class field theory (see Iwasawa's book, [I], for a more general treatment).

Let K be a local field. Lubin-Tate formal group laws (over $\mathcal{O}_K$) are certain power series $F_f(X, Y)$, which we define below. In order to give the definition, we first need some notation and a technical lemma. For each uniformizer π of K, let

$$\mathcal{F}_\pi = \{f(X) \in \mathcal{O}_K[[X]] : f(X) \equiv \pi X \pmod{\deg 2}, \text{ and } f(X) \equiv X^q \pmod{\mathcal{P}_K}\}.$$

We shall make use of the following lemma several times, for various choices of linear form $\ell(X_1, \ldots, X_m)$, for various m. It can be generalized to the case where $f \in \mathcal{F}_\pi$ and $g \in \mathcal{F}_{\pi'}$ with $\pi \neq \pi'$, but to do so involves more cumbersome notation. Since we only must address this more complicated situation once, for the sake of clarity we have opted to postpone our discussion of it until Lemma 7.6. See Iwasawa's book, [I], for the general statement.

Lemma 5.1 (Lubin, Tate). Let π be a uniformizer in a local field K, and let $\mathbb{F}_K$ have order q. Suppose $f, g \in \mathcal{F}_\pi$ and let $\ell(X_1, \ldots, X_m) = a_1 X_1 + \cdots a_m X_m$ be a linear form (with $a_i \in \mathcal{O}_K$). Then there is a unique power series $F \in \mathcal{O}_K[[X_1, \ldots, X_m]]$ such that

$$F(X_1, \ldots, X_m) \equiv \ell(X_1, \ldots, X_m) \pmod{\deg 2}$$

$$f(F(X_1, \ldots, X_m)) = F(g(X_1), \ldots, g(X_m)).$$

Proof. We construct F by defining a sequence $F_1, F_2 \ldots$ of polynomials, where $\deg F_i \leq i$, with $F_1 = \ell$ and

$$f(F_n(X_1, \ldots, X_m)) \equiv F_n(g(X_1), \ldots, g(X_m)) \pmod{\deg n + 1}.$$

It will follow that $F = \lim_{n \to \infty} F_n$ is the desired power series. Proceed recursively, beginning with $F_1 = \ell$. Suppose we have found $F_n \in \mathcal{O}_K[X_1, \ldots, X_m]$ of degree no greater than n satisfying the above congruence. To construct F_{n+1} we seek a homogeneous polynomial $h_{n+1} \in \mathcal{O}_K[X_1, \ldots, X_m]$ of degree $n + 1$ such that

$$f(F_n(X_1, \ldots, X_m) + h_{n+1}(X_1, \ldots, X_m))$$
$$\equiv F_n(g(X_1), \ldots, g(X_m)) + h_{n+1}(g(X_1), \ldots, g(X_m)) \pmod{\deg n + 2}.$$

(Note that for any such h_{n+1} we also get $F_n + h_{n+1} \equiv F_n \pmod{\deg n + 1}$ and $\deg(F_n + h_{n+1}) \leq n + 1$.) Uniqueness of the power series F will follow if we find that the homogeneous polynomial h_{n+1} is unique. If the polynomial

h_{n+1} exists, then we put $F_{n+1} = F_n + h_{n_1}$. Given the nature of f and g, we must have

$$f(F_{n+1}(X_1, \ldots, X_m)) \equiv f(F_n(X_1, \ldots, X_m)) + \pi h_{n+1}(X_1, \ldots, X_m))$$
$$F_{n+1}(g(X_1), \ldots, g(X_m)) \equiv F_n(g(X_1), \ldots, g(X_m)) + \pi^{n+1} h_{n+1}(X_1, \ldots, X_m)$$

where both congruences are (mod deg $n+2$). From this, the only possible candidate for h_{n+1} is the (homogeneous) degree $n + 1$ part of

$$\frac{f(F_n(X_1, \ldots, X_m)) - F_n(g(X_1), \ldots, g(X_m))}{\pi^{n+1} - \pi}.$$

The above has coefficients in K and has no terms of degree less than $n + 1$. We need only verify that its coefficients lie in $\mathcal{O}_K$. Since $f(X) \equiv X^q \equiv g(X) \pmod{\pi}$ and for any $a \in \mathcal{O}_K$ we have $a^q \equiv a \pmod{\pi}$, it follows that

$$F_n(g(X_1), \ldots, g(X_m)) \equiv F_n((X_1^q, \ldots, X_m^q)$$
$$\equiv F_n(X_1, \ldots, X_m)^q$$
$$\equiv f(F_n(X_1, \ldots, X_m)) \pmod{\pi}.$$

Thus π divides $f(F_n(X_1, \ldots, X_m)) - F_n(g(X_1), \ldots, g(X_m))$ and the polynomial h_{n+1} has coefficients in $\mathcal{O}_K$.   □

For $f \in \mathcal{F}_\pi$, let $F_f(X, Y)$ be the unique power series in $\mathcal{O}_K[[X, Y]]$ that satisfies

$$F_f(X, Y) \equiv X + Y \pmod{\text{deg } 2},$$
$$f(F_f(X, Y)) = F_f(f(X), f(Y)).$$

(Apply the lemma with $m = 2$, $f = g$ and $\ell(X, Y) = X + Y$.)

Exercise 7.19. Show that F_f is a commutative formal group law over $\mathcal{O}_K$. The formal group laws F_f for $f \in \mathcal{F}_{\pi_K}$ are called the *Lubin-Tate formal group laws* for π_K. (HINT: To verify the equations in the definition of formal group law, show that each side satisfies Lemma 5.1 for an appropriately chosen linear form ℓ; the uniqueness part of the lemma then gives that the two sides must be equal. For example, taking $\ell(X, Y) = X + Y$ in the lemma shows $F_f(X, Y) = F_f(Y, X)$ since both make the lemma true for this ℓ.)   ◊

Example.

5. Let $K = \mathbb{Q}_p$, $\pi = p$. It is elementary to verify that $f(X) = (1 + X)^p - 1$ is in $\mathcal{F}_p$. Now consider the multiplicative formal group law $G_m(X, Y) = X + Y + XY$. We have

$$G_m(f(X), f(Y)) = (X + 1)^p(Y + 1)^p - 1 = f(G_m(X, Y)).$$

By uniqueness, it follows that $F_f(X, Y) = G_m(X, Y)$ in this case.

Lemma 5.2 (Lubin, Tate). Suppose K is a local field, with residue field $\mathbb{F}_K$ of order q. Let π be a uniformizer in K, and let $f(X), g(X) \in \mathcal{F}_\pi$. Then for any $a \in \mathcal{O}_K$ there is a unique power series $[a]_{f,g}(X) \in \mathcal{O}_K[[X]]$ such that

$$f([a]_{f,g}(X)) = [a]_{f,g}(g(X))$$
$$[a]_{f,g}(X) \equiv aX \pmod{X^2}.$$

Proof. Apply Lemma 5.1 with $m = 1$ and $\ell(X) = aX$.   □

The power series $[a]_{f,g}(X)$ will play an important role in what is to come. By the following corollary, each power series $[a]_{f,g}(X)$ gives us a homomorphism from F_g to F_f, which will be an isomorphism precisely when its Jacobian a satisfies $a \in \mathcal{U}_K$.

Corollary 5.3 (Lubin, Tate). Let K be a local field and let π be a uniformizer in K. Suppose $f(X), g(X), h(X) \in \mathcal{F}_\pi$ and $a, b \in \mathcal{O}_K$. As is customary, we put $[a]_f = [a]_{f,f}$.

i. $[\pi]_f(X) = f(X)$.
ii. $[a]_{f,g}([b]_{g,h}(X)) = [ab]_{f,h}(X)$ for any $a, b \in \mathcal{O}_K$.
iii. $[1]_{f,g}([1]_{g,f}(X)) = X$.
iv. $[a]_{f,g}(F_g(X, Y)) = F_f([a]_{f,g}(X), [a]_{f,g}(Y))$.
v. $[a + b]_{f,g}(X) = F_f([a]_{f,g}(X), [b]_{f,g}(X))$.

Proof. For (i), note that both $F(X) = [\pi]_f(X)$ and $F(X) = f(X)$ are solutions to

$$f(F(X)) = F(f(X))$$
$$F(X) \equiv \pi X \pmod{X^2}.$$

By the uniqueness part of Lemma 5.1, it follows that $[\pi]_f(X) = f(X)$. The proofs of $(ii) - (v)$ are similar, and are left as **Exercise 7.20**.   □

As can be seen from the above corollary, the power series $[a]_f(X)$ allows us to define a formal $\mathcal{O}_K$-module structure when combined with the formal group law F_f. In turn, this allows us to define an actual $\mathcal{O}_K$-module as follows.

For $x, y \in \mathcal{P}_\Omega$, the series $F_f(x, y)$ converges to an element of $\mathcal{P}_\Omega$, which we denote $x +_f y$. It is straightforward to check that $\mathcal{P}_\Omega$ is an abelian group under this operation. Similarly, for $a \in \mathcal{O}_K$ and $x \in \mathcal{P}_\Omega$, the series $[a]_f(x)$ converges to an element of $\mathcal{P}_\Omega$, which we denote $a \cdot_f x$. Using this as our "scalar multiplication," the corollary implies that $\mathcal{P}_\Omega$ becomes an $\mathcal{O}_K$-module, which we denote $\mathcal{P}_f$.

Care must be taken to remember that the operations on $\mathcal{P}_f$ are *not* the usual ones inherited from the field Ω (hence the alternate notation for the same set). $\mathcal{P}_f$ is called the *Lubin-Tate module* for π.

The following exercise shows that, up to isomorphism, $\mathcal{P}_f$ is independent of choice of $f \in \mathcal{F}_\pi$.

Exercise 7.21. Let $f, g \in \mathcal{F}_\pi$. Show that the $\mathcal{O}_K$-modules $\mathcal{P}_f$ and $\mathcal{P}_g$ are isomorphic.◊

6 Lubin–Tate Extensions

In this section, we see how the formal group laws of Lubin and Tate lead to totally ramified extensions of the local field K. To begin, we need a lemma about polynomials.

Lemma 6.1. Let k be any field, and let $g(X) = X^n + a_{n-1}X^{n-1} + \cdots + a_0 \in k[X]$ where either char $k = 0$ or n is prime to char k. Then we may find a positive integer r and a polynomial $\tilde{g}(X) \in k[X]$ of degree less than r, such that the polynomial $h(X) = X^r g(X) + \tilde{g}(X)$ has only simple zeros.

Proof. If k has infinite cardinality, then take $r = 1$, $\tilde{g}(X) = a_0 \in k$; this yields $\frac{d}{dX}(Xg(X) + a_0) = g(X) + Xg'(X)$, which is prime to $Xg(X) + a_0$ for suitable a_0. Thus we may suppose that k has finite cardinality, say $\#k = q$. Note if g is linear, there is nothing to prove, so we assume $n > 1$. Since n is prime to char k, we know that $\deg g' = n - 1 > 0$. Let $\alpha_1, \ldots, \alpha_{n-1}$ be the zeros of $g'(X)$ and let k' be a finite extension of $k(\alpha_1, \ldots, \alpha_{n-1})$, chosen so that $\#k' = q^s > \deg g$. Take $r = q^{s+1}$ and $\tilde{g}(X) = X^q g(X) + 1$ to get

$$h(X) = X^{q^{s+1}} g(X) - X^q g(X) + 1$$
$$h'(X) = (X^{q^{s+1}} - X^q)g'(X) \qquad \text{(note char } k \text{ divides } q).$$

If β is a zero of $h'(X)$, then it is a zero of $X^{q^{s+1}} - X^q$ or of $g'(X)$. In the first case, we have $h(\beta) = 1 \neq 0$. In the second, we have $\beta \in k'$, so $\beta^{q^s} = \beta$, and again $h(\beta) = 1 \neq 0$. $\qquad\square$

We shall use the polynomials that give rise to Lubin-Tate formal group laws to construct certain totally ramified abelian extensions of a local field K. Let $q = \#\mathbb{F}_K$ and, as before, let $\mathcal{F}_{\pi_K}$ denote the set of all power series $f(X) \in \mathcal{O}_K[[X]]$ such that $f(X) \equiv \pi_K X \pmod{X^2}$ and $f(X) \equiv X^q \pmod{\pi_K}$. Suppose $f(X) \in \mathcal{F}_{\pi_K}$ is a monic polynomial of degree q. Then

$$f(X) = X^q + \pi_K(a_{q-1}X^{q-1} + \cdots + a_2 X^2) + \pi_K X, \qquad \text{where } a_j \in \mathcal{O}_K.$$

For a positive integer m, define recursively

$$f^{(1)}(X) = f(X), \ f^{(2)}(X) = f(f(X)), \ldots, f^{(m)}(X) = f(f^{(m-1)}(X)).$$

By (i) and (ii) of Corollary 5.3, we have

$$f(X) = [\pi_K]_f(X) \qquad \text{and} \qquad f^{(m)}(X) = [\pi_K^m]_f(X).$$

It follows that if $\lambda \in \mathcal{P}_f$, (the Lubin-Tate module), then

$$f^{(m)}(X) = \pi_K^m \cdot_f \lambda.$$

Exercise 7.22. Let $f^{(m)}(X)$ be as above.

a. Show, for $m \geq 2$, that $f^{(m-1)}(X) \big| f^{(m)}(X)$.

b. Prove it is possible to find $\lambda_1, \lambda_2, \ldots \in \Omega$ so that λ_1 is a zero of $f(X)$, and for each $m \geq 2$, λ_m is a zero of $f^{(m)}(X)$ but not a zero of $f^{(m-1)}(X)$. Moreover, show that the λ_m can be chosen so that $f(\lambda_m) = \lambda_{m-1}$ for $m \geq 2$.

c. Show, for $m \geq 2$, that the polynomial $\frac{f^{(m)}(X)}{f^{(m-1)}(X)}$ is Eisenstein of degree $(q-1)q^{m-1}$ and has constant term equal to π_K. $\diamond$

Let $\lambda_1, \lambda_2, \ldots$ be as in part (b) of Exercise 7.22, and put $L_m = K(\lambda_m)$. By part (c) of Exercise 7.22, the extension L_m/K is totally ramified of degree $(q-1)q^{m-1}$, the element λ_m is a uniformizer in L_m, and π_K is a norm from L_m (for any m). Since $\pi_K^m \cdot_f \lambda_m = 0$, we may view the element λ_m as a π_K^m-torsion point in the Lubin-Tate module $\mathcal{P}_f$. It is "primitive" since $\pi_K^{m-1} \cdot_f \lambda_m \neq 0$.

Example.

6. We shall find L_m for $K = \mathbb{Q}_p$, $\pi_K = p$, and $f(X) = (1+X)^p - 1$. Recall this choice of $f(X)$ qualifies as an element of $\mathcal{F}_p$. A simple computation yields $f^{(m)}(X) = (1+X)^{p^m} - 1$, so we obtain

$$\frac{f^{(m)}(X)}{f^{(m-1)}(X)} = \frac{(1+X)^{p^m} - 1}{(1+X)^{p^{m-1}} - 1)}$$
$$= \Phi_{p^m}(1+X),$$

where Φ_{p^m} is the cyclotomic polynomial. Hence we find $\lambda_m = \zeta_{p^m} - 1$ and $L_m = \mathbb{Q}_p(\zeta_{p^m})$, where ζ_{p^m} is a primitive p^m-th root of unity in Ω. If we choose the primitive p^m-th roots of unity coherently, we also get $f(\lambda_m) = \lambda_{m-1}$ for $m \geq 2$.

Proposition 6.2. Let K be a local field and fix a polynomial $f(X) \in \mathcal{F}_{\pi_K}$. For $m \in \mathbb{Z}_+$, let L_m be the field associated to $f(X)$ as discussed above. Then

$$N_{L_m/K}\mathcal{U}_{L_m} \subseteq \mathcal{U}_K^m.$$

Proof. A typical element of $\mathcal{U}_{L_m}$ has the form εu, where ε is a $(q-1)^{\text{th}}$ root of unity and $u \in \mathcal{U}_{L_m}^1$. Observe that

$$N_{L_m/K}(\varepsilon) = \varepsilon^{(q-1)q^{m-1}} = 1$$

so we must show $N_{L_m/K}(u) \in \mathcal{U}_K^m$ for any $u \in \mathcal{U}_{L_m}^1$. This is clear if $m = 1$. Assume $m \geq 2$. Now $u \in \mathcal{U}_{L_m}^1$ may be written

$$u = 1 + a_1\lambda_m + a_2\lambda_m^2 + \cdots + a_n\lambda_m^n + w$$

where $a_i \in \mathcal{O}_K$, $n = m(q-1)q^{m-1} - 1$, and the valuation of w satisfies $v(w) \geq v(\pi_K^m)$. Note this choice of n gives $(n, \text{char } \mathbb{F}_K) = 1$. Let

$$d(X) = X^n + a_1 X^{n-1} + \cdots + a_n$$

and let $g(X) \in \mathbb{F}_K[X]$ be the image of $d(X)$ modulo $\mathcal{P}_K$. Apply Lemma 6.1 with $k = \mathbb{F}_K$, and let r and $\tilde{g}(X)$ be as the lemma provides. Now lift $\tilde{g}(X)$ to $\hat{g}(X) \in \mathcal{O}_K[X]$, where $\deg \hat{g} = \deg \tilde{g}$, and let

$$h(X) = X^r d(X) + \hat{g}(X).$$

Then the image of $h(X)$ in $\mathbb{F}_K[X]$ has no multiple zeros and the zeros of $h(X)$ are in K_{ur}.

As can be seen from the proof of Lemma 6.1, we may choose the constant term of $\tilde{g}(X)$ so that the constant term of $h(X)$ is 1. Let $\alpha_1, \ldots, \alpha_t$ be the zeros of $h(X)$. Since the constant term of $h(X)$ is 1, we have $\prod \alpha_i = \pm 1$, and the zeros of $h(X)$ are all in $\mathcal{U}_{K_{\text{ur}}}$. Also,

$$(1 - \alpha_1 \lambda_m) \cdots (1 - \alpha_t \lambda_m) = 1 + a_1 \lambda_m + \cdots + a_n \lambda_m^n + w'$$

where $v(w') \geq v(\pi_K^m)$. We have

$$
\begin{aligned}
u &= 1 + a_1 \lambda_m + \cdots + a_n \lambda_m^n + w \\
&= (1 - \alpha_1 \lambda_m) \cdots (1 - \alpha_t \lambda_m) + w - w' \\
&= (1 - \alpha_1 \lambda_m) \cdots (1 - \alpha_t \lambda_m)(1 + y)
\end{aligned}
$$

where $y = \dfrac{w - w'}{\prod_i (1 - \alpha_i \lambda_m)}$. Note we have $v(y) \geq v(\pi_K^m)$ and hence $N_{L_m/K}(1 + y) \in \mathcal{U}_K^m$.
It remains to show $N_{L_m/K}\left(\prod_i (1 - \alpha_i \lambda_m)\right) \in \mathcal{U}_K^m$, which will follow if we show $N_{L_m K_{\text{ur}}/K_{\text{ur}}}\left(\prod_i (1 - \alpha_i \lambda_m)\right) \in \mathcal{U}_{K_{\text{ur}}}^m$, because $\mathcal{U}_{K_{\text{ur}}}^m \cap \mathcal{U}_K = \mathcal{U}_K^m$ and the following diagram commutes (since K_{ur}/K is unramified and L_m/K is totally ramified).

$$
\begin{array}{ccc}
L_m & \xrightarrow{\ \text{incl.}\ } & L_m K_{\text{ur}} \\
\Big\downarrow{\scriptstyle N_{L_m/K}} & & \Big\downarrow{\scriptstyle N_{L_m K_{\text{ur}}/K_{\text{ur}}}} \\
K & \xrightarrow{\ \text{incl.}\ } & K_{\text{ur}}
\end{array}
$$

Because K_{ur}/K is unramified, the Eisenstein polynomial $\dfrac{f^{(m)}(X)}{f^{(m-1)}(X)}$ remains irreducible over K_{ur}, so it is the minimum polynomial of λ_m over K_{ur}. Hence for $\alpha \in \mathcal{U}_{K_{\text{ur}}}$ we have

$$N_{L_m K_{\text{ur}}/K_{\text{ur}}}(1 - \alpha \lambda_m) = \alpha^{(q-1)q^{m-1}} \frac{f^{(m)}(\alpha^{-1})}{f^{(m-1)}(\alpha^{-1})}.$$

Thus

$$N_{L_m K_{\mathrm{ur}}/K_{\mathrm{ur}}}\left(\prod_i (1 - \alpha_i \lambda_m)\right) = \left(\prod_i \alpha_i\right)^{(q-1)q^{m-1}} \prod_i \frac{f^{(m)}(\alpha_i^{-1})}{f^{(m-1)}(\alpha_i^{-1})}$$

$$= \prod_i \frac{f^{(m)}(\alpha_i^{-1})}{f^{(m-1)}(\alpha_i^{-1})}, \quad (\text{since } \prod_i \alpha_i = \pm 1 \text{ and } m \geq 2)$$

$$= 1 + \frac{\prod_i f^{(m)}(\alpha_i^{-1}) - \prod_i f^{(m-1)}(\alpha_i^{-1})}{\prod_i f^{(m-1)}(\alpha_i^{-1})}.$$

Now α_i^{-1} is a unit, so $f^{(m-1)}(\alpha_i^{-1})$ is also a unit. Thus we need only show

$$\prod_{i=1}^{t} f^{(m)}(\alpha_i^{-1}) - \prod_{i=1}^{t} f^{(m-1)}(\alpha_i^{-1}) \equiv 0 \pmod{\pi_K^m}.$$

The Frobenius automorphism φ permutes the α_i (they are the zeros of $h(X)$), so φ also permutes the α_i^{-1}. Modulo π_K, the map φ is just $x \mapsto x^q$; also (since $f \in \mathcal{F}_{\pi_K}$) the map $x \mapsto f(x) \mod \pi_K$ is just $x \mapsto x^q$. This means $f(\alpha_i^{-1}) \equiv \alpha_j^{-1} \pmod{\pi_K}$ for some j. In general, if $a, b \in \mathcal{O}_{K_{\mathrm{ur}}}$ satisfy $a \equiv b \mod \pi_K^r$ for $r \in \mathbb{Z}_+$, then

$$a^q \equiv b^q \pmod{\pi_K^{r+1}}$$

and

$$\pi_K a^s \equiv \pi_K b^s \pmod{\pi_K^{r+1}} \qquad \text{for any } s = 1, \ldots, q - 1.$$

Hence

$$f(a) \equiv f(b) \pmod{\pi_K^{r+1}}.$$

Since $f(\alpha_i^{-1}) \equiv \alpha_j^{-1} \pmod{\pi_K}$, we may apply this result with $a = f(\alpha_i^{-1})$, $b = \alpha_j^{-1}$ and $r = 1$. A simple induction argument yields

$$f^{(m)}(\alpha_i^{-1}) \equiv f^{(m-1)}(\alpha_j^{-1}) \pmod{\pi_K^m}.$$

But then

$$\prod_{i=1}^{t} f^{(m)}(\alpha_i^{-1}) \equiv \prod_{j=1}^{t} f^{(m-1)}(\alpha_j^{-1}) \pmod{\pi_K^m}$$

as needed. $\square$

Exercise 7.23. Let f, $g \in \mathcal{F}_{\pi_K}$ be polynomials of degree q and suppose $\mu \in \mathcal{P}_\Omega$ is a zero of $g^{(m)}(X)$ but not a zero of $g^{(m-1)}(X)$. Let $a \in \mathcal{O}_K$. Show

a. $[a]_{f,g}(\mu)$ is a zero of $f^{(m)}(X)$,

b. if $u \in \mathcal{U}_K$, then $[u]_{f,g}(\mu)$ is not a zero of $f^{(m-1)}(X)$. $\Diamond$

By the exercise above, if λ_m is a zero of $f^{(m)}(X)$ but not a zero of $f^{(m-1)}(X)$, then $[u]_f(\lambda_m)$ is also a zero of $f^{(m)}(X)$ but not a zero of $f^{(m-1)}(X)$. Observe that $[u]_f(\lambda_m) \in K(\lambda_m) = L_m$ because $[u]_f(X) \in \mathcal{O}_K[[X]]$ and L_m is complete. We want to prove that the extension L_m/K is Galois; we do this by showing that varying u in $[u]_f(\lambda_m)$ yields all the conjugates of λ_m.

Lemma 6.3. Let $f(X) \in \mathcal{O}_K[[X]]$, and suppose L/K is a finite extension. If there is some $\lambda \in L$ with $v_L(\lambda) > 0$ and $f(\lambda) = 0$, then there is a power series $h(X) \in \mathcal{O}_K[[X]]$ with $f(X) = (X - \lambda)h(X)$.

Proof. We work with polynomials: For any n there is a polynomial $h_n(X) \in \mathcal{O}_K[X]$, and some constant $b_n \in \mathcal{O}_L$ such that $f(X) \equiv (X - \lambda)h_n(X) + b_n \pmod{X^n}$. Since $f(\lambda) = 0$, we have $b_n \equiv 0 \pmod{\pi_K^n}$, so $v_L(b_n) \geq n\, v_L(\lambda) > 0$. This says $\lim_{n \to \infty} v_L(b_n) = \infty$. We have

$$(X - \lambda)h_{n+1}(X) + b_{n+1} \equiv (X - \lambda)h_n(X) + b_n \pmod{X^n}$$
$$b_{n+1} \equiv b_n \pmod{\lambda^n}$$

so that

$$(X - \lambda)(h_{n+1}(X) - h_n(X)) \equiv 0 \pmod{\langle X^n, \lambda^n \rangle}.$$

Say $h_{n+1}(X) - h_n(X) = a_n X^n + a_{n-1}X^{n-1} + \cdots + a_1 X + a_0$. Then $v_L(a_0\lambda) \geq n\, v_L(\lambda)$, and $v_L(a_i\lambda - a_{i-1}) \geq n\, v_L(\lambda)$, for $i = 1, \ldots n - 1$. But this implies $v_L(a_0) \geq (n - 1)v_L(\lambda)$, $v_L(a_1) \geq (n-2)v_L(\lambda)$, $\ldots$, $v_L(a_{n-1}) \geq 0$. Hence we conclude $\lim_{n \to \infty} h_n(X)$ exists; put $h(X) = \lim_{n \to \infty} h_n(X)$. Then for any n,

$$f(X) \equiv (X - \lambda)h(X) \pmod{\langle X^n, \lambda^n \rangle}.$$

But this is only possible if $f(X) = (X - \lambda)h(X)$.   □

Lemma 6.4. Suppose $u, u' \in \mathcal{U}_K$ and let $f(X) \in \mathcal{F}_{\pi_K}$ be a polynomial of degree q. If $[u]_f(\lambda_m) = [u']_f(\lambda_m)$ then $u\,\mathcal{U}_K^m = u'\,\mathcal{U}_K^m$.

Proof. First note that $[u]_f([u']_f(X)) = [uu']_f(X)$. Because of this, it suffices to show that if $[u]_f(\lambda_m) = \lambda_m$, then $u \in \mathcal{U}_K^m$. Let $\sigma : L_m \hookrightarrow \Omega$ be any K-embedding. Then $\sigma(\lambda_m)$ is a zero of $[u]_f(X) - X$. Also, since $[u]_f(f(X)) = f([u]_f(X))$, we have that $f(\lambda_m)$ is a zero of $[u]_f(X) - X$. Applying this idea repeatedly, we see that

$f^{(r)}(\lambda_m)$ is a zero of $[u]_f(X) - X$ for any $r \le m$. But then all the zeros of $f^{(m)}(X)$ are zeros of $[u]_f(X) - X$. Now use the previous lemma to write

$$[u]_f(X) - X = f^{(m)}(X) h(X)$$

for some $h(X) \in \mathcal{O}_K[[X]]$. Note that $f^{(m)}(X) \equiv \pi_K^m X \pmod{X^2}$, and compare coefficients in the two sides of the above equation. It follows that $u - 1 = h_0 \pi_K^m$, where h_0 is the constant term of $h(X)$. Since the coefficients of $h(X)$ lie in $\mathcal{O}_K$, we have $u \in \mathcal{U}_K^m$ as desired. $\square$

Theorem 6.5. *Let K be a local field and let L_m be as above, for some polynomial* $f(X) \in \mathcal{F}_{\pi_K}$ *of degree q. Then L_m/K is Galois, and* $\mathrm{Gal}\,(L_m/K) \cong \mathcal{U}_K \big/ \mathcal{U}_K^m$.

Proof. For any $u \in \mathcal{U}_K$, we know that $[u]_f(\lambda_m)$ is a conjugate of λ_m in L_m. Thus, Lemma 6.4 implies that distinct cosets in $\mathcal{U}_K \big/ \mathcal{U}_K^m$ correspond to distinct conjugates of λ_m in L_m. But $\#\left(\mathcal{U}_K \big/ \mathcal{U}_K^m\right) = (q - 1)q^{m-1} = [L_m : K]$. It follows that all the conjugates of λ_m are in L_m, and hence that L_m/K is Galois. Now map $\mathrm{Gal}\,(L_m/K) \to \mathcal{U}_K \big/ \mathcal{U}_K^m$ by $\sigma \mapsto u\mathcal{U}_K^m$, where $u \in \mathcal{U}_K$ satisfies $[u]_f(\lambda_m) = \sigma(\lambda_m)$. It is straightforward to show that this yields a well-defined isomorphism of groups. $\square$

We have succeeded in showing that varying u in $[u]_f(\lambda_m)$ yields all the conjugates of λ_m. Considering a certain submodule of the Lubin- Tate module $\mathcal{P}_f$ leads to a similar result. To discuss this, we study the π_K^m-torsion points of $\mathcal{P}_f$ a bit more. For a degree-q polynomial $f \in \mathcal{F}_{\pi_K}$, put

$$W_{f,m} = \ker [\pi_K^m]_f = \{\lambda \in \mathcal{P}_f : \pi_K^m \cdot_f \lambda = 0\}.$$

Observe that $W_{f,m}$ is an $\mathcal{O}_K$-submodule of $\mathcal{P}_f$, and because $W_{f,m}$ is annihilated by $\pi_K^m \mathcal{O}_K$, it also can be regarded as an $\mathcal{O}_K \big/ \pi_K^m \mathcal{O}_K$-module. We have

$$\{0\} \subset W_{f,1} \subset W_{f,2} \subset \cdots \subset W_{f,m} \subset \cdots .$$

Putting $f^{(0)}(X) = X$, (so that $W_{f,0} = \{0\}$), we have (for $m \ge 1$)

$$\#W_{f,m} = \#W_{f,m-1} + (q - 1)q^{m-1},$$

from which it follows that $\#W_{f,m} = q^m$.

Now fix $\lambda_m \in W_{f,m} - W_{f,m-1}$, and define $\delta : \mathcal{O}_K \to W_{f,m}$ by $\delta(a) = a \cdot_f \lambda_m$. The map δ is easily seen to be $\mathcal{O}_K$-linear. To find $\ker \delta$, note that any non-zero $a \in \mathcal{O}_K$ may be written $a = \pi_K^t u$ for some $u \in \mathcal{U}_K$ and some $t \in \mathbb{N}$. If $\delta(a) = 0$, then

$$0 = a \cdot_f \lambda_m = [a]_f(\lambda_m)$$
$$= [\pi_K^t u]_f(\lambda_m)$$
$$= [\pi_K^t]_f([u]_f(\lambda_m))$$
$$= f^{(t)}([u]_f(\lambda_m)).$$

Since $u \in \mathcal{U}_K$, we know that $[u]_f(\lambda_m)$ is a conjugate of λ_m, so it is a zero of $f^{(m)}(X)$ and not a zero of $f^{(m-1)}(X)$. Since $f^{(t)}([u]_f(\lambda_m)) = 0$, we must have $t \geq m$. We have shown: If $a \in \ker \delta$, then $a \in \pi_K^m \mathcal{O}_K$. The converse is also clearly true. Hence $\ker \delta = \pi_K^m \mathcal{O}_K$.

By the above argument, δ gives rise to an injective $\mathcal{O}_K$-linear map

$$\bar{\delta} : {}^{\mathcal{O}_K} \big/ _{\pi_K^m \mathcal{O}_K} \hookrightarrow W_{f,m}.$$

This map is actually an isomorphism of $\mathcal{O}_K$-modules, since the cardinalities agree:
$$\#\left({}^{\mathcal{O}_K} \big/ _{\pi_K^m \mathcal{O}_K} \right) = q^m = \#W_{f,m}.$$

Now consider the map ${}^{\mathcal{O}_K} \big/ _{\pi_K^m \mathcal{O}_K} \longrightarrow \mathrm{End}_{\mathcal{O}_K}(W_{f,m})$ given by $a \mapsto [a]_f$. By Corollary 5.3, this is a ring homomorphism. Since $\mathrm{End}_{\mathcal{O}_K}\!\left({}^{\mathcal{O}_K} \big/ _{\pi_K^m \mathcal{O}_K} \right) \cong {}^{\mathcal{O}_K} \big/ _{\pi_K^m \mathcal{O}_K}$ as (finite) rings, we may use the $\mathcal{O}_K$-linear isomorphism $\bar{\delta}$ defined above to conclude

$$ {}^{\mathcal{O}_K} \big/ _{\pi_K^m \mathcal{O}_K} \cong \mathrm{End}_{\mathcal{O}_K}(W_{f,m})$$

as rings. Comparing the units in these rings, we also get

$$ {}^{\mathcal{U}_K} \big/ _{\mathcal{U}_K^m} \cong \mathrm{Aut}_{\mathcal{O}_K}(W_{f,m})$$

via the map $u \mapsto [u]_f$.

Exercise 7.24. Let $f, g \in \mathcal{F}_{\pi_K}$ be polynomials of degree q.

a. Show that $W_{f,m}$ and $W_{g,m}$ are isomorphic as $\mathcal{O}_K$-modules.

b. Show that $W_{f,m}$ is a free ${}^{\mathcal{O}_K} \big/ _{\pi_K^m \mathcal{O}_K}$-module of rank 1. ◇

We know that L_m/K is Galois; we now discover it is also independent of the choice of $f \in \mathcal{F}_{\pi_K}$.

Corollary 6.6. The extensions L_m/K depend only on the choice of uniformizer π_K; they do not depend on the choice of polynomial $f(X) \in \mathcal{F}_{\pi_K}$. The field L_m is called the mth *Lubin-Tate extension* of K associated to the uniformizer π_K.

Proof. Suppose

$$f(X) = X^q + \pi_K(a_{q-1}X^{q-1} + \cdots + a_2X^2) + \pi_K X$$
$$g(X) = X^q + \pi_K(b_{q-1}X^{q-1} + \cdots + b_2X^2) + \pi_K X,$$

where $a_j, b_j \in \mathcal{O}_K$. By Lemma 5.2, there is a unique power series $[1]_{f,g}(X) \in \mathcal{O}_K[[X]]$, such that $[1]_{f,g}(X) \equiv X \pmod{X^2}$ and $f([1]_{f,g}(X)) = [1]_{f,g}(g(X))$. Let μ_m be a zero of $g^{(m)}(X)$ that is not a zero of $g^{(m-1)}(X)$. By Exercise 7.23, $[1]_{f,g}(\mu_m)$ is a zero of $f^{(m)}(X)$ and is not a zero of $f^{(m-1)}(X)$. Hence $L_m = K([1]_{f,g}(\mu_m))$. Since $[1]_{f,g}(\mu_m) \in K(\mu_m)$, we must have $L_m \subseteq K(\mu_m)$. Since they have equal degrees over K, it follows that $L_m = K(\mu_m)$.   □

Corollary 6.7. *Let L_m be the mth Lubin-Tate extension of a local field K. Then $N_{L_m/K}\mathcal{U}_{L_m} = \mathcal{U}_K^m$.*

Proof. We have shown $N_{L_m/K}\mathcal{U}_{L_m} \subseteq \mathcal{U}_K^m$. Also, since L_m/K is totally ramified, we have $\mathrm{Gal}\,(L_m/K) \cong \mathcal{U}_K \big/ N_{L_m/K}\mathcal{U}_{L_m}$ via the map $\tilde{\theta}_{L_m/K}$. The result is then clear from Theorem 6.5.   □

We have reached the point where it is possible to study the extension K_{ab}/K, where K is a local field (so $\mathbb{F}_K$ is finite).

Theorem 6.8. *Let K be a local field. There are isomorphisms $\mathrm{Gal}\,(K_{ab}/K)_{ram} \cong \mathcal{U}_K$ and $\mathrm{Gal}\,(K_{ab}/K) \cong \mathcal{U}_K \times \hat{\mathbb{Z}}$.*

Proof. We have shown that if L/K is a finite abelian extension, then there is an isomorphism $\tilde{\theta}_{L/K} : \mathcal{U}_K \big/ N_{L/K}\mathcal{U}_L \xrightarrow{\cong} \mathrm{Gal}\,(L/K)_{ram}$. Now we take the limit over all finite abelian extensions L to obtain an isomorphism

$$\theta_{K_{ab}/K} : \varprojlim \mathcal{U}_K \big/ N_{L/K}\mathcal{U}_L \xrightarrow{\cong} \mathrm{Gal}\,(K_{ab}/K)_{ram}.$$

For any such L, the group $\mathcal{U}_L$ is compact and the map $N_{L/K}$ is continuous. It follows that $N_{L/K}\mathcal{U}_L$ is a compact subgroup in $\mathcal{U}_K$. But it also has finite index in $\mathcal{U}_K$, since $\mathrm{Gal}\,(L/K)_{ram}$ is finite. Thus $N_{L/K}\mathcal{U}_L$ is both open and closed in $\mathcal{U}_K$. In particular, there is some $m \in \mathbb{Z}_+$ such that $\mathcal{U}_K^m \subseteq N_{L/K}\mathcal{U}_L$. We have shown that $\mathcal{U}_K^m = N_{L_m/K}\mathcal{U}_{L_m}$ where L_m is the mth Lubin-Tate extension of K. Thus, we must have

$$\varprojlim \mathcal{U}_K \big/ N_{L/K}\mathcal{U}_L = \mathcal{U}_K.$$

This allows us to consider $\theta_{K_{ab}/K}$ as an isomorphism $\mathcal{U}_K \longrightarrow \mathrm{Gal}\,(K_{ab}/K)_{ram}$.
    For the second assertion, it suffices to recall that

$$\text{Gal}(K_{ab}/K) \cong \text{Gal}(K_{ab}/K)_{ram} \times \hat{\mathbb{Z}}$$

by Exercise 7.5. □

Corollary 6.9. Let K be a local field with uniformizer π_K. Put $L_{\pi_K} = \cup L_m$ where the L_m are the Lubin-Tate extensions of K associated to the uniformizer π_K. Then $K_{ab} = L_{\pi_K} K_{ur}$.

Proof. Since $L_{\pi_K} K_{ur}/K$ is abelian, we have $L_{\pi_K} K_{ur} \subseteq K_{ab}$. For the reverse containment, let $\alpha : \text{Gal}(K_{ab}/K) \longrightarrow \text{Gal}(L_{\pi_K} K_{ur}/K)$ be the natural homomorphism and consider the following commutative diagram with exact rows.

$$
\begin{array}{ccccccccc}
1 & \longrightarrow & \text{Gal}(K_{ab}/K)_{ram} & \longrightarrow & \text{Gal}(K_{ab}/K) & \longrightarrow & \text{Gal}(K_{ur}/K) & \longrightarrow & 1 \\
 & & \alpha' \downarrow & & \alpha \downarrow & & = \downarrow & & \\
1 & \longrightarrow & \text{Gal}(L_{\pi_K} K_{ur}/K)_{ram} & \longrightarrow & \text{Gal}(L_{\pi_K} K_{ur}/K) & \longrightarrow & \text{Gal}(K_{ur}/K) & \longrightarrow & 1
\end{array}
$$

(Here α' is the homomorphism induced by α.) Taking projective limits as in the theorem, we also have homomorphisms

$$\theta_{K_{ab}/K} : \mathcal{U}_K \longrightarrow \text{Gal}(K_{ab}/K)_{ram}$$
$$\theta_{L_{\pi_K} K_{ur}/K} : \mathcal{U}_K \longrightarrow \text{Gal}(L_{\pi_K} K_{ur}/K)_{ram}$$

where in the first case the limit is over all the finite abelian extensions L/K, and in the second case the limit is over finite extensions L/K, with $L \subseteq L_{\pi_K} K_{ur}$. By the theorem, $\theta_{K_{ab}/K}$ is an isomorphism. Similarly $\theta_{L_{\pi_K} K_{ur}/K}$ is an isomorphism. Moreover, we have

$$\alpha' \circ \theta_{K_{ab}/K} = \theta_{L_{\pi_K} K_{ur}/K}$$

so that α', (and hence α), is also an isomorphism. □

Example.

7. Let $K = \mathbb{Q}_p$. Using $\pi = p$ and $f(X) = (X+1)^{p^m} - 1$, we have computed $\lambda_m = \zeta_{p^m} - 1$, so that $L_m = \mathbb{Q}_p(\zeta_{p^m})$. The extension $L_\pi/\mathbb{Q}_p$ will then be obtained by adjoining all p-power roots of unity to $\mathbb{Q}_p$. The maximal unramified extension of $\mathbb{Q}_p$ can be obtained by adjoining all the roots of unity whose orders are prime to p. By Corollary 6.9, we now have that the maximal abelian extension of $\mathbb{Q}_p$ is $\mathbb{Q}_p(\mu_\infty)$, where μ_∞ is the set of all roots of unity in Ω.

Exercise 7.25. Give $K^\times \cong \mathcal{U}_K \times \mathbb{Z}$ the topology of open subgroups H of finite index (where H is open with respect to the topology of the valuation v_K). Show that the completion of $K^\times$ in this topology is $\mathcal{U}_K \times \hat{\mathbb{Z}}$. ◇

7 The Local Artin Map

Observe that Exercise 7.25 gives us an embedding $K^\times \hookrightarrow \mathcal{U}_K \times \hat{\mathbb{Z}}$. Also, since we have $\mathcal{U}_K \times \hat{\mathbb{Z}} \cong \mathrm{Gal}(K_{\mathrm{ab}}/K)$, we have an embedding $K^\times \hookrightarrow \mathrm{Gal}(K_{\mathrm{ab}}/K)$. Our aim, however, is to find a *canonical* isomorphism — we want to be able to choose the isomorphism $\mathcal{U}_K \times \hat{\mathbb{Z}} \cong \mathrm{Gal}(K_{\mathrm{ab}}/K)$ so that for any finite abelian extension L/K the kernel of the map given by the composition

$$K^\times \xrightarrow{\text{incl.}} \mathcal{U}_K \times \hat{\mathbb{Z}} \xrightarrow{\cong} \mathrm{Gal}(K_{\mathrm{ab}}/K) \xrightarrow{\text{nat.}} \mathrm{Gal}(L/K)$$

will be $N_{L/K}L^\times$. Thus, if we succeed in finding such an isomorphism, we shall have obtained a local analogue of Artin Reciprocity.

Lemma 7.1. If L/K is a finite abelian extension, then $[K^\times : N_{L/K}L^\times] = [L : K]$.

Proof. Let K_L/K denote the maximal unramified subextension of L/K. Then Theorem 3.4 implies that $[L : K_L] = \#\left(\mathcal{U}_K \big/ N_{L/K}\mathcal{U}_L\right)$. Consider the following commutative diagram with exact rows,

$$
\begin{array}{ccccccccc}
1 & \longrightarrow & \mathcal{U}_L & \longrightarrow & L^\times & \xrightarrow{v_L} & \mathbb{Z} & \longrightarrow & 1 \\
& & \big\downarrow{\scriptstyle N_{L/K}} & & \big\downarrow{\scriptstyle N_{L/K}} & & \big\downarrow{\scriptstyle \mu} & & \\
1 & \longrightarrow & \mathcal{U}_K & \longrightarrow & K^\times & \xrightarrow{v_K} & \mathbb{Z} & \longrightarrow & 1
\end{array}
$$

where $\mu : \mathbb{Z} \longrightarrow \mathbb{Z}$ is multiplication by $[K_L : K]$. It follows that $[K^\times : N_{L/K}L^\times] = [K_L : K][\mathcal{U}_K : N_{L/K}\mathcal{U}_L] = [K_L : K][L : K_L] = [L : K]$. $\qquad\Box$

Now suppose L/K is a finite totally ramified abelian extension, and choose the uniformizer π_K to be a norm from L. Let $K_{(t)}/K$ be an unramified extension of K of degree t. We want to use the map $\theta_{L/K}$ that was defined previously for finite totally ramified abelian extensions of K to construct a map for $LK_{(t)}$. We let $\rho_{LK_{(t)}/K} : K^\times \longrightarrow \mathrm{Gal}(LK_{(t)}/K)$ be the unique homomorphism that satisfies

$$
\begin{aligned}
\rho_{LK_{(t)}/K}(u) &= \theta_{L/K}(u^{-1}) && \text{for } u \in \mathcal{U}_K \\
\rho_{LK_{(t)}/K}(\pi_K) &= \varphi && \text{Frobenius in } \mathrm{Gal}(LK_{(t)}/L).
\end{aligned}
$$

Note that $\mathrm{Gal}(L/K)_{\mathrm{ram}} = \mathrm{Gal}(L/K)$ so that $\theta_{L/K}$ is onto $\mathrm{Gal}(L/K)$ by Proposition 3.2. Since we also have $\mathrm{Gal}(L/K) \cong \mathrm{Gal}(LK_{(t)}/K_{(t)})$, the above definition of $\rho_{LK_{(t)}/K}(u)$ for $u \in \mathcal{U}_K$ should be interpreted by identifying $\mathrm{Gal}(L/K)$ and $\mathrm{Gal}(LK_{(t)}/K_{(t)})$. Since $LK_{(t)}/L$ is finite and unramified, we have $\mathrm{Gal}(LK_{(t)}/L) = \langle\varphi\rangle$, so the map $\rho_{LK_{(t)}/K}$ is surjective. To be useful, we want the map $\rho_{LK_{(t)}/K}$ to depend only on the extension $LK_{(t)}/K$ and not on the totally ramified subextension

L/K used to define it. Our first task is to show that if E is another finite totally ramified abelian extension of K such that $EK_{(t)} = LK_{(t)}$, then the maps $\rho_{LK_{(t)}/K}$ and $\rho_{EK_{(t)}/K}$ agree.

Lemma 7.2. Let $K_{(t)}/K$ be a finite unramified extension of K of degree t. Let L/K and E/K be finite totally ramified abelian extensions of K such that $LK_{(t)} = EK_{(t)}$. Consider the composition

$$K^\times \xrightarrow{\ \rho_{LK_{(t)}/K}\ } \operatorname{Gal}(LK_{(t)}/K) \xrightarrow{\ \text{nat.}\ } \operatorname{Gal}(E/K).$$

The kernel of this composition is $N_{E/K}E^\times$.

Proof. By the previous lemma it suffices to prove that $N_{E/K}E^\times$ is contained in the kernel. Since we already know that $N_{E/K}\mathcal{U}_E$ is contained in the kernel, it remains to show that $N_{E/K}(\pi_E)$ is in the kernel, where π_E is a uniformizer in E.

Note that $\rho_{LK_{(t)}/K}(\mathcal{U}_K) = \operatorname{Gal}(LK_{(t)}/K)_{\text{ram}}$, so there is some $u \in \mathcal{U}_K$ such that $\langle \rho_{LK_{(t)}/K}(u)\,\varphi \rangle$ has fixed field E. Let π_L be a uniformizer in L, chosen so that $N_{L/K}(\pi_L) = \pi_K$ and let $\pi_E = \varepsilon \pi_L$. We leave it as **Exercise 7.26** to show that $N_{LK_{(t)}/K_{(t)}}(\varepsilon) \in \mathcal{U}_K$. By our choice of u, we have $\big(\rho_{LK_{(t)}/K}(u)\,\varphi\big)(\pi_E) = \pi_E$ so

$$\frac{\big(\theta_{L/K}(u^{-1})\big)(\pi_L)}{\pi_L} = \frac{\big(\rho_{LK_{(t)}/K}(u)\big)(\pi_L)}{\pi_L}$$

$$= \frac{\big(\rho_{LK_{(t)}/K}(u)\,\varphi\big)(\pi_L)}{\pi_L} \qquad \text{since } \varphi(\pi_L) = \pi_L$$

$$= \frac{\big(\rho_{LK_{(t)}/K}(u)\,\varphi\big)(\varepsilon^{-1})}{\varepsilon^{-1}} \qquad \text{since } \varepsilon^{-1}\pi_E = \pi_L$$

$$= \frac{\big(\rho_{LK_{(t)}/K}(u)\,\varphi\big)(\varepsilon^{-1})}{\varphi(\varepsilon^{-1})}\,\frac{\varphi(\varepsilon^{-1})}{\varepsilon^{-1}}$$

$$\in \frac{\varphi(\varepsilon^{-1})}{\varepsilon^{-1}}\,\mathcal{V}(\hat{L}_{\text{ur}}/\hat{K}_{\text{ur}}).$$

We have shown that $i\big(\theta_{L/K}(u^{-1})\big)$ and $(\varphi - 1)\big(\varepsilon^{-1}\mathcal{V}(\hat{L}_{\text{ur}}/\hat{K}_{\text{ur}})\big)$ are equal. From the diagram (∗), it follows that $N_{LK_{(t)}/K_{(t)}}(\varepsilon) \in u\,N_{L/K}\mathcal{U}_L$. But then

$$\rho_{LK_{(t)}/K}(N_{E/K}(\pi_E)) = \rho_{LK_{(t)}/K}(u\pi_K) = \rho_{LK_{(t)}/K}(u)\,\varphi.$$

Since this fixes the elements of E, the result follows.   □

Lemma 7.2 tells us that the definition of $\rho_{LK_{(t)}/K}$ does not depend on the choice of finite totally ramified abelian extension L/K. For, if E/K is another such extension, with $EK_{(t)} = LK_{(t)}$, then our definition of $\rho_{EK_{(t)}/K}$ would require

$$\rho_{EK_{(t)}/K}(u) = \theta_{E/K}(u^{-1}) \qquad \text{for } u \in \mathcal{U}_K$$
$$\rho_{EK_{(t)}/K}(\pi'_K) = \varphi \qquad \text{Frobenius in Gal}(EK_{(t)}/E),$$

where π'_K is a uniformizer in K that is a norm from E.

Exercise 7.27. Use Lemma 7.2 to verify that $\rho_{LK_{(t)}/K}$ and $\rho_{EK_{(t)}/K}$ are equal on $K^\times$. $\qquad\qquad\qquad\qquad\qquad\qquad\qquad\qquad\qquad\qquad\qquad\qquad\qquad\qquad \Diamond$

With what we have done, we are able to define a local reciprocity homomorphism, which plays a role comparable to the Artin map in the global class field theory. (It is called the *local Artin map* or the *local norm residue map*, denoted ρ_K.) To do so, we choose a uniformizer π_K of K, and use the union L_{π_K} of the Lubin-Tate extensions of K, recalling that $K_{\mathrm{ab}} = L_{\pi_K} K_{\mathrm{ur}}$, and we may identify $\mathrm{Gal}(K_{\mathrm{ab}}/K_{\mathrm{ur}}) = \mathrm{Gal}(L_{\pi_K}/K)$. Define $\rho_K : K^\times \longrightarrow \mathrm{Gal}(K_{\mathrm{ab}}/K)$ to be the unique homomorphism that satisfies

$$\rho_K(u) = \theta_{K_{\mathrm{ab}}/K}(u^{-1}) \in \mathrm{Gal}(L_{\pi_K}/K) \qquad \text{for } u \in \mathcal{U}_K$$
$$\rho_K(\pi_K) = \varphi \qquad\qquad\qquad\qquad \text{Frobenius in Gal}(K_{\mathrm{ab}}/L_{\pi_K}).$$

Since $\pi_K \in N_{L_m/K} L_m^\times$ for all the Lubin-Tate extensions L_m/K, it follows that ρ_K agrees with $\rho_{L_m K_{(t)}/K}$. Thus the definition of ρ_K is independent of choice of uniformizer π_K of K by Lemma 7.2. Recall $\mathcal{U}_K \times \hat{\mathbb{Z}} \cong \mathrm{Gal}(K_{\mathrm{ab}}/K)$. By Exercise 7.28 below, the homomorphism ρ_K is the restriction to $K^\times$ of an isomorphism $\mathcal{U}_K \times \hat{\mathbb{Z}} \longrightarrow \mathrm{Gal}(K_{\mathrm{ab}}/K)$.

Exercise 7.28. Give $\mathrm{Gal}(K_{\mathrm{ab}}/K)$ the Krull topology.

a. Is ρ_K continuous? Prove that your answer is correct.

b. Is the image $\rho_K(K^\times)$ dense in $\mathrm{Gal}(K_{\mathrm{ab}}/K)$? Prove that your answer is correct.

c. Show ρ_K can be extended to an isomorphism $\mathcal{U}_K \times \hat{\mathbb{Z}} \longrightarrow \mathrm{Gal}(K_{\mathrm{ab}}/K)$. $\qquad \Diamond$

Lemma 7.3. Let L/K be a finite abelian extension and let K_L be the maximal unramified subextension of L/K, where $[K_L : K] = n$. Then the following diagram commutes.

$$
\begin{array}{ccc}
K_L^\times & \xrightarrow{\;N_{K_L/K}\;} & K^\times \\
\rho_{K_L}\big\downarrow & & \big\downarrow\rho_K \\
\mathrm{Gal}(L/K_L) & \xrightarrow{\;\text{incl.}\;} & \mathrm{Gal}(L/K)
\end{array}
$$

Proof. Choose a totally ramified abelian extension L'/K and an associated positive integer t as in the Decomposition Theorem. Then $L'K_{(t)} = LK_{(t)}$, where as usual $K_{(t)}$ is the unramified extension of K of degree t. We also have $L'_{\mathrm{ur}} = L_{\mathrm{ur}}$ and $K_L \subseteq K_{(t)}$ so that $n \leq t$.

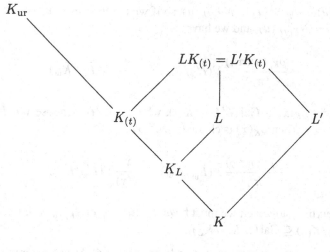

$$K_L^\times \xrightarrow{\;\;N_{K_L/K}\;\;} K^\times$$

$$\rho_{K_L} \downarrow \qquad\qquad\qquad \rho_K \downarrow$$

$$\mathrm{Gal}\,(L'K_{(t)}/K_L) \xrightarrow{\;\;\text{incl.}\;\;} \mathrm{Gal}\,(L'K_{(t)}/K)$$

It suffices to show that the following diagram commutes.

Since any element of $K_L^\times$ can be expressed as a product of an element of $\mathcal{U}_{K_L}$ times a power of some fixed uniformizer, we shall check the commutativity of this diagram by first considering where the mappings send a uniformizer and then where they send units.

Note that if $\varphi \in \mathrm{Gal}\,(L'K_{(t)}/L')$ is the Frobenius automorphism, then φ^n is the Frobenius automorphism in $\mathrm{Gal}\,(L'K_{(t)}/L'K_L)$. Choose a uniformizer π_K so that it is a norm from L', (note π_K is a uniformizer in K_L too); then $\rho_K(N_{K_L/K}(\pi_K)) = \rho_K(\pi_K^n) = \varphi^n = \rho_{K_L}(\pi_K)$. Hence it remains to check commutativity on elements of $\mathcal{U}_{K_L}$.

We want to show for $u \in \mathcal{U}_{K_L}$, that $\rho_{K_L}(u) = \rho_K(N_{K_L/K}(u))$. Choose $v \in \mathcal{U}_{\hat{L}'_{ur}}$ so that $N_{\hat{L}'_{ur}/\hat{K}_{ur}}(v) = u$. (The unit v exists because the extension is unramified, so the norm is surjective on units.) Recall that in $\mathrm{Gal}\,(L'K_L/K_L) = \mathrm{Gal}\,(L'K_{(t)}/K_{(t)})$, the element $\rho_{K_L}(u)$ is characterized by

$$\frac{\rho_{K_L}(u)(\pi_{L'})}{\pi_{L'}}\, \mathcal{V}(\hat{L}'_{ur}/\hat{K}_{ur}) = \frac{v}{\varphi^n(v)}\, \mathcal{V}(\hat{L}'_{ur}/\hat{K}_{ur}),$$

where $\pi_{L'}$ is any uniformizer in L'. But then

$$\frac{\rho_{K_L}(u)(\pi_{L'})}{\pi_{L'}}\, \mathcal{V}(\hat{L}'_{ur}/\hat{K}_{ur}) = \frac{v}{\varphi(v)} \cdot \frac{\varphi(v)}{\varphi^2(v)} \cdots \frac{\varphi^{n-1}(v)}{\varphi^n(v)}\, \mathcal{V}(\hat{L}'_{ur}/\hat{K}_{ur}).$$

Now $u \cdot \varphi(u) \cdots \varphi^{n-1}(u) = N_{K_L/K}(u)$, so if we put $w = v \cdot \varphi(v) \cdots \varphi^{n-1}(v)$, then $N_{\hat{L}'_{ur}/\hat{K}_{ur}}(w) = N_{K_L/K}(u)$, and we have

$$\frac{\rho_{K_L}(u)(\pi_{L'})}{\pi_{L'}} \mathcal{V}(\hat{L}'_{ur}/\hat{K}_{ur}) = \frac{w}{\varphi(w)} \mathcal{V}(\hat{L}'_{ur}/\hat{K}_{ur}).$$

Now consider $\rho_K(x) \in \text{Gal}(L'K_{(t)}/K_{(t)})$, where $x \in \mathcal{U}_K$. Choose $y \in \mathcal{U}_{\hat{L}'_{ur}}$ so that $N_{\hat{L}'_{ur}/\hat{K}_{ur}}(y) = x$. Then $\rho_K(x)$ is characterized by

$$\frac{\rho_K(x)(\pi_{L'})}{\pi_{L'}} \mathcal{V}(\hat{L}'_{ur}/\hat{K}_{ur}) = \frac{y}{\varphi(y)} \mathcal{V}(\hat{L}'_{ur}/\hat{K}_{ur}).$$

Putting everything together, we must have $\rho_{K_L}(u) = \rho_K(N_{K_L/K}(u))$, (an element of $\text{Gal}(L'K_{(t)}/K_{(t)}) \subseteq \text{Gal}(L'K_{(t)}/K_L)$). $\qquad\square$

The following theorem amounts to a local version of Artin Reciprocity.

Theorem 7.4. *Let L/K be a finite abelian extension. Consider the composition*

$$K^\times \xrightarrow{\rho_K} \text{Gal}(K_{ab}/K) \xrightarrow{restr.} \text{Gal}(L/K).$$

The kernel of this composition is $N_{L/K}L^\times$.

Proof. We have shown $[K^\times : N_{L/K}L^\times] = [L : K]$ so it suffices to show $N_{L/K}L^\times$ is contained in the kernel. Suppose K_L is the maximal unramified subextension of L/K, where $[K_L : K] = n$. Consider ρ_{L/K_L}. Since L/K_L is totally ramified, $\ker \rho_{L/K_L} = N_{L/K_L}L^\times$ by Lemma 7.2. Also $N_{L/K}L^\times = N_{K_L/K}(N_{L/K_L}L^\times)$. By commutativity of the diagram in the previous lemma, the theorem follows. $\qquad\square$

Corollary 7.5. The open subgroups of finite index in $K^\times$ are precisely the subgroups of the form $N_{L/K}L^\times$, for L/K finite abelian. Indeed, any open subgroup of finite index in $K^\times$ is the kernel of the composition

$$K^\times \xrightarrow{\rho_K} \text{Gal}(K_{ab}/K) \xrightarrow{nat.} \text{Gal}(L/K)$$

for some finite abelian extension L/K.

Proof. **Exercise 7.29.** $\qquad\square$

Lubin and Tate have given an explicit construction of the reciprocity homomorphism ρ_K using formal groups. We discuss this construction next. To begin, we need a technical lemma about formal power series. It provides a generalization of Lemma 5.1.

Lemma 7.6. Let π and π' be uniformizers in a local field K, say with $\pi' = u\pi$, where $u \in \mathcal{U}_K$. Let $q = \#\mathbb{F}_K$ and suppose $f(X)$, $g(X)$ are polynomials of degree q such that $f(X) \in \mathcal{F}_\pi$ and $g(X) \in \mathcal{F}_{\pi'}$ We use φ to denote the Frobenius automorphism in $\text{Gal}(K_{ur}/K)$ and also its extension to $\hat{K}_{ur}$.

i. There is a power series $\vartheta(X) \in \mathcal{O}_{\hat{K}_{ur}}[[X]]$ such that $\vartheta^{\varphi}(X) = \vartheta([u]_f(X))$, and $\vartheta(X) \equiv \varepsilon X \pmod{X^2}$ for some $\varepsilon \in \mathcal{U}_{\hat{K}_{ur}}$. (Here $\vartheta^{\varphi}(X)$ is the power series obtained by applying φ to the coefficients of $\vartheta(X)$.)

ii. There is a power series $\vartheta'(X)$ that satisfies (i), and also satisfies

$$\vartheta'([a]_f(X)) = [a]_g(\vartheta'(X)) \qquad \text{for any } a \in \mathcal{O}_K.$$

Proof. i. Choose $\varepsilon \in \mathcal{U}_{\hat{K}_{ur}}$ so that $u = \varphi(\varepsilon)\varepsilon^{-1}$, (possible since $\varphi - 1$ is surjective on $\mathcal{U}_{K_{ur}}$ by Lemma 3.1). Put $\vartheta_1(X) = \varepsilon X$. We have

$$\vartheta_1^{\varphi}(X) = \varphi(\varepsilon)X = \varepsilon u X \equiv \vartheta_1([u]_f(X)) \pmod{X^2}.$$

Continuing recursively, suppose we have constructed $\vartheta_m(X)$ satisfying

$$\vartheta_m(X) \equiv \varepsilon X \pmod{X^2},$$
$$\vartheta_m^{\varphi}(X) \equiv \vartheta_m([u]_f(X)) \pmod{X^{m+1}}.$$

It follows that there is some $c \in \mathcal{O}_{\hat{K}_{ur}}$ such that

$$\vartheta_m^{\varphi}(X) - \vartheta_m([u]_f(X)) \equiv -cX^{m+1} \pmod{X^{m+2}}.$$

By the surjectivity of $\varphi - 1$ on $\mathcal{O}_{\hat{K}_{ur}}$, there is $b \in \mathcal{O}_{\hat{K}_{ur}}$ such that

$$\varphi(b) - b = (\varepsilon u)^{-(m+1)}c = \varphi(\varepsilon)^{-(m+1)}c.$$

Thus

$$\varphi(b\varepsilon^{m+1}) = \left(b + \varphi(\varepsilon)^{-(m+1)}c\right)\varphi(\varepsilon)^{m+1} = b\varphi(\varepsilon)^{m+1} + c.$$

Now put

$$\vartheta_{m+1}(X) = \vartheta_m(X) + b\varepsilon^{m+1}X^{m+1}.$$

Then $\vartheta_{m+1}(X) \equiv \vartheta_m(X) \equiv \varepsilon X \pmod{X^2}$ and

$$\vartheta_{m+1}^{\varphi}(X) - \vartheta_{m+1}([u]_f(X))$$
$$\equiv \left(\vartheta_m^{\varphi}(X) + (b\varphi(\varepsilon)^{m+1} + c)X^{m+1}\right) - \left(\vartheta_m([u]_f(X)) + b\varphi(\varepsilon)^{m+1}X^{m+1}\right)$$
$$\equiv \vartheta_m^{\varphi}(X) - \vartheta_m([u]_f(X)) + cX^{m+1}$$
$$\equiv 0 \pmod{X^{m+2}}.$$

The desired series is then $\vartheta(X) = \lim_{m} \vartheta_m(X)$.

ii. Consider the power series $\vartheta(X)$ from (i). There is a power series in $\mathcal{O}_{\hat{K}_{ur}}[[X]]$, which we shall denote $\vartheta^{-1}(X)$, such that $\vartheta(\vartheta^{-1}(X)) = X = \vartheta^{-1}(\vartheta(X))$. Put

$$h(X) = \vartheta^{\varphi}(f(\vartheta^{-1}(X))).$$

Applying (i) and noting that $f(X) = [\pi]_f(X)$, we get

$$h(X) = \vartheta([u]_f(f(\vartheta^{-1}(X)))) = \vartheta([u]_f[\pi]_f(\vartheta^{-1}(X))) = \vartheta([\pi']_f(\vartheta^{-1}(X))).$$

We leave it as **Exercise 7.30** to show that $h^{\varphi}(X) = h(X)$ so that $h(X)$ has coefficients in $\mathcal{O}_K$. If we examine the series $h(X)$ more closely, we find

$$h(X) \equiv \varphi(\varepsilon)\pi\varepsilon^{-1}X \equiv u\varepsilon\pi\varepsilon^{-1}X \equiv u\pi X \equiv \pi'X \pmod{X^2},$$
$$h(X) \equiv \vartheta^{\varphi}((\vartheta^{-1}(X))^q) \equiv \vartheta^{\varphi}((\vartheta^{-1})^{\varphi}(X^q)) \equiv X^q \pmod{\pi}.$$

This means (by Lemma 5.2) there is a power series $[1]_{g,h}(X) \in \mathcal{O}_K[[X]]$ that satisfies $[1]_{g,h}(X) \equiv X \pmod{X^2}$ and $g([1]_{g,h}(X)) = [1]_{g,h}(h(X))$. From this we can define the series

$$\vartheta'(X) = [1]_{g,h}(\vartheta(X)).$$

Because the coefficients of $[1]_{g,h}(X)$ are in $\mathcal{O}_K$, it follows that $\vartheta'(X)$ satisfies (i). It remains to show $\vartheta'([a]_f(X)) = [a]_g(\vartheta'(X))$ for any $a \in \mathcal{O}_K$. Equivalently, we shall show $\vartheta'([a]_f((\vartheta')^{-1}(X))) = [a]_g(X)$. Let $r(X) = \vartheta'([a]_f((\vartheta')^{-1}(X)))$. Then

$$\begin{aligned}
g(r(X)) &= g(\vartheta'([a]_f((\vartheta')^{-1}(X)))) \\
&= g([1]_{g,h}(\vartheta([a]_f(\vartheta^{-1}([1]_{h,g}(X)))))) \\
&= [1]_{g,h}(h(\vartheta([a]_f(\vartheta^{-1}([1]_{h,g}(X)))))) \\
&= [1]_{g,h}(\vartheta([\pi']_f(\vartheta^{-1}(\vartheta([a]_f(\vartheta^{-1}([1]_{h,g}(X)))))))) \\
&= [1]_{g,h}(\vartheta([\pi']_f([a]_f(\vartheta^{-1}([1]_{h,g}(X)))))) \\
&= [1]_{g,h}(\vartheta([a]_f([\pi']_f(\vartheta^{-1}([1]_{h,g}(X)))))) \\
&= [1]_{g,h}(\vartheta([a]_f(\vartheta^{-1}(h([1]_{h,g}(X)))))) \\
&= [1]_{g,h}(\vartheta([a]_f(\vartheta^{-1}([1]_{h,g}(g(X)))))) \\
&= \vartheta'([a]_f((\vartheta')^{-1}(g(X)))) \\
&= r(g(X)).
\end{aligned}$$

But this says that the power series $r(X)$ satisfies the definition of $[a]_g(X)$. Since $[a]_g(X)$ is the unique power series that behaves this way, we conclude $r(X) = [a]_g(X)$ as needed.   □

Exercise 7.31. Let π and π' be uniformizers in a local field K, and suppose $f(X)$, $g(X)$ are polynomials of degree q, where $q = \#\mathbb{F}_K$, such that $f(X) \in \mathcal{F}_\pi$ and $g(X) \in \mathcal{F}_{\pi'}$. What does the result of Lemma 7.6 tell us about the relationship between the formal group laws $F_f(X, Y)$ and $F_g(X, Y)$?   ◊

Let K be a local field and fix the uniformizer $\pi = \pi_K$. Recall the union L_π of the Lubin-Tate extensions L_m/K satisfies $L_\pi K_{\mathrm{ur}} = K_{\mathrm{ab}}$. Let λ_m be as in Exercise 7.22 (so $L_m = K(\lambda_m)$). Let $f(X) \in \mathcal{F}_\pi$. We have seen, for $u \in \mathcal{U}_K$, that there is an automorphism $\sigma_u \in \mathrm{Gal}\,(L_\pi/K)$ such that $\sigma_u(\lambda_m) = [u]_f(\lambda_m)$ for any m, (see Theorem 6.5).

With what we have done, we are able to define explicitly a homomorphism $\gamma_\pi : K^\times \longrightarrow \mathrm{Gal}\,(L_\pi K_{\mathrm{ur}}/K)$, which we shall show is just the reciprocity homomorphism ρ_K. To determine γ_π completely, we need only give the image of π and of an arbitrary element u of $\mathcal{U}_K$. We put

$$\gamma_\pi(u) = \sigma_{u^{-1}} \quad \text{in } \mathrm{Gal}\,(L_\pi K_{\mathrm{ur}}/K_{\mathrm{ur}}) \cong \mathrm{Gal}\,(L_\pi/K),$$

$$\gamma_\pi(\pi) = \varphi \quad \text{Frobenius in } \mathrm{Gal}\,(L_\pi K_{\mathrm{ur}}/L_\pi).$$

Theorem 7.7. *The homomorphism γ_π defined above does not depend on the choice of uniformizer π. Moreover, γ_π is the local Artin map, i.e., $\gamma_\pi = \rho_K$.*

Proof. Let π and π' be uniformizers in K, say $\pi' = u\pi$. Note that $\gamma_\pi(\pi')$ and $\gamma_{\pi'}(\pi')$ induce the Frobenius automorphism on K_{ur}, while on $L_{\pi'}$, by definition $\gamma_{\pi'}(\pi')$ is the identity. Hence to deduce that $\gamma_\pi(\pi') = \gamma_{\pi'}(\pi')$, we want to show that $\gamma_\pi(\pi')$ is the identity on $L_{\pi'}$. To get this, it suffices to show that $\gamma_\pi(\pi')(\lambda'_m) = \lambda'_m$ for all m. (Here λ'_m generates $L_{\pi'} = K(\lambda'_m)$.) Recall, from Exercise 7.22, the element λ'_m is a zero of $\frac{g^{(m)}(X)}{g^{(m-1)}(X)}$, where $g(X) \in \mathcal{F}_{\pi'}$ is a monic polynomial of degree q.

Now let λ_m be a zero of $\frac{f^{(m)}(X)}{f^{(m-1)}(X)}$, where $f(X)$ is a monic polynomial of degree q in $\mathcal{F}_\pi$ so that $L_\pi = K(\lambda_m)$. Consider a power series $\vartheta'(X) \in \mathcal{O}_{\hat{K}_{\mathrm{ur}}}$ as in (ii) of the previous lemma. It follows that $\vartheta'(\lambda_m)$ is a zero of $\frac{g^{(m)}(X)}{g^{(m-1)}(X)}$. Thus we may put $\vartheta'(\lambda_m) = \lambda'_m$. Let φ denote the Frobenius automorphism in $\mathrm{Gal}\,(L_\pi K_{\mathrm{ur}}/L_\pi)$. Then

$$
\begin{aligned}
\gamma_\pi(\pi')(\lambda'_m) &= \gamma_\pi(u)\,\varphi(\vartheta'(\lambda_m)) \\
&= \gamma_\pi(u)(\vartheta'([u]_f(\lambda_m))) \\
&= \vartheta'([u]_f(\gamma_\pi(u)(\lambda_m))) \\
&= \vartheta'([u]_f([u^{-1}]_f(\lambda_m))) \\
&= \vartheta'(\lambda_m) \\
&= \lambda'_m.
\end{aligned}
$$

Finally, we note that γ_π and ρ_K agree, since at π each is the Frobenius automorphism on K_{ur} and the identity on L_π.   □

Exercise 7.32. Let K be a local field, and let L_m be the mth Lubin-Tate extension of K. Factor $a \in K^\times$ as $a = u\pi_K^t$, where $u \in \mathcal{U}_K$ and $t \in \mathbb{Z}$. Show that $a \in \ker \rho_{L_m/K}$ if and only if $u^{-1} \in \mathcal{U}_K^m$.   ◊

Exercise 7.32 tells us $\ker \rho_{L_m/K} = \langle \pi_K \rangle \times \mathcal{U}_K^m$, a subgroup of $K^\times$. But we know that $\ker \rho_{L_m/K} = N_{L_m/K} L_m^\times$ by Theorem 7.4. Hence L_m is the class field over K of the group $\langle \pi_K \rangle \times \mathcal{U}_K^m$.

Exercise 7.33. Suppose E/K is unramified, and let $a \in K^\times$. Find, as explicitly as possible, $\rho_{E/K}(a)$.   ◊

Bibliography

[AT]   Artin, E. and Tate, J., *Class Field Theory*, W. A. Benjamin, New York, 1967.
[CF]   Cassels, J. and Fröhlich, A., eds., *Algebraic Number Theory*, Academic Press, London and New York, 1967.
[FT]   Fröhlich, A. and Taylor, M., *Algebraic Number Theory*, Cambridge University Press, 1991.
[G]   Gras, G., *Class Field Theory*, 2nd edition, Monographs in Mathematics, Springer–Verlag, Berlin, Heidelberg, 2005.
[Haz1]   Hazewinkel, M., *Formal Groups and Applications*, Academic Press, New York, 1978.
[I]   Iwasawa, K., *Local Class Field Theory*, Oxford University Press, Clarendon Press, New York, Oxford, 1986.
[J]   Janusz, G., *Algebraic Number Fields*, Academic Press, New York, 1973.
[JR]   Jones, J. and Roberts, D., *Database of Local Fields*, http://math.asu.edu/˜jj/localfields/.
[L1]   Lang, S., *Algebraic Number Theory*, Addison–Wesley, Reading, MA, 1970, 2nd edition, Graduate Texts in Mathematics, 110, Springer–Verlag, New York, 1994.
[L2]   Lang, S., *Elliptic Functions*, Addison–Wesley, Reading, MA, 1973.
[L3]   Lang, S., *Cyclotomic Fields I and II*, Graduate Texts in Mathematics, 121, Springer–Verlag, New York, 1990.
[Lo]   Long, R., *Algebraic Number Theory*, Marcel Dekker, New York, 1977.
[Ma]   Marcus, D., *Number Fields*, Universitext, Springer–Verlag, New York, 1977.
[Mil]   Milne, J., *Class Field Theory*, http://www.jmilne.org/math/.
[N]   Neukirch, J., *Class Field Theory*, Grundlehren, 280, Springer–Verlag, Berlin, Heidelberg, 1986.
[Se1]   Serre, J-P., *A Course in Arithmetic*, Graduate Texts in Mathematics, 7, Springer–Verlag, New York, 1973.
[Se2]   Serre, J-P., *Local Fields*, Graduate Texts in Mathematics, 67, Springer–Verlag, New York, 1979.
[Sh]   Shimura, G., *Introduction to the Arithmetic Theory of Automorphic Functions*, Iwanami Shoten and Princeton University Press, Princeton , 1971.
[Si]   Sinnott, W., *Class Field Theory* , course notes, Ohio State University, 1985.
[Wa]   Washington, L., *Introduction to Cyclotomic Fields*, Graduate Texts in Mathematics, 83, Springer–Verlag, New York, 1982.
[Wei1]   Weil, A., *Basic Number Theory*, Grundlehren, 144, Springer–Verlag, Berlin, Heidelberg, 1967.

CHRONOLOGY OF SOME IMPORTANT HISTORICAL REFERENCES

Beginnings of the Kronecker-Weber Theorem
[K1]   Kronecker, L., *Über die algebraisch auflösbaren Gleichungen I*, Sber. preuss. Akad. Wiss. (1853), 365–374.

Beginnings of the Principal Ideal Theorem (first complete statement was later given by Hilbert), and early notion of class field
[K2]   Kronecker, L., *Grundzüge einer arithmetischen Theorie der algebraischen Grössen*, J. reine angew. Math. **92** (1882), 1–122.
Beginnings of class field theory for imaginary quadratic fields
[K3]   Kronecker, L., *Zur Theorie der elliptischen Funktionen I–XXII*, Sber. preuss. Akad. Wiss. (1883–1890).
First complete proof of Kronecker-Weber Theorem
[We1]   Weber, H., *Theorie der Abel'schen Zahlkörper I,II*, Acta Math. Stockh. **8,9** (1886, 1887).

Definition of class field, Uniqueness Theorem, Universal Norm Index Inequality, generalization of Dirichlet's Theorem on Primes in Arithmetic Progressions, beginnings of Existence Theorem, etc.
[We2]   Weber, H., *Über Zahlengruppen in algebraischen Körpen I, II, III*, Math. Annln. **48, 49, 50** (1897–1898).

Hilbert class fields, notion of class field theory as a study of relatively abelian extensions of number fields, etc.
[Hi1]   Hilbert, D., *Der Theorie der algebraischer Zahlkörper*, Jahresber. der Deutsch. Math. Ver. **4** (1897), 177–546; English transl. by Adamson, I. with an introduction by Lemmermeyer, F. and Schappacher, N. (1998), Springer–Verlag, Berlin.
[Hi2]   Hilbert, D., *Über die Theorie der relativ-Abel'schen Zahlkörper*, Nachr. Ges. Wiss. Göttingen (1898), 377–399.

The famous problems
[Hi3]   Hilbert, D., *Mathematische Probleme. Vertrag auf internat. Math. Kongr. Paris, 1900*, Nachr. Ges. Wiss. Göttingen (1900), 253–297.

Existence of Hilbert class fields
[Fur1]   Furtwängler, Ph., *Allgemeiner Existenzbeweis für den Klassenkörper eines beliebigen algebraischen Zahlkörpers*, Math. Annln. **63** (1907), 1–37.

Hilbert class fields of imaginary quadratic fields and the relation to complex multiplication
[We3]   Weber, H., *Lehrbuch der Algebra III*, Braunschweig, 1908.
[Fue]   Fueter, R., *Abel'sche Gleichungen in quadratisch-imaginären Zahlkörpern*, Math. Annln. **75** (1914), 177–255.

Proofs of Existence, Completeness, Ordering, Isomorphy Theorems, full realization of Kronecker's ideas on abelian extensions of imaginary quadratic fields
[T]   Takagi, T., *Über eine Theorie des relativ-Abel'schen Zahlkörpers*, J. Coll. Sci. Imp. Univ. Tokyo **41** (1920), Nr. 9, 1–133.

Beginnings of Artin Reciprocity
[A1]   Artin, E., *Über eine neue Art von L-Reihen*, Abh. Math. Semin. Univ. Hamburg **3** (1924), 89–108.

The Density Theorem
[Ch]   Chebotarev, N., *Die bestimmung der Dichtigkeit einer Menge von Primzahlen welche zu einer gegebenen Substitutionsklasse gehören*, Math. Ann. **95** (1926), 191–228.

Proof of Artin Reciprocity
[A2]   Artin, E., *Beweis des allgenmeinen Reziprozitätsgesetzes*, Abh. Math. Semin. Univ. Hamburg **5** (1927), 353–363.

First proof of the Principal Ideal Theorem
[Fur2] Furtwängler, Ph., *Beweis des Hauptidealsatzes für Klassenkörper algebraischen Zahlkörpen*, Abh. Math. Semin. Univ. Hamburg **7** (1930), 14–36.

Reduction of the Principal Ideal Theorem to group theory
[A3] Artin, E., *Idealklassen in Oberkörpern und allgemeines Reziprozitätsgesetz*, Abh. Math. Semin. Univ. Hamburg **7** (1930), 46–51.
[Iy] Iyanaga, S., *Zum Beweise des Hauptidealsatzes*, Abh. Math. Semin. Univ. Hamburg **10** (1934), 349–357.

Main theorems of local class field theory
[Has] Hasse, H., *Die Normenresttheorie relativ-Abel'sches Zahlkörper als Klassenkörpertheorie im Kleinen*, J. reine angew. Math. **162** (1930), 145–154.
[Sc] Schmidt, F. K., *Zur Klassenkörpertheorie im Kleinen*, J. reine angew. Math. **162** (1930), 155–168.
[Ch1] Chevalley, C., *Sur la théorie du corps de classes dans les corps finis et les corps locaux*, J. Fac. Sci. Tokyo Univ. **2** (1933), 365–476.

Use of idèles for class field theory, avoided L-functions in the proof of the Universal Norm Index Inequality, etc.
[Ch2] Chevalley, C., *La théorie du corps de classes*, Ann. of Math **41** (1940), 394–417.

Origin of formal groups
[Bo] Bochner, S., *Formal Lie groups*, Ann of Math. **47** (1946), 192–201.

Use of idèles in their modern form
[Wei2] Weil, A., *Sur la théorie du corps de classes*, J. Math. Soc. Japan **3** (1951), 1–35.

Use of cohomology
[HN] Hochschild, G. and Nakayama, T., *Cohomology in Class Field Theory*, Ann. of Math. **55** (1952), 348–366.
[Tat] Tate, J., *The higher dimensional cohomology groups of class field theory*, Ann. of Math. **56** (1952), 294–297.

The group theoretic Principal Ideal Theorem revisited
[Wi] Witt, E., *Verlagerung von Gruppen und Hauptidealsatz*, Proc. Int. Congr. of Math. Amsterdam, Ser. II **2** (1954), 70–73.

Interpretation of formal groups via power series
[Laz1] Lazard, M., *Sur les groupes de Lie formels à un paramètre*, Bull. Soc. Math. France **83** (1955), 251–274.
[Laz2] Lazard, M., *Lois de groupes et analyseurs*, Ann. Ecole Norm. Sup. **72** (1955), 299–400.

Existence of infinite Hilbert class field towers
[GS] Golod, E., and Shafarevich, I., *Über Klassenkörpertürme*, Izv. Adad. Nauk. SSSR **28** (1964), 261–272 (Russian); English transl. in AMS Transl. (2) **48**, 91–102.

Use of formal groups in local class field theory
[LT] Lubin, J. and Tate, J., *Formal Complex Multiplication in Local Fields*, Annals of Math. **81** (1965), 380–387.
[Haz2] Hazewinkel, M., *Local class field theory is easy*, Adv. in Math. **18** (1975), 148–181.
[Col] Coleman, R., *Division Values in Local Fields*, Inv. Math. **53** (1979), 91–116.

Use of algebraic K-theory
[Ka] Kato, K., *A generalization of local class field theory by using K-groups I, II, III*, J. Fac. Sci. Univ. Tokyo **26,27,29** (1979, 1981, 1982), 303–376, 603–683, 31–43.
[Bl] Bloch, S., *Algebraic K-theory and class field theory for arithmetic surfaces*, Ann. of Math. **114** (1981), 229–265.

[KS]   Kato, K. and Saito, S., *Global class field theory of arithmetic schemes*, Contemp. Math.
         55 (1986), AMS, 255–331.
[R]   Raskind, W., *Abelian class field theory of arithmetic schemes*, Proc. Symp. Pure Math.
         58 (1995), AMS, 85–187.

The Principal Ideal Theorem revisited
[Su]   Suzuki, H., *A generalization of Hilbert's Theorem 94*, Nagoya Math J. **121** (1991),
         161–169.

Computational class field theory
[Coh]   Cohen, H., *Advanced Topics in Computational Number Theory*, Graduate Texts in Math-
         ematics, 193, Springer-Verlag, New York, 2000.

A modern look at class field theory
[Miy]   Miyake, K., ed., *Class Field Theory – Its Centenary and Prospect*, Adv. Studies in Pure
         Math. 30, Math. Soc. Japan , Tokyo, 2001.

Index

A

Abel's Lemma, 31–32
Absolute value
  Archimedean, 9–11, 63, 64
  equivalence of, 10, 12, 64
  non-Archimedean, 9, 11
  p-adic, 9, 11–12, 46–47, 63–64, 70
Admissible ideal, 107, 123
$\mathcal{A}_{K/F}$, 108, 121
Approximation Theorem, 45–47, 51, 71
Artin automorphism, 5, 7, 108, 125
Artin map, 61, 108–109, 112, 114, 115, 122,
    126, 127, 128, 130–133, 135–136,
    138, 145, 148–153, 160, 167, 182,
    194, 210–218
  global, on ideals, 109, 111, 115, 122, 126,
    127, 133
  global, on idèles, 126, 127, 128, 130, 135,
    146, 153
  local, 182, 195, 210–218
Artin Reciprocity, 54, 57, 60, 101, 105–147,
    151, 152–153, 156, 159–160, 182,
    186, 210, 214
Artin's Lemma, 118–121
Artin symbol, 105–109, 111, 132, 159, 167

B

Bernoulli number, 172

C

$\mathcal{C}_F$, 2, 14, 45, 49, 50, 53, 61, 70, 71, 141–142,
    153–156, 158–159, 179
C_F, 85, 86, 87, 101, 107, 127, 167–168
Character, 17–19, 20–24, 25–30, 33, 36, 38,
    45, 49, 54, 55, 56, 57, 108, 114,
    124–125, 140, 145, 154
  field associated to, 22–25, 29, 30, 140, 202
  induced, 9, 20–21, 61, 106, 164, 186, 189,
    191, 209

  trivial, 17, 21
    see also Dirichlet character, Weber
      character
Character group, 17, 28, 29, 30
  field associated to, 29, 30, 140, 202
Chebotarev Density Theorem, 125, 145, 160
Chinese Remainder Theorem, 46, 117, 165
Class field, 1, 7, 8, 14, 15, 45, 50, 53, 54, 55,
    59, 61, 74, 101, 105, 108, 124, 126,
    133, 135, 137, 138, 139, 144, 145,
    146, 147, 148, 149, 150, 151, 153,
    154, 155, 156, 157, 158, 159, 160,
    169, 177, 178
Class formation, 63
Class number, 2, 14, 37, 50, 133, 154, 157–158,
    169, 170
  relative, 169
CM-field, 156, 170
1-cocycle, 85
Completeness Theorem, 61
Complete Splitting Theorem, 136, 146, 149,
    152
Completion, 1, 8–14, 47, 66, 69, 70, 83, 103,
    143, 148, 150, 166, 176, 182, 183,
    194, 209
Complex multiplication, 155, 156
Conductor (of an abelian extension), 151, 153
Conductor Discriminant Formula, 24
Consistency Property, 109–110, 112, 122, 132,
    159, 160, 167
Crossed homomorphism, 85, 86
Cyclotomic units, 170

D

Decomposition field, 5, 119
Decomposition group, 4–5, 7, 28–29, 38, 83,
    108, 112, 117, 119, 124, 131, 132
Decomposition Theorem, 184, 212
Dedekind-Kummer Theorem, 2

Dedekind zeta function, 36–37, 38, 42
Diagonal embedding, 69, 179
Dirichlet character, 17, 20, 21, 23–24, 26, 33,
      45, 49, 54
  conductor of, 20, 24
  even, 22, 25
  generalized, 49, 54
  odd, 20, 21, 24, 27, 170
  order of, 21
  primitive, 21
  quadratic, 21, 27
Dirichlet density, 40–41, 43, 53, 54, 55, 108,
      124, 161–162
Dirichlet L–function, 30, 33, 35, 49, 54, 170
Dirichlet series, 4, 30–35
Dirichlet's Theorem on Primes in Arithmetic
      Progressions, 17–44, 49, 61, 145
Dirichlet Unit Theorem, 8
Discriminant, 3, 14, 24, 37
Divisor, 21, 47, 48, 60, 61, 107, 149, 152

E
$\mathcal{E}_F$, 68–69, 71, 72, 73, 92, 105, 107, 133, 151
$\mathcal{E}_{F,\mathrm{m}}^+$, 71
Elliptic curves, 155–156
Euler product, 33, 49, 54
Exact Hexagon Lemma, 77
Existence Theorem, 61, 125, 133, 135–179
Exponent of an abelian extension, 139

F
Fermat's Last Theorem, 171
First Inequality, 101
$F_f(X, Y)$, 198, 199–200
F_{m}^+, 51–52, 71–72, 92, 93–94, 106, 126
$\mathcal{F}_\pi$, 198
Formal group law, 195–196, 199, 200
  additive, 196
  homomorphism of, 4, 18, 49, 64, 69, 126,
      187
  Jacobian of, 197, 200
  Lubin-Tate, 181, 198, 199–200, 201, 202,
      206, 207, 208, 209, 212, 217, 218
  multiplicative, 196, 199
Fractional ideal, 1–4, 9, 13, 14, 45, 47, 48, 50,
      59, 65, 69, 110, 122, 127, 133, 141,
      142, 157, 171
Frobenius automorphism, 4, 5, 28, 165–166,
      183, 185, 191, 194, 204, 213, 214,
      217
Frobenius element, 5–6, 125
$F_\mathbf{S}$, 141, 142, 143, 144, 145, 147, 155
Fundamental system of units, 8

G
G_a, 196, 197
G^{ab}, 156, 159
Genera of quadratic forms, 102
Global Cyclic Norm Index Inequality, 60,
      101–102
G_m, 196–197, 199–200, 317
G_{ram}, 182

H
Hasse-Minkowski Theorem, 103
Hasse's Norm Principle, 103
Hensel's Lemma, 11, 13
Herbrand quotient, 63, 75–82, 85, 87–88, 94,
      96, 100, 101
Hilbert class field, 30, 61, 105, 133, 136,
      153–157, 169, 177
Hilbert class field tower, 157
Hilbert p-class field, 154, 177–178
Hilbert's Twelfth Problem, 53, 156
Hilbert Theorem 90, 86

I
Ideal class group, 1–2, 14, 48, 49, 71, 87, 141,
      153, 154, 177, 179
  generalized, 45, 48
  strict, 49
Idèle, 63–103
  content of, 69
  principal idèle, 71
Idèle class group, 63, 85, 87, 94, 96, 101, 127
$\mathcal{I}_F(\mathrm{m})$, 48, 50–60, 61, 72, 74, 75, 87, 91,
      92, 93–94, 101, 106, 108, 110,
      111–112, 114, 115, 121, 123–124,
      126
Inclusion Theorem, 126
Inertia field, 5
Inertia subgroup, 5, 28, 132, 150, 175–176
Infinite Galois extension, 63, 162, 166, 167,
      175
Isomorphy Theorem, 15, 54, 61, 74, 105, 112,
      135, 153
Iwasawa theory, 175, 177, 178

J
$J_{F,\mathrm{m}}^+$, 71–72, 92, 93, 94, 105–106, 126, 127,
      131
J_F, 68–75, 84–87, 91–94, 101–103, 105, 106,
      107, 126, 127, 130, 131, 133, 135,
      136–137, 138, 139, 141, 142, 143,
      144, 145, 146, 147, 148, 149, 151,
      153, 159, 160, 167, 168
$J_{F,\mathbf{S}}$, 141–142, 144–145, 146, 147
J-invariant, 155

K

K_{ab}, 183, 186, 194, 208, 209, 210, 212, 214, 217

Kronecker-Weber Theorem, 155, 181

Krull topology, 162, 166, 212

$K(t)$, 183, 184, 185, 212

Kummer extension, 136, 141, 146

K_{ur}, 183–185, 191, 192, 203–204, 209, 211, 212, 214, 217

L

Langlands philosophy, 15

Layer Theorem, 5

Legendre symbol, 20, 128

Leopoldt's Conjecture, 179

L_m, 49–50, 54–59, 61, 124–125, 145, 202, 205, 206, 208, 209, 217, 218

Local global principle, 63

Lubin-Tate extension, 201–218

Lubin-Tate module, 200, 201, 202, 206

M

Modulus, 17, 20, 36, 38, 45, 47, 49

N

$\mathcal{N}_{K/F}(\mathfrak{m})$, 59, 110

Normal Basis Theorem, 82

Norm (of an idèle), 74, 84

Norm (of an ideal), 3–4, 14

Norm residue symbol, 127, 167, 168, 188

Number field, 1–15, 24, 29–30, 37, 43, 45, 47, 49, 54, 59, 61, 63, 64–65, 66, 68–74, 82, 83, 89, 91, 93, 94, 96, 98, 101, 103, 105–109, 110, 111, 113, 114, 115, 117, 118, 123, 124, 125, 127, 130, 132, 133, 136, 138, 139, 141, 144, 145, 148, 150, 151, 153, 154, 156, 157–158, 159–162, 166, 167, 171, 175, 176, 177, 178, 182

O

Ordering Theorem, 136–137, 160, 168

Ord_p, 9

Ord_v, 65

Orthogonality Relations, 19–20, 36

Ostrowski's Theorem, 10, 64

P

p-adic expansion, 10

p-adic integer, 10–11, 67

$\mathcal{P}_{F,m}^+$, 47, 48–49, 51–52, 53, 54, 55, 57, 59, 60, 61, 72, 74–75, 87, 91, 92, 93, 94, 101, 106, 107–108, 111, 112, 114, 115, 121, 123, 124, 125, 126, 127

$\mathcal{P}_F$, 2, 45, 47, 49, 50, 51, 53, 60, 61, 70, 71, 106, 156

Place, 64–65

    Archimedean, 64, 176

    discrete, 64

    finite, 64–65, 66, 69, 70, 73, 83, 89, 93–94, 127, 131, 132, 143, 144, 148, 149, 150, 151

    imaginary, 65–66, 71, 96, 98, 131

    infinite, 65, 66, 70–71, 107, 131, 132, 141, 150, 176

    non-Archimedean, 9, 11–12, 64

    ramified infinite, 107, 131, 132

    real, 65, 66, 70–71, 94, 127

    unramified, 131, 151

Prime

    completely split, 1, 7, 53, 108, 111, 124, 160, 162, 145, 154

    imaginary, 47, 107

    inert, 2

    infinite, 30, 45–47, 107, 132, 152, 154, 169

    irregular, 172

    ramified, 1, 27, 42, 93, 101, 107, 110, 111, 121, 124, 130, 151, 152, 177, 184

    ramified infinite, 107, 131

    real, 47, 107, 155

    regular, 172

    totally ramified prime, 2, 13, 14, 26, 93, 129, 159, 169, 177, 178, 181–182, 184–185, 186, 188, 190, 191, 192, 193–194, 201, 202, 203, 208, 210, 211, 212, 214

    unramified, 7, 107, 111, 125, 131, 151, 160

Principal Divisor Theorem, 105

Principal Ideal Theorem, 105, 156, 157

Product formula, 65, 66, 145, 148

Profinite completion, 166

Profinite group, 163–164, 166

Projective limit, 164, 167, 209

Projective system, 164–165

Q

Quadratic Reciprocity, 17, 105, 128–130

R

Ramification index, 2, 13, 26, 66, 88, 89, 91, 175, 176, 185

Ray, 45–46, 47–60

    modulo a divisor, 48, 60

Ray class group, 45–61, 63, 152
    modulo a divisor, 48, 60
    strict, 48–49, 50
Ray class number, 50
    strict, 50
Reduction Lemma, 136–137, 138, 139
Regulator, 8, 37
Residue field, 2, 5, 10–11, 13–14, 38, 109–110,
        181, 182–183, 186, 191, 200
Residue field degree, 2, 13–14, 38, 109, 183
Restricted topological product, 68, 69
$\mathcal{R}_{F,\mathfrak{m}}^{+}$, 48–49, 53, 60
Riemann zeta function, 17, 30, 33, 35

S
$\mathcal{S}_{K/F}^{1}$, 160–162
$\mathcal{S}_{\sigma}$, 125
Second Inequality, 101
Shapiro's Lemma, 80, 82, 88–89, 95, 100
S-idèle classes, 71, 85, 142
$\mathcal{S}_{K/F}$, 43, 53, 54, 57, 58, 59, 60, 108, 124, 125,
        145, 146, 160, 161, 162
Snake Lemma, 102, 193
Spl(f), 126
S-units, 141, 142, 155
Sup-norm, 96–97

T
Tate cohomology groups, 76
Topological group, 66–67, 68–69, 162–164
Totally positive element, 47
Transfer, 157

U
$\mathcal{U}_K$, 87, 182
$\mathcal{U}_K^{\mathfrak{m}}$, 182
$\mathcal{U}_v$, 66

Uniformizer, 10, 89, 131, 143, 146, 150, 182,
        198, 200, 202, 207, 209
Unit circle group, 95
Universal Norm Index Inequality, 45, 47–60,
        63, 74–75, 101, 112

V
Valuation, 9–10, 182, 186–187, 203, 209
    discrete, 9, 10, 182, 186–187
    normalized, 182
Vandiver's Conjecture, 172
Verlagerung, 157
V_F, 64
$\mathcal{V}(L/K)$, 187, 188

W
Weber character, 49
Weber function, 155
Weber L–function, 49, 54, 124, 145
$W_{f,\mathfrak{m}}$, 206, 207

Z
$\hat{\mathbb{Z}}$, 165
$\mathbb{Z}_p$-extension, 166, 175, 176, 177, 179
    cyclotomic, 166

η, 69
$\eta_{\mathfrak{m}}$, 72
$\theta_{Kab/K}$, 208
ι, 70
$\iota_{\mathfrak{p}}, \iota_v$, 13,69
$\lambda_{\mathfrak{m}}$, 202
ρ_K, 212
$\rho_{K/F}$, 126
φ, 191
φ_v, 130
ω_F, 167